Open-Shell Coordination Compounds based on Cyanide and Scorpionate Ligands

Zur Erlangung des akademischen Grades eines

DOKTORS DER NATURWISSENSCHAFTEN

(Dr. rer. nat.)

Fakultät für Chemie und Biowissenschaften

Karlsruher Institut für Technologie (KIT) - Universitätsbereich

genehmigte

DISSERTATION

von

M.Sc. Delphine Garnier

aus

Paris

Dekan: Prof. Dr. P. Roesky

Referent: Prof. Dr. F. Breher

Korreferent: Prof. Dr. R. Lescouëzec

Tag der mündlichen Prüfung: 17.07.2015

Bibliografische Information der Deutschen Nationalbibliothek

Die Deutsche Nationalbibliothek verzeichnet diese Publikation in der Deutschen Nationalbibliografie; detaillierte bibliografische Daten sind im Internet über http://dnb.d-nb.de abrufbar.
Bibliographic information published by the Deutsche Nationalbibliothek

The Deutsche Nationalbibliothek lists this publication in the Deutsche Nationalbibliografie; detailed bibliographic data are available on the Internet at http://dnb.d-nb.de .

ISBN 978-3-8325-4091-3

Logos Verlag Berlin GmbH
Comeniushof, Gubener Str. 47,
10243 Berlin
Tel.: +49 (0)30 42 85 10 90
Fax: +49 (0)30 42 85 10 92
INTERNET: http://www.logos-verlag.de

Université Pierre et Marie Curie

Karlsruhe Institute of Technology

ED 406

Institut Parisien de Chimie Moleculaire (IPCM) – UMR 8232

Equipe ERMMES

Open-shell Coordination Compounds based on Cyanide and Scorpionate Ligands

Par Delphine Garnier

Thèse de doctorat de Chimie Moléculaire

Dirigée par Prof. Dr. Frank Breher et Prof. Dr. Rodrigue Lescouëzec

Présentée et soutenue publiquement le 10/07/15

Devant un jury composé de : Prof. Dr. Frank Breher

Prof. Dr. Rodrigue Lescouëzec

Dr. Cyril Ollivier

Prof. Dr. Rolf Schuster

Prof. Dr. Jan Paradies

This work was produced between October 2011 and June 2015 at the Institute of Inorganic Chemistry of Karlsruhe Institute of Technology (KIT, Karlsruhe, Germany) under the supervision of Prof. Dr. Frank Breher and at the Institut Parisien de Chimie Moléculaire (IPCM) – UMR 8232 – ERMMES work group (University Pierre and Marie Curie (UPMC), Paris, France) under the supervision of Prof. Dr. Rodrigue Lescouëzec.

« Pourquoi faire simple quand on peut faire compliqué ? »

Proverbe Shadok

À Mamie, à Mimie et à Jo

À mes parents

À Joël

Table of content

1 Introduction

This work focuses on molecular systems based on cyanide and scorpionate ligands. We are particularly interested in the (photo)magnetic properties of polymetallic species obtained from self-assembly of building blocks complexes of the art $[Fe(L)(CN)_3]^{n-}$ because of their potential as switchable molecular materials. The first section of this chapter will introduce the chemistry of scorpionate ligands. The second part will focus on molecular cyanide-based systems, and in particular, the presentation of two photomagnetic phenomena that may occur in our targeted compounds: the Light-Induced Excited Spin-State Trapping (LIESST) effect and the Electron Transfer Coupled with a Spin Transition (ETCST).

Scorpionates ligand systems

Since they were first reported by Trofimenko in 1966,[1–3] tris(pyrazolyl)borate ligands Tp^R, and more generally scorpionate ligands, have been widely used to coordinate a great deal of metal across the periodic table.[4–9] The adjective "scorpionate" describes tripodal ligands systems, which are able to coordinate a metal ion with two identical donor moieties like the pincers of a scorpion. Depending on the nature of the X moiety (Figure 1.1) and that of the metal ion, scorpionates may arch above the plane to "sting" and coordinate the metal ion in a *fac*-manner (see Figure 1.1). Thus, they may display two interchangeable coordination modes. If the X moiety is identical to the two first "claws", the ligand is referred to as "homoscorpionate", while it is referred to as "heteroscorpionate" if the third donor group is different. This introduction will restrain itself to the first category. The most emblematic member of this widely used family of ligands is the Tp ligand (Tp = tris(pyrazolyl)borate), with a {BH} as bridgehead entity YR.

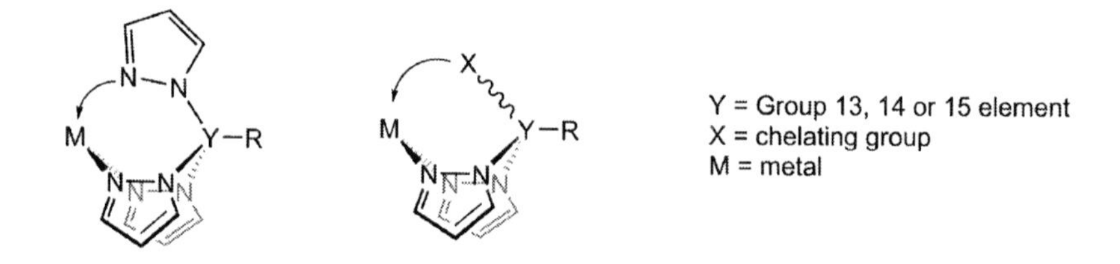

Figure 1.1: Schematic representation of homo- (left) and heteroscorpionates (right) with pyrazolyl donor moieties.

Tridentate ligands of this form, also called "scorpionate ligands of the first generation", usually behave as six electron donors, which holds a striking analogy with a wide array of "sandwich" and "half-sandwich" compounds. Yet, under certain conditions, they are able to display κ^1-*N* or κ^2-*N* coordination modes.[6] Because of this analogy with sandwich compounds, scorpionate ligands are often compared with ligands from the cyclopentadienyl family (Cp^R); this comparison is nowadays more and more disputed, because of the fundamental differences in energy and symmetry of the relevant orbitals between the two ligand systems.[6,10]

By modifying the nature, number and position of substituents of the heterocycle rings, a wide array of new Tp-based ligands with different electronic and steric properties can be prepared and used to tune the coordinated metal electronic properties. These ligands are noted Tp^R, with R being the substituent(s) at the 3, 4 or 5-position. Aside from the 3,5-methylation of the pyrazolyl rings to form the Tp* ligand, the most widespread modification is the introduction of sterically demanding substituents at the 3-position. In absence of such bulky groups, the ligands form octahedral sandwich complexes with many transition metal ions,[3,6] while their presence can lead to tetrahedral geometry around the metal ion.[6,11,12] Introduction of additional donor moieties as substituent at the 3-position allows the increase of the denticity of the ligands from 3 to 6.

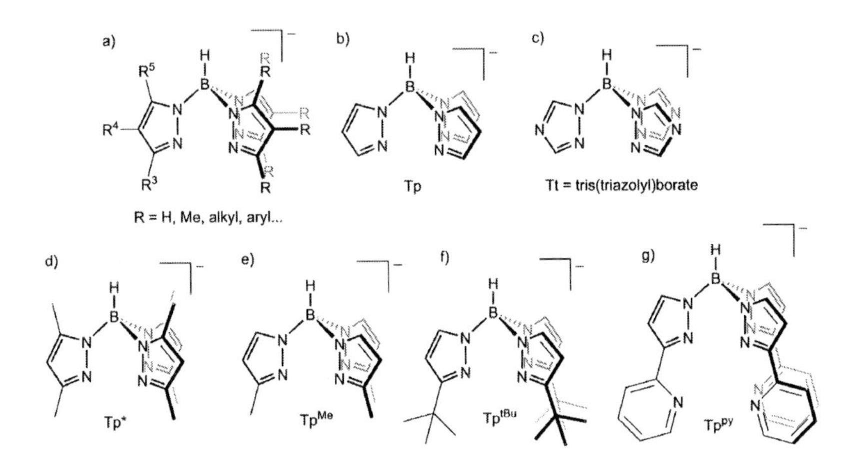

Figure 1.2 – Selected boron-based scorpionate ligands: a) general representation, b) and c): examples for ligands of the first scorpionate generation, d) to g): examples for ligands of the second scorpionate generation.

The replacement of the bridgehead boron atom in the Tp^R ligands by elements of the carbon atom column led to the development of the carbon-based tris(pyrazolyl)methane (Tpm) family of homoscorpionate ligands based on the same structural scheme. The neutral Tpm ligand was first reported by Trofimenko in the 1970[13] but remained little used before the improved synthesis published by Reger *et al.* in 2001.[14] Like their parent Tp and Tp* ligands, the Tpm and Tpm* ligands were used to coordinate a wide array of metal ions across the periodic table: coordination compounds range from classical sandwich complexes to complicated coordination polymers.[4,7,15–18]

While the scorpionate ligands of the second generation bear ring substituents which tune their electronic and steric properties, scorpionates of the third generation possess a supplementary function in apical position which will influence the ligand properties. This apical chemical function can, for instance, increase the solubility of the ligand in a given solvent, introduce an anchor function to graft complexes on surfaces or offer a supplementary donor moiety for coordination purposes. For instance, the introduction of a SO_3^- moiety by Kläui *et al.*[19] to form the Tpms ligand drastically improves the water solubility of the complexes thereof in respect to analogous Tpm complexes. Tpms was found to exhibit κ^3-*N* coordination mode,[20] as well as κ^2-*N*, κ^2-*N* / κ^1-*O* and κ^1-N / κ^1-*O* coordination modes[21–25] (see Figure 1.3) depending on the coordinating metal fragment and the ring substituents.

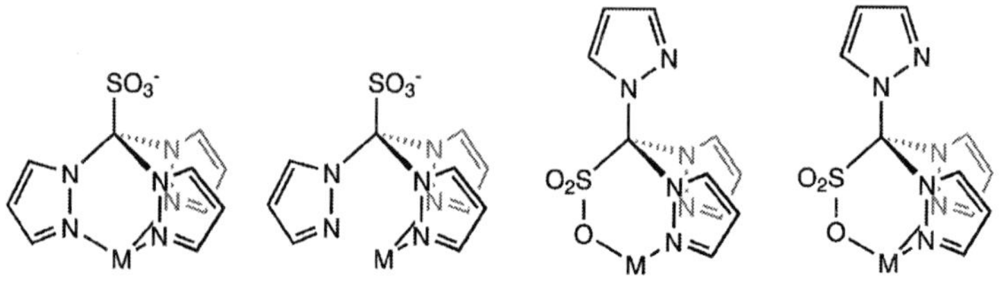

κ^3-*N* κ^2-*N* κ^1-*N* / κ^1-*O* κ^2-*N* / κ^1-*O*

Figure 1.3: Different coordination modes exhibited by the Tpms ligand.

The reaction of the apical carbon atom of the Tpm ligand with paraformaldehyde to produce the tris(pyrazolyl)ethanol (Tpe) ligand was reported by Reger *et al.*[26] It opened the route to facile functionalisation of the apical position in order to synthesise bitopic ligand systems, as illustrated by Figure 1.4.[4,5,26–36] In particular, the pyridine-functionalised ligand TpmPy exhibits different coordination modes depending on the nature of the coordinated metal. As reported for Tpe (Figure 1.4.a),[10,37] the TpmPy ligand produces the classical κ^3-*N* sandwich iron(II) sandwich complex b). However, in presence of *cis*-[$PdCl_2(CH_3CN)_2$], c) is formed.[38] In 2011, the group of Schatzschneider succeeded for the first time in anchoring a Tpm derivative tricarbonyl molybdenum complex at the surface of SiO_2 nanoparticles.[39] The complex showed light-induced release of CO, but was proven stable in solution if kept in the dark. The synthesis occurred by functionalising the hydroxyl moiety by a terminal alkyne function which was able to react to triazine by click-chemistry reaction with azide-functionalised SiO_2 nanoparticles. Another interesting example is the dendritic-like molybdenum complex e) reported by Reger *et al.* in 2002.[34]

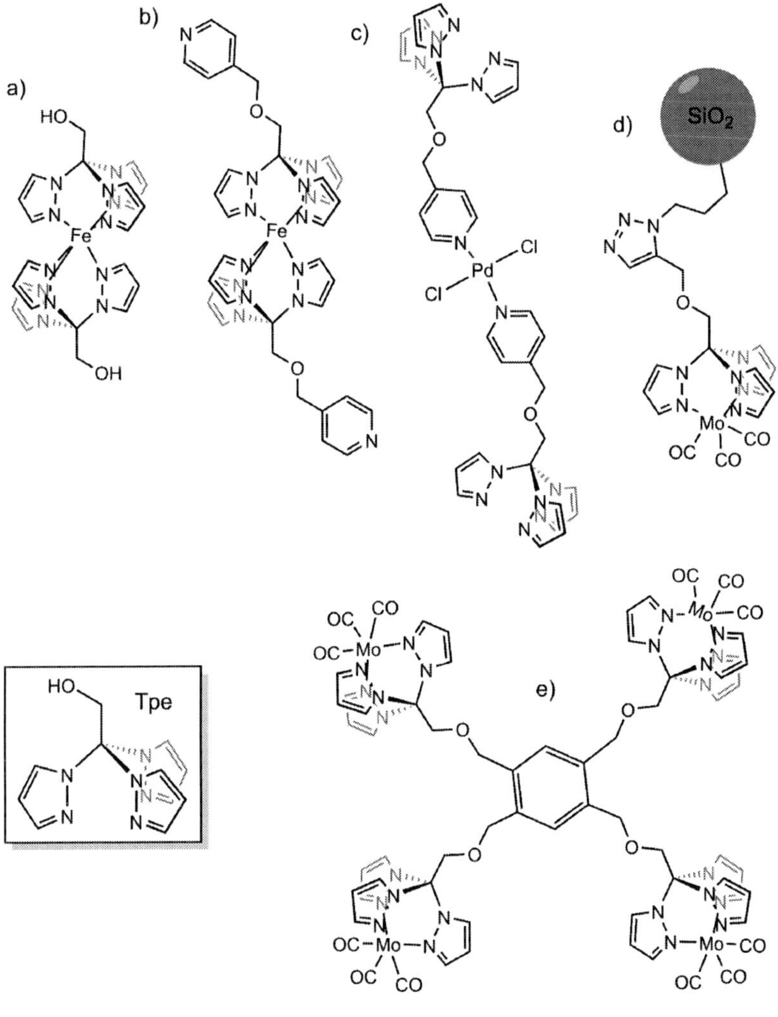

Figure 1.4: Examples of complexes of Tpe-based ditopic ligands.

Direct deprotonation of Tpm derivatives with a strong base produces the carbanionic tris(pyrazolyl)methanide Tpmd ligand. This ambidentate ligand features three nitrogen donors and a "nake" formally sp^3 hybridised pyramidal carbanion, facing in the opposite direction. Such ligand systems have been coined as Janus ligands, in reference to the Roman god of doors and gates.[5] 4- and 6-coordinated monomeric sandwich and half sandwich transition metal complexes of Tpmd derivatives were reported.[4,5,17,18,40–44] Some examples are depicted in Figure 1.5. Compounds a) to e) exhibit κ^3-*N* coordination modes towards the metal ion M (M = Mg, Zn, Cd, Fe, Co).[42,44] It has been showed that the carbanion can also act alternatively or simultaneously as *C*-donor or Lewis base.[5,42,45] Notably, Tpmd can act as κ^1-*C* or κ^3-*N* donor moiety towards coinage metals (M = Au, Ag, Cu).[43] In f) the gold(I) ion forms covalent C–Au bond. Despite the coordination of the nitrogen donors to form a sandwich complex e), the lone pair of the carbanion can still act as a donor moiety towards Lewis acids (compound c), here towards tri(ethyl)aluminium).

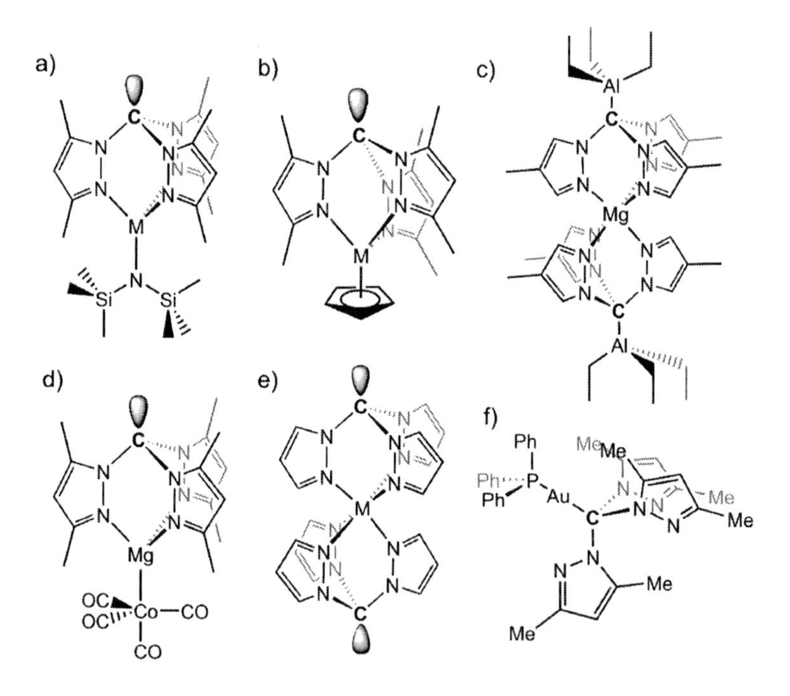

Figure 1.5: Some examples of complexes based on Tpmd derivatives.

In the framework of this work, it is noteworthy that the neutral iron(II) sandwich complex $[Fe^{II}(Tpmd^*)_2]$ exhibits a spin transition featuring a small hysteresis loop with $T_{1/2} = 270$ K.[41,42]

Photomagnetic systems

Magnetic properties arise from the presence of unpaired electrons.[46] In order to produce a photomagnetic system, the easiest way is a reorganisation of the valence electrons within the compound. There are several possibilities,[47–49] but the two most important are likely the spin crossover systems (reorganisation of the electronic configuration between the orbitals of one metal centre) and the photo-induced electron transfer (electronic configuration reorganisation between two metal centres). This corresponds to the light-induced phenomena called Light-Induced Excited Spin-State Trapping (LIESST) and Electron Transfer Coupled with a Spin Transition (ETCST).

According to the crystal field theory, the metal centred d-orbitals in octahedral geometry are split into two e_g and t_{2g} subsets, whose energy difference corresponds to the ligand field Δ. For octahedral complexes of first-row transition metals with $[Ar]3d^4$–$3d^7$ electronic configuration, two possible electronic ground states are possible.

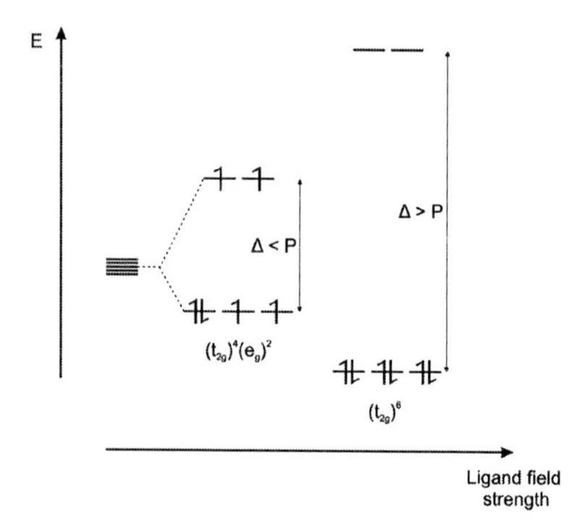

Figure 1.6: Electronic configuration of an iron(II) (d^6) octahedral complex in function of the ligand field strength Δ: low-spin and high-spin states.

When the ligand field, Δ, is greater than the interelectronic repulsion energy, *P*, it is energetically more favourable to fully occupy the t_{2g} orbitals than to promote one or more electrons in the higher energy e_g orbitals: the metal adopts a low-spin state (LS). If Δ is smaller than *P*, it becomes energetically more favourable to follow Hund's first rule and the e_g orbital are filled: the metal adopts a high-spin state (HS) (see Figure 1.6 for a d^6 configuration). Since the e_g orbitals possess an antibonding character, the metal-ligand bond lengths are longer in high-spin complexes than in low-spin complexes.

Most of the d^4-d^7 octahedral complexes exhibit either a high-spin or a low-spin ground state. However, if *P* is of the same order of magnitude as Δ, the difference in energy between the high-spin and low-spin states (ΔE – Figure 1.7) is in the order of magnitude of the thermal energy, k_BT.[50] In such a case, minor external stimuli (temperature change, pressure or light irradiation) can induce a spin-state change. This phenomenon is called spin crossover (SCO).[51–59]

The first spin-state transition was observed by Cambi *et al.*[60,61] in an iron(III) complex in the 1930s. Since then, this phenomenon was reported for iron, cobalt, nickel and chromium(II) complexes.[62–64] The iron(II) spin crossover complexes are of particular

The conversion of a low-spin state into a metastable high-spin state at low temperature by laser light irradiation was first described by Decurtins *et al.* in 1984.[72] It proceeds as follows: under laser light irradiation at a wavelength λ_1, and at low temperature, the complex undergoes a spin-allowed electronic transition from the $^1A_1$ low-spin (LS) ground state to the $^1T_1$ excited state (with a lifetime of the order of the nanosecond). The complex then undergoes a fast non-radiative relaxation process over two intersystem crossing steps to lead to the metastable $^5T_2$ high-spin state. The return to the $^1A_1$ ground state is prevented by the energy activation barrier. The system can fall back to the ground state by quantum tunnelling even if this process is slow. As a result, at low enough temperature, the metastable state exhibits a long lifetime (the photo-induced high-spin (HS) state is "trapped"). The return to the ground state will proceed by increasing the temperature near the so-called T_{LIESST} (or T_{relax}). It is when $k_B T_{\text{LIESST}} = k_B T$. The reverse conversion from the metastable high-spin state into the low-spin ground state can also be triggered by laser light irradiation at low temperature (reverse LIESST). It was first described by Hauser two years after the seminal paper of Decurtins *et al.*[72] In that case, the irradiation of the compound in its metastable state by another laser light wavelength λ_2 triggers the spin-allowed transition to the excited state ^5E. Rapid relaxation into the ground low-spin state $^1A_1$ occurs through two non-radiative intersystem crossing steps over the $^3T_1$ excited state. The LIESST effect was first observed for $[Fe(ptz)_6](BF_4)_2$ single crystals.[72] However, it was soon observed in functionalised molecular materials such as spin crossover complexes trapped in polymer films, KBr pellets or for spin crossover complexes grafted at the surface of nanoparticles.[73–75] It is noteworthy that the LIESST effect does not always convert a low-spin ground state into a high-spin metastable state. The inverse situation, converting a high-spin ground state into a low-spin metastable state has also been observed.[76]

Photo-induced electron rearrangement leading to reversible change of the magnetic properties can also be obtained by photo-induced electron transfer, in some mixed-valence compounds (of the class II). In photomagnetic compounds, the electron transfer can be triggered by laser light irradiation of the IVCT band (InterValence Charge Transfer), also named Metal to Metal Charge Transfer (MMCT) band. The photomagnetic effect due to photo-induced electron transfer has only been observed for the moment in cyanide-bridged compounds. In 1996, O. Sato, K. Hashimoto *et al.* reported for the first time that the magnetisation of the {FeCo} Prussian Blue Analogue

(PBA) $K_{0.2}Co_{1.4}[Fe(CN)_6] \cdot 6.9\,H_2O$ increases at 5 K under light irradiation.[77] When irradiated by red light, the magnetisation of the compound increases and ferrimagnetic interactions appear, so that the material becomes a magnet with an ordering temperature of T_C = 16 K. Irradiation in near infrared allows the decrease of the photo-induced magnetisation and a partial return to the initial state. Full conversion was observed on thin {FeCo} films by the same authors.[78]

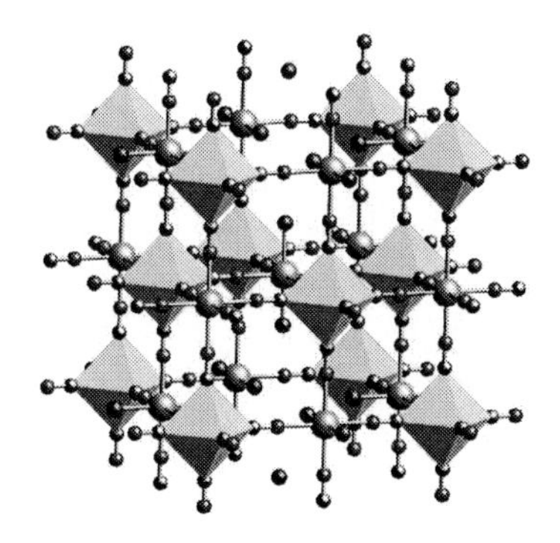

Figure 1.8: Molecular structure of a non-stoichiometric PBA. Iron(II) ions are represented as yellow octahedra.[79,80]

The photomagnetic effect observed in some {FeCo} PBAs can be explained by the occurrence of a photo-induced Electron Transfer Coupled with a Spin Transition (ETCST). The first wavelength promotes an electron transfer from the iron(II) ion to the cobalt(III) ion, which also undergoes a spin-state transition, thus converting diamagnetic $\{Fe^{II}_{LS}–C{\equiv}N–Co^{III}_{LS}\}$ $((t_{2g})^6$ and $(t_{2g})^6$ electronic configurations, $S = 0)$ pairs into $\{Fe^{III}_{LS}–C{\equiv}N–Co^{II}_{HS}\}$ $((t_{2g})^5$ and $(t_{2g})^5(e_g)^2$ electronic configurations, S = 1/2 and 3/2) paramagnetic ones.[81] As for the LIESST effect, the photo-induced state is metastable, but the respective T_{relax} are generally higher.

Since the seminal article of O. Sato *et al.*,[77] many {FeCo} PBAs have been studied but only some of them show photomagnetic properties (with more or less strong effect).

Indeed, it soon appeared that the magnetic properties of these non-stoichiometric materials were strongly dependent on their chemical formula. PBAs are non-stoichiometric cyanide-bridged 3D coordination polymers (see Figure 1.8) whose cubic structure (volume of the cavities ≈ 125 Å^3) can accommodate a wide variety of alkali ions.[79,80] Depending on their chemical formula, they contain various amounts of vacancies and inserted alkali ions. This leads to the coexistence in the same material of numerous different cobalt environments, and thus, of numerous different {Fe–C≡N–Co} pairs with not necessarily the same magnetic properties, and possible interactions with each other. For instance, it was shown by NMR that the stoichiometric {Fe_2Cd_3} PBA exhibits three different cadmium(II) sites, each of them with several isomers.[82] In non-stoichiometric PBAs, the number of possible coexisting {FeCo} configurations increases exponentially.

Over the last years, numerous efforts have been devoted to identify the critical parameters, which play a key role in the occurrence of the photomagnetic properties.[83–85] It has been shown that, (i) the ligand field on the cobalt and the iron ions have to be well-adjusted. If the ligand field on the cobalt ions is too strong, low-spin cobalt(III) oxidation state is too stabilised compared to high-spin cobalt(II) state for a photomagnetic effect to be observed. *A contrario*, a too weak ligand field on the cobalt ions stabilises the high-spin cobalt(II) state too much compared to low-spin cobalt(III) state, and the compound remains paramagnetic over the whole temperature range;[83,84] (ii) the 3D network must be flexible enough to allow dilatation that is concomitant with the ETCST phenomenon: indeed, the Co–N distances increase by 0.2 Å during the conversion of low-spin cobalt(III) in high-spin cobalt(II) ions.

These parameters critically depend on various chemical and structural factors such as: (i) the nature of the coordination sphere of the cobalt ion; (ii) the nature and amount of inserted alkali ions; (iii) the amount of {$Fe(CN)_6$} vacancies and (iv) the geometry of the cyanide bridges {Fe–C≡N–Co}. As these factors are interdependent, the rationalisation of the magnetic properties may be complicated and it requires the study of series of compounds where selected parameters are varied in order to shed light on their specific influence on the magnetic properties.

Magnetic properties are known to be strongly dependent on the structural geometry of the observed species. Because of their sophisticated local structure, it is very difficult to rationalise the (macroscopically measured) magnetic properties of the PBAs with respect to particular electronic / structural parameters, especially when the respective parameters are intricated. In order to better understand the parameters governing the ETCST phenomenon in the {Fe–C≡N–Co} bridges, the synthesis and characterisation of molecular models are of great interest. Indeed, the linear geometry of the cyanide bridge and the generally octahedral coordination sphere of the involved metal ions allows the nature of the interactions (ferro- or antiferromagnetic) to be predicted by Kahn's model.[86]

Such molecules can be prepared by self-assembly of preformed, carefully chosen metal complexes, the so-called "building blocks". Substituted cyanidometallates of the type $[Fe^{II/III}(L)_x(CN)_y]^{n-}$, where L are polydentate ligands preventing polymerisation, are reacted with partially blocked $[M^{II}(L')_z(S)_a]^{2+}$ units (for an example, see Figure 1.9).[87,88] By controlling the nature of the L and L' ligands and the coordination sphere of the metal ions, it is possible –in some extend– to control the electronic and structural properties of the resulting material.

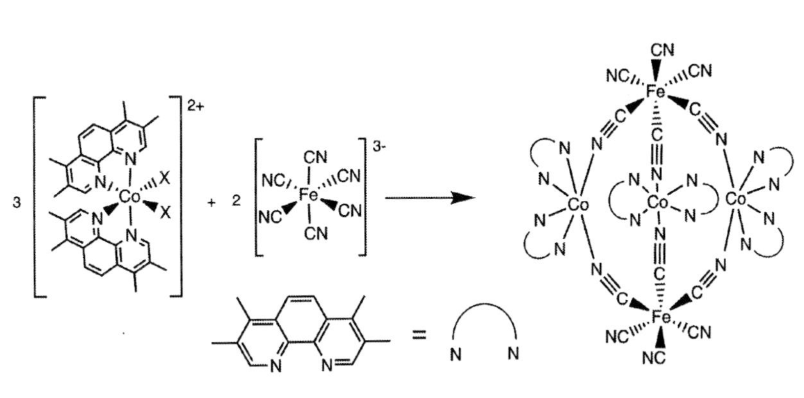

Figure 1.9: Synthesis of the $\{Fe_2Co_3\}$ trigonal-based pyramidal complex of Dunbar *et al.*[89–92]

The first complex being considered as a molecular model of Prussian Blue Analogue was reported by Dunbar *et al.* in 2004,[89–92] and consists of a trigonal based pyramidal complex obtained by self-assembly of hexacyanidoferrate(III) and partially blocked

$[Co(tmphen)_2(S)_2]^{2+}$ units (see Figure 1.9). This complex undergoes a thermally induced ETCST but is not photomagnetic. Of particular interest is an octanuclear cyanide-bridged $\{Fe_4Co_4\}$ complex with a cubic core reported by Holmes *et al.,* which also exhibits thermally induced ETCST and photomagnetic properties.[93] It is worth noticing its $T_{relax} = 200$ K, which is the highest reported up to date.

Since then, a wide array of $\{Fe_2Co_2\}$ molecular squares with photomagnetic properties were also reported by Holmes *et al.*, Lescouëzec *et al.* and Oshio *et al.* They all exhibit both thermally and photo-induced ETCST.[94–102] Furthermore, the Parisian group in which this work was partly performed reported that the magnetism of such squares can be reversibly and quantitatively switched on and off.[97,98]

The $\{Fe^{III}Co^{II}\}$ pair is also known to lead to interesting magnetic properties such as magnetic bistability at low temperature (nano-magnet).[103,104] The most known examples are the so-called single chain magnets (SCM) which behave as magnets at low temperature. This is due to the efficient magnetic exchange interactions through the cyanide bridges and the magnetic anisotropy of both metal ions that exhibit first order orbital moment and significant spin-orbit coupling.

Taking profit of this property, Sato *et al.* reported first an original multifunctional photomagnetic SCM.[104] Since then, other systems have been prepared. It is worth noticing the triple switch[105,106] chiral molecular chain exhibiting magnetic and electric bistability as well as photomagnetic behaviour.

Finally, it is noteworthy that the $\{Fe_xCo_y\}$ systems are not the only pairs showing photomagnetic properties. While some $\{Fe_2Fe_2\}$[49,98,107–109] mixed valence compounds undergo thermo- and photo-induced spin transition, ETCST phenomenon was also reported for $\{WCo\}$[91,103–105] $\{MoCu\}$, $\{FeMn\}$ and $\{OsM\}$ (M = Fe, Co) pairs.[83,110–113]

2 Overall aims

This PhD thesis was carried out within the framework of a collaboration between the workgroup ERMMES (Équipe de Recherche en Magnétisme Moléculaire Et Spectroscopie) under the supervision of Prof. Dr. Rodrigue Lescouëzec, at the Institut Parisien de Chimie Moléculaire (IPCM) of the Pierre et Marie Curie University (UPMC), in Paris, France, and the workgroup of Prof. Dr. Frank Breher, under his supervision, at the Institute of Inorganic Chemistry, at the Karlsruhe Institute of Technology (KIT) in Karlsruhe, Germany. The target of this PhD work, at the frontier between the research themes of the two research groups, was to synthesise new iron complexes based on cyanide and carbon-based scorpionate ligands, which can either be further functionalised at the apical position to introduce a satellite donor atom/moiety, or form multimetallic species through *N*-coordination of the three cyanide ligands. The properties of these new building blocks were extensively characterised (electronic and structural properties) and are compared to the already literature-known $[Fe^{III}(L)(CN)_3]^{n-}$ complexes (L = scorpionate ligand) in chapter 3:

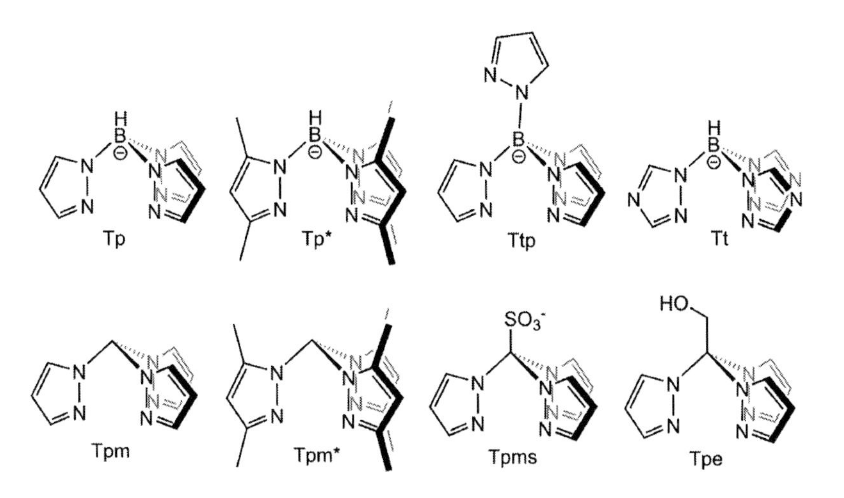

Figure 2.1: Capping ligands L used in this work for $[Fe^{II/III}(L)(CN)_3]^{n-}$ complexes.

In chapter 4 to 6, their reactivity towards metal ions and partially blocked subunits $\{M^{II}(L')_x(S)_y\}^{2+}$ (L' = blocking ligands, S = solvent molecules) known to exhibit photomagnetic properties in the right electronic and structural conditions ($\{Co(bik)_2\}$,[96–98] $\{Fe(bik)_2\}$[98,107] and $\{Co(Tpe)\}$[93]) were studied in order to obtain magnetic molecular materials of low dimensionality. A particular interest was shown for the {FeCo} systems because of their potential bistability and/or photomagnetic properties.

3 Mononuclear iron(II) and iron(III) building blocks

In the self-assembly approach, the geometric and electronic properties of tailored building block complexes of "blocked topology" allow the orientation of their self-assembly towards polymetallic materials of controlled architecture and physical properties.

In this chapter, we prepared new iron(II) and iron(III) building blocks, whose structure and electronic properties are characterised in depth. These building blocks, of the family of the $[Fe^{III}(Tp)(CN)_3]^-$ ([**1**]$^-$), represent interesting starting materials for the design of functional molecular materials such as Single Chain Magnets (SCM), Single Molecule Magnet (SMM) and photomagnetic molecules.

The molecular structures are obtained by X-ray diffraction on single crystals. Of particular interest will be the distortions from the octahedral ideal symmetry, which play a key role in the electronic / magnetic properties (structural distortions have a key role in the magnetic anisotropy). Here, we will have a close look at the distortion from the C_{3v} symmetry of the *fac* species: at the iron atom, with the octahedral distortion, and at the scorpionate ligand, with the pyrazolyl torsion angles.

As explained in chapter 1, matching redox potentials is a necessary condition to obtain reversible electron transfer systems. Studying the redox properties of the $[Fe(L)(CN)_3]^{n-}$ building blocks is therefore crucial for the synthesis of charge transfer systems.

Electronic properties will be probed by EPR spectroscopy and SQUID magnetometry. These techniques allow to look at the magnetic properties including the magnetic anisotropy. Since the $[Fe^{III}(L)(CN)_3]^{n-}$ complexes exhibit a $^2T_2$ ground term, which shows first order spin-orbit coupling, these systems are expected to exhibit significant magnetic anisotropy, which plays a key role in SMM and SCM materials.

Finally, spin density extension along the cyanide bridges will be probed by NMR and compared when possible to DFT calculations. Indeed, spin density plays an important role in all polynuclear paramagnetic materials as it governs the magnetic exchange.

3.1 Tricyanido iron (II) and iron (III) complexes of scorpionate ligand L

3.1.1 Syntheses of tricyanido iron complexes

The syntheses of $[Fe^{II}(L)(CN)_3]^{n-}$ anionic complexes (L = Tp, Tp*, Ttp, Tt, Tpm, Tpm*, Tpms, Tpe) follow two different routes, depending on the nature of the substituents at the 3-position of the binding rings. In case of hydrogen atoms (L = Tp, Ttp, Tt, Tpm, Tpms and Tpe),[98,114–116] a two-step synthesis is mandatory to produce the iron(II) complexes (see Scheme 3.1). The direct reaction of $Fe^{II}Cl_2$ with *ca* 3 equivalents of alkali metal cyanide and one equivalent of the desired scorpionate ligand in methanol at room temperature always leads to the formation of a mixture of the corresponding sandwich iron(II) complex and ferrocyanides. However, replacing one of the scorpionate ligands by three cyanides in the coordination sphere of the sandwich complex to produce the corresponding tricyanido iron(II) complex of scorpionate ligand is possible under heating and exclusion of light, in methanol for L = Tp and Ttp and in isopropanol for all the others.

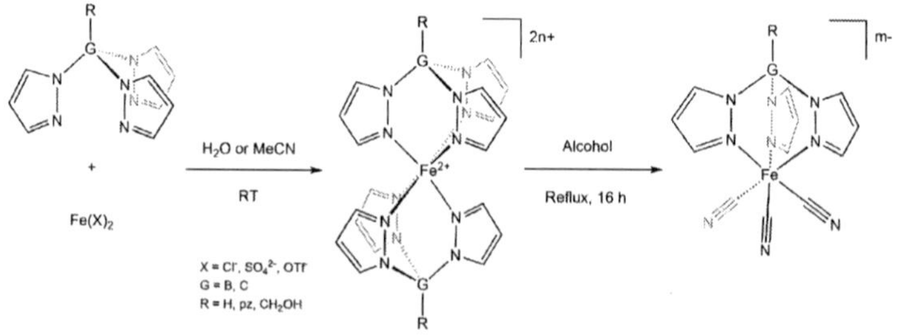

Scheme 3.1: Two-step synthesis of the $[Fe^{II}(L)(CN)_3]^{n-}$ anion if the 3-position of the ring bears an hydrogen.

Depending on the redox potential, the oxidation of the iron(II) $[Fe^{II}(L)(CN)_3]^{n-}$ complexes into iron(III) usually takes an extra step to produce the respective $[Fe^{III}(L)(CN)_3]^{(n-1)-}$. This oxidation usually occurs with a mild oxidant in water (I_2, H_2O_2) or acetonitrile ($[Fc][PF_6]$). It is noteworthy that $PPh_4[Fe^{III}(Tp)(CN)_3]$ can be synthesised in only one step, using the literature protocol of Kim *et al.*[117]

$PPh_4[Fe^{III}(Tp)(CN)_3]$ (PPh_4[**1**]), $PPh_4[Fe^{III}(Ttp)(CN)_3]$ (PPh_4[**9**]) and $(PPh_4)_2[Fe^{II}(Tpms)(CN)_3]$ ($(PPh_4)_2$[**5**]) are literature-known compounds and were synthesised using this method without any change in the literature protocol.[114,116]

If the pyrazolyl rings of the scorpionate ligand are 3,5-dimethylated, the steric hindrance induced by the methyl moieties at the metal centre allows a certain control over its coordination behaviour and may prevent the formation of the bis-scorpionate iron(II) complexes using adequate conditions.[118,119] The formation of $[Fe^{II}(L)(CN)_3]^{n-}$ (L = Tp* or Tpm*) can thus take place as a one-pot synthesis, by pre-coordination of the scorpionate ligand to the iron(II) ion in methanol (Tpm*) or acetonitrile (Tp*), then slow dropwise addition of the resulting methanolic or acetonitrile solution to a stirred cyanide methanolic solution under exclusion of light. Syntheses for $Et_4N[Fe^{III}(Tp^*)(CN)_3]$ (Et_4N[**7**]), $PPh_4[Fe^{II}(Tpm^*)(CN)_3]$ (PPh_4[**3**]) and $[Fe^{III}(Tpm^*)(CN)_3]$ (**8**) are already described in the literature[115,120] but modifications of these protocols were used in this work.

$X[Fe^{II}(Tpm)(CN)_3]$ (X[2], X = $[PPh_4]^+$, Na^+)

PPh_4[**2**] is a pale yellow solid produced in poor yield after a two-step synthesis. In D_2O, the ^1H NMR spectrum of [**2**]$^-$ (as sodium salt) exhibits one set of signals for the three pyrazolyl heterocycles, which confirms the C_{3v} symmetry of [**2**]$^-$ in solution. The proton at the heterocyclic 4-position appears at $\delta = 6.41$ ppm as a doublet of doublet, that is at the same position as it appears in the free ligand with coupling constants $J_{HH} = 2.2$ and 2.9 Hz to protons at the heterocyclic 5- and 3-positions respectively. These two are

~0.6 ppm shifted toward higher frequency compared to the free ligand and appear at $\delta = 8.11$ and 8.18 ppm respectively. They couple with each other very weakly ($J_{HH} = 0.6 - 0.7$ Hz). The peak corresponding to the apical proton appears in freshly dissolved samples at 9.14 ppm. However, its intensity decays rapidly and no signal is found in a matter of minutes, while the chemical shift of the three other signals remains unchanged. This is typical of a proton/deuterium chemical exchange in H_2O, which indicates that the apical proton is sufficiently acidic to be easily deprotonated. The complex can thus form either the corresponding Tpmd dianionic species, or undergo an apical functionalisation directly on the coordinated ligand. The functionalisation is complicated by the poor yield of the complex (*ca* 8%), its non-solubility in any solvent except water and, as a tetraphenylphosphonium salt, methanol. Furthermore, the aqueous crystallisation step of the synthesis makes getting the necessary anhydrous PPh_4[**2**] compound extremely difficult. Although the $[Fe^{II}(Tpm)(CN)_3]^-$ ([**2**]$^-$) complex offers a route to functionalisation on the carbon apical atom (in contrast with the borate derivative), its poor yield and poor solubility make it a poor candidate for building block functionalisation.

As a side-effect of the rapid proton/deuterium exchange in D_2O, the apical carbon, whose intensity is usually very weak in ^{13}C NMR, does not benefit anymore from the NOE signal enhancement provided by its attached proton during the usual *zgpg30* pulse sequence. As a result, it cannot be detected by a ^{1}H, ^{13}C gHMQC or a gHMBC experiment and therefore remains invisible. This chemical exchange phenomenon also takes place for PPh_4[**2**] in deuterated methanol but at a much slower rate: the apical carbon was detected at $\delta = 74.9$ ppm. Finally, the cyanide quaternary carbons come at $\delta = 175.5$ ppm.

X[Fe^{II}(Tpm*)$(CN)_3$] (X[3], X = $[PPh_4]^+$, Na^+)

Na[**3**] is a yellow brownish solid obtained in good yield (*ca* 78%) in a one-pot synthesis in methanol, followed, for crystallisation purposes, by a cation metathesis to produce

PPh_4[**3**] from water. Depending on unidentified parameters, PPh_4[**3**] crystallises either as red tetrahedra or orange rods. In methanol-*d*$_4$, the coordination of the Tpm* ligand has no impact on the chemical shift of the 4-pz-C*H* proton (δ = 6.03 ppm). However, the protons of the methyl groups at the 3- and 5-positions of the pyrazolyl heterocycles are ~0.6 ppm shifted toward lower frequency compared to free ligand and come at δ = 2.76 and 2.57 ppm respectively. The apical proton appears at slightly lower frequency, at 7.83 ppm (8.15 ppm for Tpm* in CD_3OD). All carbon atoms were detected in the ^{13}C NMR spectrum. Of special interest are the cyanide quaternary carbon atoms at δ = 171.9 ppm.

X[Fe^{II}(Tpe)$(CN)_3$] (X[4], X = $[PPh_4]^+$, Na^+)

Na[**4**] is pale yellow, slightly light-sensitive solid obtained after a two-step synthesis in high yield (95%). Even if the first step, the formation of the amaranth red literature-known sandwich compound [$Fe^{II}$$(Tpe)_2$]$(OTf)_2$,[10] is quite air-sensitive and should be carried out in air-free conditions, the replacement of one of the Tpe ligands by three *C*-binding cyanides can be performed under lab atmosphere. $[\mathbf{4}]^-$ is relatively unstable in methanol at room temperature so the cyanuration reaction was performed in refluxing isopropanol under light exclusion conditions to avoid side reactions at high temperatures.

A cation metathesis with one equivalent of PPh_4Cl in water precipitates $[\mathbf{4}]^-$ as tetraphenylphosphonium salt. It is worth noting that PPh_4[**4**] is not soluble in pure acetonitrile, but soluble in acetonitrile/isopropanol or acetonitrile/water, even for very small proportions of water/alcohol. It is therefore very important to dry the solid completely between the two washings in order to get PPh_4[**4**] in good yields. The ^1H NMR spectrum of PPh_4[**4**] exhibits the expected set of signals for a C_{3v}-symmetric complex, indicating that the Tpe is κ^3-*N* coordinated to the iron ion. It is worth noting that the signal appearing at 8.33 ppm in the ^1H NMR spectrum and corresponding to the proton at the heterocyclic 3-position is unusually broad with a width at half height of 22 Hz. While the 4-pz-C*H* signal is not affected by the coordination (6.42 ppm for

PPh_4[**4**] in methanol, 6.41 ppm for Tpe in the same solvent), the 3- and 5-pz-C*H* signals are both significantly shifted toward higher frequency (8.33 ppm and 8.24 ppm respectively, to be compared with 7.34 ppm and 7.68 ppm for Tpe). The coordination of the Tpe has a significant influence on the CH_2OH sharp singlet: it is shifted of +0.50 ppm compared to the free ligand (5.55 ppm *vs* 5.05 ppm for Tpe). The 4-pz-*C*H carbon in the ^{13}C NMR spectrum gives a signal at relatively lower frequency (108.5 ppm) while the 5-pz-*C*H position is more deshielded (149.2 ppm in case of PPh_4[**4**]). The 3-pz-*C*H carbon is not detected. This is probably due to a weaker than usual coordinative behaviour. However, the detection of the two ^{15}N signals in ^{1}H, ^{15}N gHMBC at 212.9 ppm (trivalent 1-pz-*N*) and 253.5 ppm (imine-like 2-pz-*N*) indicates that this intramolecular motion is limited and does not consist of chemical exchange with the hydroxyl moiety: in case of chemical exchange, the detection of the 2-pz-*N* nucleus would be compromised at room temperature. The 1-pz-*N* signal position is merely influenced by coordination, and can be found at a typical chemical shift; 2-pz-*N* is shifted of about 50 ppm toward lower frequency compared to free Tpe, but lies at a rather high, but not unseen, chemical shift for a coordinated pyrazolyl heterocycle.[10,121] It may be an indication for a higher than usual metal to ligand backdonation for the cyanides (*trans* effect). Even if no signal could be detected in the ^{15}N NMR spectrum direct measurement for the natural abundance cyanide nitrogens, a signal corresponding to the cyanide moieties could be detected in the $^{13}C\{^{1}H\}$ spectrum at 172.1 ppm, that is, at about the same chemical shift as it is observed for PPh_4[**2**] and PPh_4[**3**].

$(PPh_4)_2[Fe^{II}$(Tpms)$(CN)_3]$ ($(PPh_4)_2$[5])

The tris(pyrazolyl)methanesulfonate (Tpms) ligand can exhibit different coordination modes: along with the "classical" κ^3-(*N*, *N*, *N*) coordination mode, κ^3-(*N*, *N*, *O*) and κ^2-(*N*, *N*) coordination configurations have been reported and/or suggested.[15,16,20-25,122-131] Tpms is, like Tp, a monoanionic ligand. It also features an additional, albeit weak compared to the three nitrogens, anionic SO_3^- donor moiety, which can possibly undergo further reactivity. A protocol for the synthesis of tricyanido iron(II) complex $(PPh_4)_2$[**5**] was reported by Gu *et al.* in 2005, but, surprisingly, neither crystal structure, cyclic voltammetry nor NMR data were published for the tricyanido iron(II) complex itself. Slow evaporation of acetonitrile, water and acetonitrile/water was unsuccessful due to the relatively high solubility of the complex in those two polar solvents, which is attributed to the two tetraphenylphosphonium countercations (acetonitrile), the sulfonate group and the dianionic charge (water). Suitable crystals of $(PPh_4)_2$[**5**] for X-ray diffraction analysis were obtained in a few days by acetonitrile/diethyl ether layering, but slow decomposition of the compound was also observed.

In deuterated acetonitrile, and apart from the $[PPh_4]^+$ multiplets, freshly dissolved $(PPh_4)_2$[**5**] exhibits one set of pyrazolyl signals: while the 5-pz-C*H* provides a sharp doublet at $\delta = 8.71$ ppm ($J_{HH} = 2.5$ Hz), the 3-pz-C*H* signal is broad ($\Delta\nu_{1/2} = 40$ Hz) and shifted towards lower frequency at $\delta = 8.00$ ppm. As usual, the 4-pz-C*H* signal is found at clearly lower frequency ($\delta = 6.25$ ppm), this time in the form of an undefined multiplet. The broadness of the 3-pz-C*H* signal is indicative of a pyrazolyl molecular motion in acetonitrile solution at room temperature; indeed, even in darkness, new peaks appear in the spectrum after a few hours, while a brownish mirror is found on the NMR tube walls.

$PPh_4[Fe^{III}(Tt)(CN)_3]$ (PPh_4[6])

The synthesis of PPh_4[**6**] follows the same synthetic pathway as for $PPh_4[Fe^{III}(Tp)(CN)_3]$ (PPh_4[**1**]), but two crucial changes are necessary to obtain pure PPh_4[**6**] in moderate yields (*ca* 50%). The use of methanol as a solvent leads to the decomposition of the reagent and the formation of coordination polymers with no trace of the iron(II) product. However, the same reaction in isopropanol works fine and leads to the desired iron(II) compound. It was oxidised in water, in presence of PPh_4Cl with hydrogen peroxide to form a green precipitate (like PPh_4[**1**]) but was recrystallised in pure acetonitrile to produce green crystals. Recrystallisation in different acetonitrile/water mixtures as was performed for PPh_4[**1**] led to bluish green oils.

$PPh_4[Fe^{III}(Tp^*)(CN)_3]$ (PPh_4[7])

The synthesis of [**7**]$^-$ was already reported as ammonium salt by Li *et al.*[115] It consists of three synthetic steps, while the product is obtained after only one step following the new synthesis reported in this work. Even though the reported yield (58%) is higher than that reported in this work (32%), the instability of the first intermediate product (that needs to be immediately used) and the potential hazards related to the partial solvent removal from a hydrogen peroxide contaminated acetonitrile solution and its layering by diethyl ether are major drawbacks. The degased $\{Fe^{II}(Tp^*)\}$ solution is to be added to the cyanide solution shortly after its formation as $[Fe^{II}(Tp^*)_2]$ tends to precipitate over time, and the resulting suspension needs to be kept oxygen free by bubbling inert gas into the solution. As shown afterwards in this chapter, the redox potential of PPh_4[**7**] allows a clean oxidation by the oxygen of air in acetonitrile making an oxidation step not necessary.

$[Fe^{III}(Tpm^*)(CN)_3]$ (8)

While Tp, Tt and Ttp iron(II) complexes can be oxidised by a 30% wt solution of hydrogen peroxide in water at 50°C, **8** is best prepared using milder oxidants. For small quantities (0.1 mmol), best yields were obtained with PPh_4[**3**] as starting material. The oxidation step takes place in dry acetonitrile under inert atmosphere with $[Fc][PF_6]$ as the oxidation reagent. Due to its low solubility in dry acetonitrile, **8** is collected by filtration, washed with acetonitrile and recrystallised from an acetonitrile/water 4:1 mixture to afford orange rods (92%). For bigger scale reactions, it is advisable to add dropwise a concentrated ethanolic solution of 0.6 equivalents of iodine to a stirred aqueous Na[**3**] solution. **8** is collected by filtration in moderate yields (*ca* 45%), which is somewhat compensated by the absence of the metathesis step and the short duration of the overall synthesis. This low yield can be explained to some extend by interactions between the iodine species and **8** in solution, which drastically increases the solubility of **8** in the ethanol/water mixture. This is supported by the isolation of X-ray diffraction suitable dark red crystals with a metallic sheen of co-crystallised **8** and HI_5 from the red mother liquor by slow evaporation after a few weeks. **8** gives a paramagnetic ^{1}H NMR spectrum, whose apical proton signal intensity also decreases over time in methanol-d_4. However, the much slower exchange rate compared to [**2**]$^-$ can be explained either by a lesser acidity of the corresponding proton, or by the spatial hindrance induced by the three methyl moieties at the 5-position of the pyrazolyl moieties.

3.1.2 Structural analyses

$PPh_4[Fe^{II}(Tpm)(CN)_3] \cdot 2\ H_2O$ (PPh_4[2])

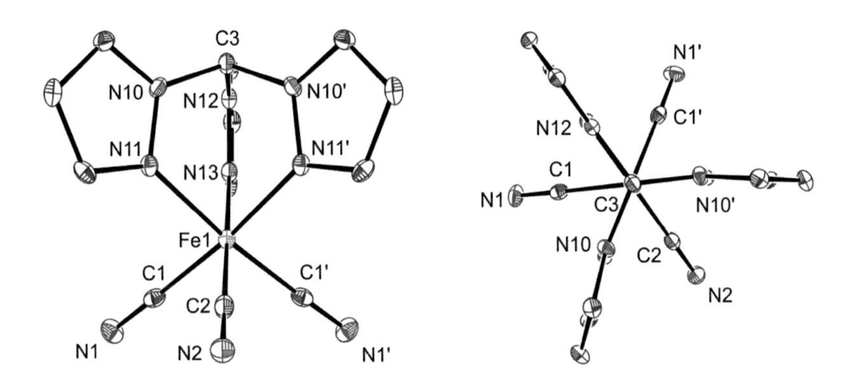

Figure 3.1: Perspective view of the molecular structure of the anion in PPh_4[**2**]. Side (left) and top (right) view (along the Fe···C3 axis). Atoms are displayed as 30% probability ellipsoids. Hydrogen atoms, solvent molecules and tetraphenylphosphonium countercation are omitted for clarity. Equivalent atoms (noted with apostrophe) are generated by the following symmetry operations: +x, +y, ½-z. Selected bond lengths (Å) and angles (°) for PPh_4[**2**] · 2 H_2O: Fe1–C1 1.910(6), Fe1–C2 1.908(9), Fe–N11 2.005(5), Fe–N13 2.000(7), C1-Fe1-C2 91.4(3), C1-Fe1-C1' 92.0(4), N11-Fe1-N13 86.5(2), N11-Fe1-N11' 83.2(3), C1-Fe1-N11 92.4(2), C1-Fe1-N13 90.5(2), C2-Fe1-N11 91.4(2), Fe1-C1-N1 177.9(6), Fe1-C2-N2 179.1(8), Fe1-N11-N10-C3 3.5, Fe1-N13-N12-C3 0.0.

PPh_4[**2**] crystallises in the orthorhombic space group *Pbcm*. Its crystal structure consists of a negatively charged tricyanido tris(pyrazolyl)methane iron(II) complex, its tetraphenylphosphonium countercation and two water lattice molecules. A perspective view of the iron(II) anionic unit [**2**]$^-$ is depicted in Figure 3.1 and selected bond lengths and angles are listed in the caption. [**2**]$^-$ is placed at a special crystallographic position, half of the molecule being the mirror image of the other half through a σ_v symmetry plane containing the following atoms : Fe1, C2, N2, C3, and the five N12-N13 pyrazole ring atoms. The iron(II) ion is in a slightly distorted octahedral C_3N_3 environment formed by three imine moieties from the pyrazolyl rings of the *fac*-coordinating Tpm ligand and the carbon atoms of three cyanides. Its octahedral distortion (defined as the sum of the deviations to 90° of the twelve angles of an octahedron) amounts to 27.2°. Viewed along the Fe···C axis, each cyanide ligand points between two pyrazole rings in a C_3-symmetric fashion. The Fe–C bonds lengths are identical (1.909 Å). These values are slightly over 1.900 Å, which usually indicates an iron(III) species in the $[Fe(L)(CN)_3]^-$ borate family. However here, the overall charge states unambiguously the +II nature of the iron oxidation state. Fe–N bonds are also identical (Fe–N = 2.003 Å) and longer than the Fe–C ones, as usual. They are quite long compared to other previously reported low-spin iron(II) Tpm species that do not contain cyanide (average 1.97 Å)[69] but not unusual for $[Fe(L)(CN)_3]^{n-}$ units.[98,114,115] It is also worth mentioning that the coordination of the cyanides to the iron(II) ion can be considered as linear (Fe-C-N > 177.9(6)°). The intramolecular distance between the bridging atom C3 and the iron atom amounts to 3.06 Å.

The shortest intermolecular Fe···Fe distance is 7.34 Å because of hydrogen bonding from one layer to the other mediated by water molecules. The lattice water molecules are involved in hydrogen bonds with the N_{CN} of the cyanide ligands leading to two-dimensional hydrogen bond networks.

$PPh_4[Fe^{II}(Tpe)(CN)_3] \cdot 2\ H_2O$ (PPh_4[4])

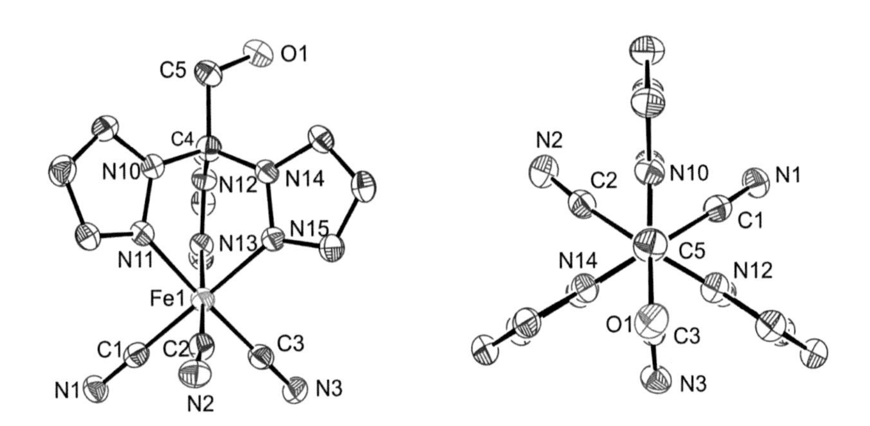

Figure 3.2: Perspective view of the anionic unit in PPh_4[**4**]. Side (left) and top (right) view, (along the Fe···C4 axis). Atoms are displayed as 30% probability ellipsoids. Hydrogen atoms, solvent molecules and tetraphenylphosphonium countercation are omitted for clarity. Selected bond lengths (Å) and angles (°) for PPh_4[**4**] · 2 H_2O: Fe1–C1 1.900(6), Fe1–C2 1.899(6), Fe1–C3 1.890(7) Fe1–N11 1.981(5), Fe1–N13 1.994(4), Fe1–N15 1.990(4), C4–C5 1.527(8), C1-Fe1-C2 88.5(2), C1-Fe1-C3 88.6(2), C2-Fe1-C3 94.1(3), C1-Fe1-N11 93.3(2), C1-Fe1-N13 94.3(2), C2-Fe1-N11 88.5(2), C2-Fe1-N15 92.4(2), C3-Fe1-N13 91.2(2), C3-Fe1-N15 91.9(2), N11-Fe1-N13 86.17(19), N11-Fe1-N15 86.15(18), N13-Fe1-N15 84.79(18), Fe1-C1-N1 177.5(5), Fe1-C2-N2 173.3(5), Fe1-C3-N3 176.6(5), C4-C5-O1 112.0(5), Fe1-N11-N10-C4 -0.21, Fe1-N13-N12-C4 +0.76, Fe1-N15-N14-C4 -2.18.

PPh_4[**4**] crystallises in the triclinic space group $P\bar{1}$ ($Z = 2$). It consists of a monoanionic tricyanido iron(II) complex, a tetraphenylphosphonium as countercation and two water lattice molecules. A perspective view of the iron(II) anionic unit of PPh_4[**4**] is depicted in Figure 3.2, with selected bond lengths and angles listed in the caption. In the metal complex unit, a tris(pyrazolyl)ethanol ligand (Tpe) κ^3-*N* *fac*-coordinates the iron ion, leading to the usual C_3N_3 environment. This environment exhibits a quite important octahedral distortion (34.5°) for such a complex. PPh_4[**4**] exhibits almost identical Fe–C (average: 1.896 Å) and Fe–N_{pz} (average: 1.988 Å) bond lengths. The Fe–C bond lengths values are consistent with a low-spin iron(II) oxidation state, which is confirmed by the overall charge of the metal complex. While two cyanides bind the metal ion almost linearly, the third is slightly shifted from linearity (Fe1-C2-N2 = 173.3(5)°). The three pyrazole "arms" of the Tpe ligand exhibit only minimal torsion. The hydroxyl group points exactly between two pyrazole rings and, when viewed along the iron-C4-C5 axis (see Figure 3.5 right), eclipses one of the cyanide ligands.

Within the unit cell and along the *b* axis, the two neighbouring iron complex units are positioned head-to-tail, which gives rise to parallel displaced π-π stacking between the pyrazole rings of pairs of neighbours [$C_{pz}\cdots C_{pz}$ = 3.54 Å], and iron-iron intermolecular distances of 7.98 Å. This interaction does not take place between iron complexes of neighbouring unit cells. The anionic metal complex and its countercation pile up in a segregated fashion along the *a* axis. The cohesion within a pile is promoted by intermolecular head-to-tail hydrogen bridges between the hydroxyl moiety of one metal complex, and one of the cyanides of its neighbour (OH···N distance: 1.974 Å).

The lattice water molecules are involved in hydrogen bonds with one of the nitrogen atoms of the cyanide ligands leading to a two-dimensional hydrogen bond networks.

$(PPh_4)_2[Fe^{II}(Tpms)(CN)_3] \cdot 2\ MeCN \cdot H_2O$ ($(PPh_4)_2$[5])

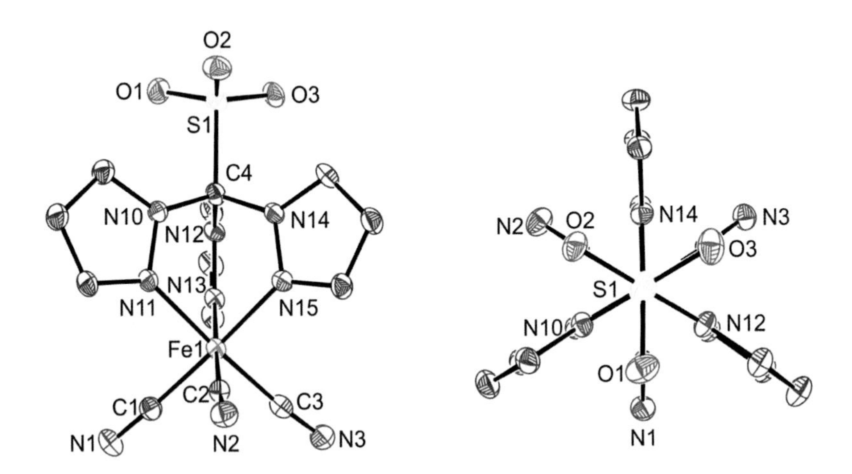

Figure 3.3: Perspective view of molecular structure of the anionic unit in $(PPh_4)_2$[5]. Side (left) and top (right) view, (along the Fe···C4 axis). Atoms are displayed as 30% probability ellipsoids. Hydrogen atoms, solvent molecules and tetraphenylphosphonium countercations are omitted for clarity. Selected bond lengths (Å) and angles (°) for $(PPh_4)_2$[5] · 2 MeCN · H_2O: Fe1–C1 1.907(3), Fe1–C2 1.901(3), Fe1–C3 1.901(4), Fe1–N11 1.981(3), Fe1–N13 1.975(3), Fe1–N15 1.980(2), C4–S1 1.884(3), S1–O1 1.436(2), S1-O2 1.436(2), S1–O3 1.441(2), C1-Fe1-C2 92.32(13), C1-Fe1-C3 91.26(14), C2-Fe1-C3 90.83(14), N11-Fe1-N13 84.41(11), N11-Fe1-N15 86.98(10), N13-Fe1-N15 85.30(10), C1-Fe1-N11 91.22(12), C1-Fe1-N13 92.21(12), C2-Fe1-N11 92.66(12), C2-Fe1-N15 90.09(12), C3-Fe1-N13 91.91(13), C3-Fe1-N15 90.39(12), Fe1-C1-N1 177.6(3), Fe1-C2-N2 175.4(3), Fe1-C3-N3 175.8(3), O1-S1-O2 115.85(15), O1-S1-O3 114.24(15), O2-S1-O3 115.24(14), C4-S1-O1 103.16(14), C4-S1-O2 102.98(14), C4-S1-O3 102.78(13), Fe1-N11-N10-C4 4.75, Fe1-N13-N12-C4 0.05, Fe1-N15-N14-C4 -1.45.

$(PPh_4)_2$[5] · 2 MeCN · H_2O crystallises in the triclinic space group $P\bar{1}$ ($Z = 2$). Its crystal structure consists of a dianionic tricyanido tris(pyrazolyl)methanesulfonate iron(II) complex, two tetraphenylphosphonium countercations, two acetonitrile and one (between two positions disordered) lattice water molecules. A perspective view of the iron(II) dianionic unit $[\mathbf{5}]^{2-}$ is depicted in Figure 3.3 and selected bond lengths and angles are listed in the caption. The Tpms ligand exhibits a κ^3–N coordination mode, leading to a classical C_3N_3 environment with an octahedral distortion of 26.2°. The Fe–C bond lengths do not confirm or infirm the attributed formal +II oxidation state (average: 1.903 Å).

However the presence of the two countercations confirms the charge of [**5**]$^{2-}$ and the iron(II) ion oxidation state. The Fe–N bonds are also about the same length (mean Fe–N distance: 1.979 Å); this is a slightly smaller than found values in PPh_4[**1**], PPh_4[**4**] and the literature-known chain {[Fe^{II}(Tpms)$(CN)_3$][$Mn^{II}$$(H_2O)_2$$(DMF)_2$]} · DMF,[116] but remains in the normal range for low-spin Tpm derivatives.[69] The three cyanide ligands *C*-bind the iron atom almost linearly. All three oxygen atoms exhibit similar bond lengths to the sulphur atom (mean value: 1.438 Å), similar C4-S1-O ≈ 103° angles and similar O-S1-O ≈ 115° angles indicating that the negative charge is equally delocalised over the three atoms. The iron, apical C4 and sulphur atoms are aligned (Fe···C4–S1 = 179.1°), so that the whole complex is (almost) C_{3v} symmetric around this axis. Each oxygen points in an alternated manner between two pyrazolyl rings (see Figure 3.3), and therefore eclipses a cyanide ligand. It is noteworthy that the pyrazolyl ring torsion in the {Fe(Tpms)} moiety remains small with a maximal twist of 4.75°. Finally, tetraphenylphosphonium ions and iron complex units are piled up in a segregated fashion along the *b* axis. In each pile, the iron(II) units are well spatially isolated from each other, with the smallest iron-iron intermolecular distance of 10.44 Å. The shortest Fe···Fe intermolecular distances is 8.61 Å. The lattice water molecule interacts with the nitrogen atom of the cyanide ligands but is not involved in further 3D network.

PPh_4[Fe^{II}(Tpm*)$(CN)_3$] (PPh_4[3])

Two very different-looking crystal phases of PPh_4[**3**] were grown from aqueous solutions at room temperature. PPh_4[**3**] · 12 H_2O (a) crystallises in a few days as orange rod-like crystals, whereas PPh_4[**3**] · 7 H_2O (b) phase appears more slowly as red tetrahedral crystals. A third phase of PPh_4[**3**], PPh_4[**3**] · CH_3CN (c), was previously obtained by layering an acetonitrile solution with diethyl ether.[120] The two new crystal phases reported in this work crystallise in the triclinic space group $P\bar{1}$ ($Z = 2$) while the third phase (c) crystallises in the monoclinic space group $P2_1/c$. They consist of an anionic tricyanido tris(3,5-dimethylpyrazolyl)methane iron(II) complex, a tetraphenyl phosphonium cation as counterion and a variable amount of lattice water molecule (12 for (a) and only 7 for (b)). The average iron-carbon bond length in the heptahydrate (b) amounts to 1.900 Å, which is slightly longer than in the dodecahydrate (a) (mean Fe–C bond length: 1.885 Å). However, it remains in the usual iron-carbon bond length range reported for comparable iron(II) tricyanido complexes. It is noteworthy that the literature known phase of $PPh_4[Fe^{II}(Tpm^*)(CN)_3]$ (c)[120] reports much longer iron(II)-carbon bond lengths at 200 K, with a mean value of 1.913 Å. In this type of complexes, iron-carbon bond lengths over 1.900 Å are usually associated with an oxidation state of the iron ion of +III, while bond lengths smaller than 1.900 Å are usually found in iron(II) complexes. In the three crystallographic phases, the Fe–N_{pz} bond lengths are all above 2.000 Å. In the dodecahydrate (a), the heptahydrate (b) and the acetonitrile solvate (c) phases, two Fe–N_{pz} bonds are about the same length (mean values: 2.035 Å, 2.034 Å and 2.038 Å, respectively). The dodecahydrate (a) and the acetonitrile solvate (c) phases exhibit a smaller third Fe–N_{pz} bond than the two first (2.016(2) Å and 2.017(5) Å, respectively). The third bond of the heptahydrate is clearly longer than the two others and amounts to 2.055(7) Å. The octahedral distortion of the hydrates amount to 27.1° and 25.0° (dodeca- and heptahydrate, respectively), which are comparable to the values found for PPh_4[**2**] and $(PPh_4)_2$[**5**]. The coordination sphere in the acetonitrile solvate phase (c) is slightly less distorted (distortion of 23.3°). The most striking difference between the three crystallographic phases lies in the Tpm* binding configuration: while the acetonitrile solvate (c) exhibits pyrazolyl torsion angles of maximum 3.9° (in average 1.9°), the pyrazolyl torsion angle mean value reaches 6.9° for the dodecahydrate (a) and 15.5° for the heptahydrate (b) phase, underlying the strong influences that weak intermolecular interactions can exert on the coordination sphere distortion (and therefore on the electronic/magnetic properties). The iron(II) complexes in each phase are well separated, with smallest iron-iron distances of 8.74 Å, 8.83 Å and 9.82 Å for (b) , (a) and (c), respectively. The latter forms segregate piles of anions and cations along its *b* axis. In the dodecahydrate (a), the molecules also form piles along the *b* axis, but in an alternate manner. Finally, the heptahydrate phase (b) is constituted from alternating {*0*, *a*, *b*} plans of anion and cations. In the two latter cases, the N_{CN} atoms are involved in hydrogen bonds with lattice H_2O molecules leading to extended hydrogen-bonded networks.

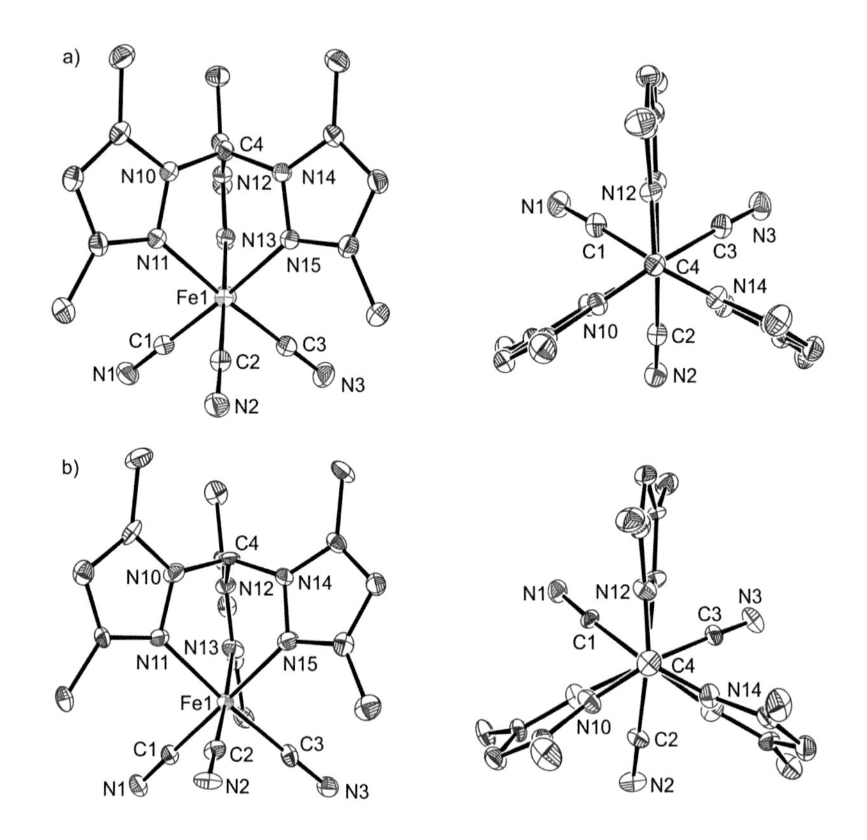

Figure 3.4: Perspective view of the molecular structure of the anionic unit in two different crystal phases of PPh_4[**3**]: a) PPh_4[**3**] · 12 H_2O and b) PPh_4[**3**] · 7 H_2O. Side (left) and top (right) view (along the C4···Fe1 axis). Atoms are displayed as 30% probability ellipsoids. Hydrogen atoms, solvent molecules and tetraphenylphosphonium countercations are omitted for clarity.

a) Selected bond lengths (Å) and angles (°) for PPh_4[**3**] · 12 H_2O: Fe1–C1 1.887(3), Fe1–C2 1.876(3), Fe1–C3 1.892(3), Fe1–N11 2.032(2), Fe1–N13 2.016(2), Fe1–N15 2.037(2), C1-Fe1-C2 87.80(13), C1-Fe1-C3 91.70(12), C2-Fe1-C3 88.97(13), N11-Fe1-N13 86.56(10), N11-Fe1-N15 86.32(10), N13-Fe1-N15 86.77(10), C1-Fe1-N11 91.32(11), C1-Fe1-N13 91.31(11), C2-Fe1-N11 92.74(12), C2-Fe1-N15 94.09(11), C3-Fe1-N13 91.77(12), C3-Fe1-N15 90.62(11), Fe1-C1-N1 178.1(3), Fe1-C2-N2 177.6(3), Fe1-C3-N3 178.2(3), Fe1-N11-N10-C4 6.72, Fe1-N13-N12-C4 7.24, Fe1-N15-N14-C4 6.69.

b) Selected bond lengths (Å) and angles (°) for PPh_4[**3**] · 7 H_2O: Fe1–C1 1.897(9), Fe1–C2 1.899(8), Fe1–C3 1.905(9), Fe1–N11 2.031(7), Fe1–N13 2.055(7), Fe1–N15 2.037(7), C1-Fe1-C2 89.6(4), C1-Fe1-C3 90.5(4), C2-Fe1-C3 90.4(4), N11-Fe1-N13 86.0(3), N11-Fe1-N15 86.6(3), N13-Fe1-N15 86.1(3), C1-Fe1-N11 93.7(3), C1-Fe1-N13 91.5(3), C2-Fe1-N11 90.6(3), C2-Fe1-N15 92.8(4), C3-Fe1-N13 92.9(3), C3-Fe1-N15 89.13(3), Fe1-C1-N1 177.2(8), Fe1-C2-N2 179.8(8), Fe1-C3-N3 178.6(9), Fe1-N11-N10-C4 15.97, Fe1-N13-N12-C4 15.70, Fe1-N15-N14-C4 14.81.

$PPh_4[Fe^{III}(Tp^*)(CN)_3] \cdot MeCN$ (PPh_4[7])

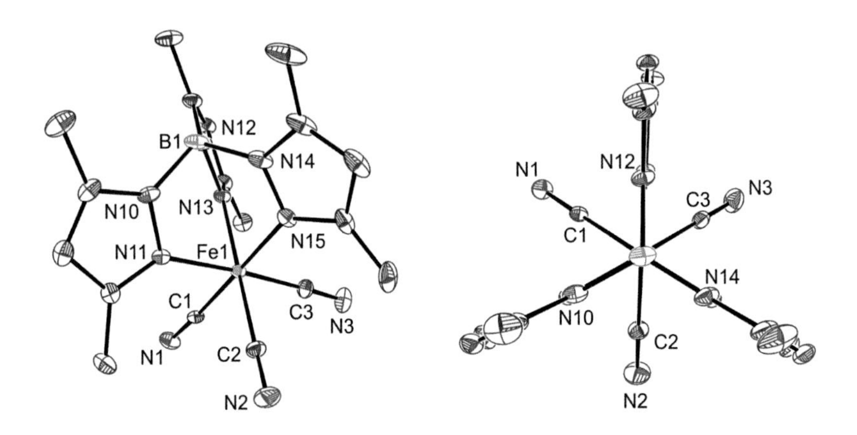

Figure 3.5: Perspective view of the molecular structure of the anion in PPh_4[**7**]. Side (left) and top (right) view (along the Fe···B axis). Atoms are displayed as 30% probability ellipsoids. Hydrogen atoms, solvent molecules and tetraphenylphosphonium countercation are omitted for clarity. Selected bond lengths (Å) and angles (°) for PPh_4[**7**] · MeCN: Fe1–C1 1.922(3), Fe1–C2 1.930(3), Fe1–C3 1.917(3), Fe1–N11 2.025 (2), Fe1–N13 1.990(2), Fe1–N15 2.014(2), C1-Fe1-C2 89.66(11), C1-Fe1-C3 87.19(12), C2-Fe1-C3 86.56(11), N11-Fe1-N13 89.96(9), N11-Fe1-N15 89.33(9), N13-Fe1-N15 88.36(9), C1-Fe1-N11 90.84(10), C1-Fe1-N13 90.38(9), C2-Fe1-N11 90.58(11), C2-Fe1-N15 91.60(11), C3-Fe1-N13 92.27(11), C3-Fe1-N15 93.32(11), Fe1-C1-N1 176.8(3), Fe1-C2-N2 178.8(3), Fe1-C3-N3 178.4(3), Fe-N11-N10-B1 0.20, Fe-N13-N12-B1 4.28, Fe1-N15-N14-B1 -0.48.

PPh_4[7] crystallises in the monoclinic space group $P2_1/c$, ($Z = 4$). Its crystal structure consists of a negatively charged tricyanido tris(3,5-dimethylpyrazolyl)borate iron(III) complex, its tetraphenyl phosphonium countercation and an acetonitrile molecule. It is isostructural to the literature-known $PPh_4[Fe^{II}(Tpm^*)(CN)_3] \cdot MeCN$,[120] where the change in oxidation state of the iron ion compensates the replacement of the neutral Tpm* by the negatively charged Tp*. A perspective view of the anion is depicted in Figure 3.5 and selected bond lengths and angles are listed in the caption. The Fe–C bond lengths exhibit a mean distance of 1.923 Å, which is typical for cyanido iron(III) complexes. The Fe–N bonds are about the same length with a mean value of 2.010 Å. This difference contributes to the elongation along the pseudo C_3 symmetry axis (Fe···B). The octahedral distortion around the iron(III) ion amounts to 17.9°, which is low compared to the values obtained for the other tricyanido scorpionate iron(III) and (II) complexes. Two of the pyrazolyl rings of the Tp* are aligned with the Fe···B axis with a torsion angle of about 0°, while the third is significantly more bent (Fe-N13-N12-B1 = 4.28°). The three cyanide ligands bind the iron ion in an almost linear way (Fe-C-N = 176.8(3)–178.8(3)°).

The shortest Fe···Fe intermolecular distance is 9.85 Å, indicating that the iron(III) complexes are spatially well-isolated from each other.

$[Fe^{III}(Tpm^*)(CN)_3]$ (8)

Depending on the crystallisation method, two different crystal structures of **8** could be obtained during this PhD thesis: crystals of **8** were obtained by slow evaporation of water (Figure 3.6.a) and **8** · 2 MeCN was obtained by slow evaporation of an acetonitrile/water 4:1 mixture (Figure 3.6.b). A third one, **8** · DMF, obtained by layering DMF with diethyl ether, is literature-known (not pictured).[120] Surprisingly, a fourth, with HI_5 co-crystallised type of **8** could be produced by slow evaporation of the ethanol/water 1:1 filtrate from the big-scale synthesis of **8** (Figure 3.6.c). In water and in DMF/diethyl ether, **8** crystallises in the orthorhombic space group *Pbca* (Z = 1). **8** · 2 MeCN and **8** · 0.5 HI_5 · 0.5 H_2O both crystallise in monoclinic $P2_1/n$ (one formula unit per asymmetric unit, Z = 4) and *C2/c* (one formula unit per asymmetric unit, Z = 8) space group, respectively. While **8** · 2 MeCN and **8** · DMF crystallise with various amounts of respective lattice solvents in the unit cell (two acetonitrile and one DMF per iron complex, respectively), no water molecule is detected by X-ray diffraction for the crystals grown from an aqueous solution. In **8** · 0.5 HI_5 · 0.5 H_2O, two complexes of **8** crystallise with an HI_5 molecule and one, between two sites disordered water molecule.

8 is a neutral, iron(III) complex, featuring a κ^3-*N* imine-like *fac*-coordinating Tpm* ligand and three *C*-bound terminal cyanides. This leads, as for every tricyanido complex presented in this work, to an octahedral C_3N_3 iron(III) environment. It is only slightly distorted in the solvent-free phase (octahedral distortion: 16.2°), while the solvent-containing phases exhibit octahedral distortions ranging from 25.1° to 29.8°. The latter values are comparable with the octahedral distortion found for the three PPh_4[**3**] phases, while the first value compares well with the boron-capped iron(III) analogue PPh_4[**7**].

In all crystal structures, **8** exhibits Fe1–C bond lengths longer than 1.900 Å, which is consistent with the overall charge and the iron(III) oxidation state. The $Fe–N_{pz}$ distances range from 1.964 Å to 2.009(3) Å. The cyanides *C*-bind the iron(III) ion almost linearly in the following three phases: the acetonitrile solvate, the DMF solvate and the HI_5 co-crystallised phases. In the solvent-free phase, the cyanide are slightly bent (174.9(7)°–176.5(6)°).

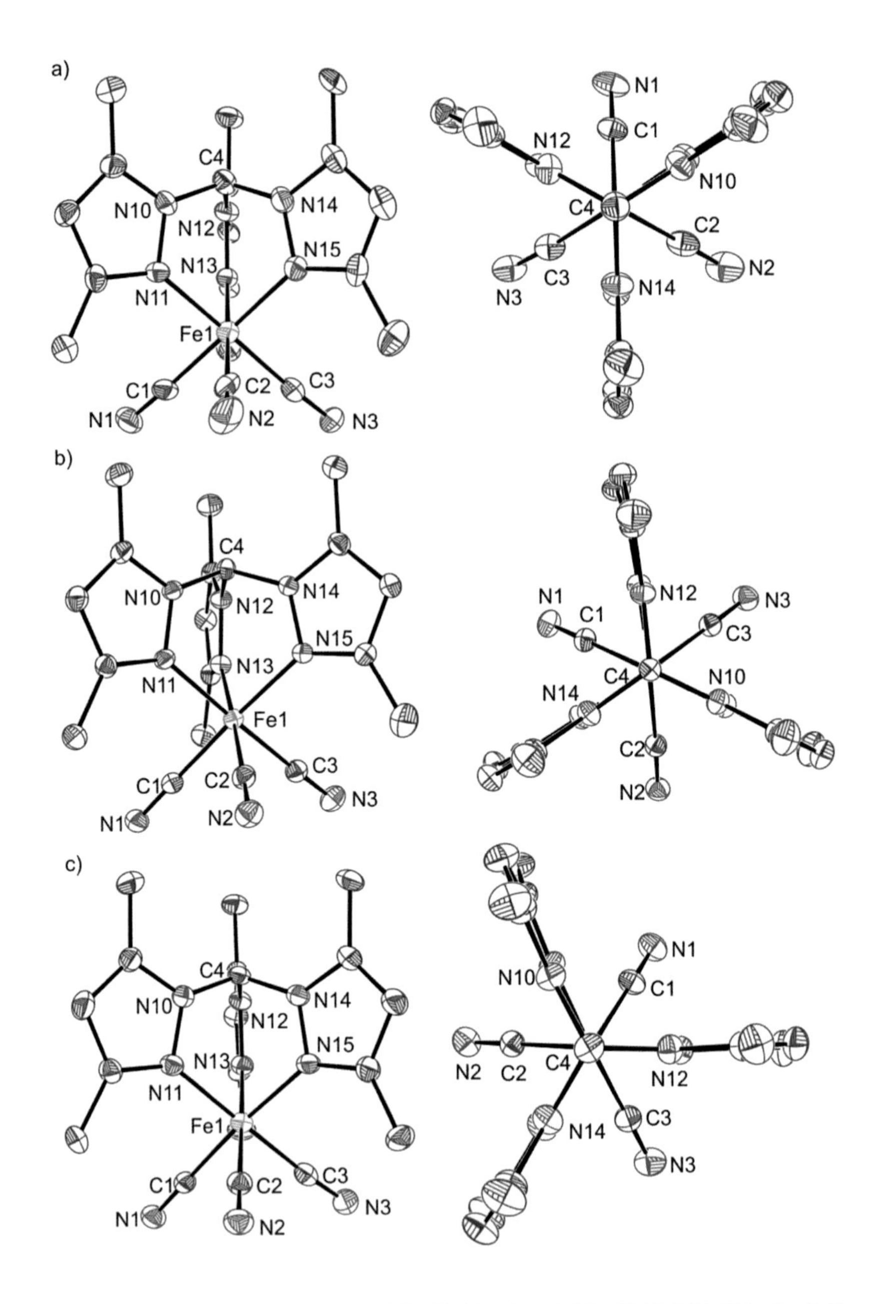

Figure 3.6: Perspective view of **8** (a), **8** · 2 MeCN (b) and **8** · 0.5 HI_5 · 0.5 H_2O (c). Side view (left) and view from above, along the Fe···C4 axis (right). Atoms are displayed as 30% probability

ellipsoids. Hydrogen atoms, eventual solvent lattice molecules and co-crystallised HI_5 are omitted for clarity.
a) Selected bond lengths (Å) and angles (°) for **8**: Fe1–C1 1.911(8), Fe1–C2 1.907(8), Fe1–C3 1.920(8), Fe1–N11 1.967(5), Fe1–N13 1.974(5), Fe1–N15 1.987(5), C1-Fe1-C2 91.1(3), C2-Fe1-C3 89.8(3), C1-Fe1-C3 88.4(3), C2-Fe1-N11 89.9(2), C1–Fe1–N11 90.7(2), C1-Fe1-N13 91.0(2), C3-Fe1-N13 93.1(3), N11-Fe1-N13 87.2(2), C2-fe1-N15 90.6(3), C3-Fe1-N15 91.6(3), N11-Fe1-N15 89.3(2), N13-Fe1-N15 87.3(2), Fe1-C1-N1 175.5(6), Fe1-C2-N2 174.9(7), Fe1-C3-N3 176.5(6), C1-N10-N11-Fe1 1.4(7), C1-N12-N13-Fe1 -2.0(6), C1-N14-N15-Fe1 -4.7(7).
b) Selected bond lengths (Å) and angles (°) for **8** · 2 MeCN: Fe1-C1 1.910(3), Fe1-C2 1.920(3), Fe1-C3 1.914(3), Fe1-N11 2.009(3), Fe1-N13 1.998(3), Fe1-N15 2.000(3), C3-Fe1-C2 89.43(14), C3-Fe1-C1 87.30(14), C2-Fe1-C1 85.97(14), C3-Fe1-N13 92.24(13), C1-Fe1-N13 93.23(12), C2-Fe1-N11 93.00(12), C1-Fe1-N11 93.12(13), N13-Fe1-N11 85.33(11), C3-Fe1-N15 91.35(13), C2-Fe1-N15 92.03(13), N13-Fe1-N15 88.80(11), N11-Fe1-N15 88.31(11), Fe1-C1-N1 179.1(3), Fe1-C2-N2 178.1(3), Fe1-C3-N3 177.5(3), C1-N10-N11-Fe1 3.0, C1-N12-N13-Fe1 -4.2, C1-N14-N15-Fe1 0.1.
c) Selected bond lengths (Å) and angles (°) for **8** · 0.5 HI_5 · 0.5 H_2O: Fe1–C1 1.924(7), Fe1–C2 1.904(7), Fe1–C3 1.901(7), Fe1–N11 1.979(5), Fe1–N13 2.001(5), Fe1–N15 1.993(5), C1-Fe1-C2 86.5(3), C1-Fe1-C3 85.7(3), C2-Fe1-C3 89.6(3), C1-Fe1-N11 94.0(2), C1-Fe1-N13 93.6(2), C2-Fe1-N11 92.6(2), C3-Fe1-N13 91.5(2), N11-Fe1-N13 86.3(2), C2-Fe1-N15 91.0(2), C3-Fe1-N15 91.0(2), N11-Fe1-N15 89.3(2), N13-Fe1-N15 88.9(2), Fe1-C1-N1 177.9(6), Fe1-C2-N2 179.2(6), Fe1-C3-N3 176.5(6), C1-N10-N11-Fe1 7.2, C1-N12-N13-Fe1 1.3, C1-N14-N15-Fe1 2.7.

The iron to bridgehead carbon distance in **8** is smaller compared to the other scorpionate-based tricyanido iron complexes. They range from 2.99 Å (HI_5 co-crystallised phase) to 3.03 Å (DMF solvate). In each case, the iron complexes do not exhibit π-π interactions with nearby complexes. By far, the smallest iron-iron intermolecular distance is reported for solvent-free **8**, and amounts to 7.35 Å, because of CH-π interactions between the cyanide moieties and the methyl groups of the Tpm* ligand.

3.1.3 Fourier Transform InfraRed spectroscopy

Fourier Transform InfraRed spectroscopy (FT-IR) is a particularly efficient and valuable tool in cyanide chemistry, which justifies its widespread use as a characterisation method in this work. The C≡N functional groups typically absorb in the triple bond region of IR spectra, leading to at least one sharp absorption band of variable intensity between 2000 cm^{-1} and 2300 cm^{-1}, which corresponds to the stretching of the triple C≡N bond.[132]

In the IR spectra of the compounds presented in this work, this triple-bond region is mostly unpopulated and allows facile identification of the presence (or absence) of cyanide moieties in the corresponding compounds.

The frequency of the absorptions also bears structural and electronic information: (i) whether the cyanide moiety is free or coordinated to one or two metal ions; (ii) its bonding mode (terminal or bridging); (iii) the oxidation state of the *C*-coordinated metal; (iv) if it is involved in weak interactions.[133]

Thus, it is literature-known that ferricyanides always absorb over 2100 cm^{-1}, that is at higher frequency than ferrocyanides (< 2100 cm^{-1}) and free cyanide (~ 2080 cm^{-1}), but far below organic nitriles (2260 – 2222 cm^{-1}).[132] Bridging coordination mode usually shifts absorption bands to higher frequencies, while weak interactions have the same but much smaller effect.

Solid-state FT-IR spectra were recorded for tricyanido iron(II) and iron(III) complexes bearing scorpionate ligands at room temperature. In order to make the comparison easier, selected absorption frequencies are reported in Table 3.1, in cm^{-1}. When the compound is available as tetraphenylphosphonium salt, it further absorbs at the following characteristic frequencies: 525, 721, 995, 1108, 1438 and 1483 cm^{-1}. The aromatic C–H of the phenyl ring also absorb above 3000 cm^{-1} but under 3100 cm^{-1}; however, these absorptions are very weak and are often not visible when the sample is freshly filtered.

Table 3.1: Selected FT-IR (ATR) frequencies of tricyanido iron(II) and iron(III) complexes (4 cm^{-1} resolution). All values are given in cm^{-1}.

No	Name	Ref.	ν_{CN}	ν_{BH}	ν_{CH}	ν_{CH2}	Ring stretch
–	$K_2[Fe^{II}(Tp)(CN)_3]$	[98,114]	2016, 2037, 2056	2472	3105, 3126, 3154	–	1514
–	$K_2[Fe^{II}(Tt)(CN)_3]$	[98], this work	2048, 2066	2530	3132, 3113	–	1499
–	$K_2[Fe^{II}(Ttp)(CN)_3]$	[98]	2018, 2035, 2057	–	3142 (br)	–	1511
–	$(Et_4N)_2[Fe^{II}(Tp^*)(CN)_3]$	[115]	2060, 2043	2507	–[a]	–[a]	1544
[2]$^-$	$PPh_4[Fe^{II}(Tpm)(CN)_3]$	this work	2045, 2054, 2064	–	3114, 3134, 3159	3001	1514
[3]$^-$	$PPh_4[Fe^{II}(Tpm^*)(CN)_3]$	this work, [120]	2042, 2048, 2064	–	3136	2975, 2924	1566
[4]$^-$	$PPh_4[Fe^{II}(Tpe)(CN)_3]$	this work	2047, 2054, 2068	–	3109, 3133, 3148	2996	1519
[5]$^{2-}$	$(PPh_4)_2[Fe^{II}(Tpms)(CN)_3]$	[116], this work	2051, 2061, 2073	–	3110, 3128, 3166	–	1520
[1]$^-$	$PPh_4[Fe^{III}(Tp)(CN)_3]$	[98,114]	2123	2502	3115, 3133, 3150	–	1501
–	$Et_4N[Fe^{III}(Tp^{Me})(CN)_3]$	[134]	2121	2481	3121, 3138	–[a]	1504
[7]$^-$	$PPh_4[Fe^{III}(Tp^*)(CN)_3]$	[115], this work	2119	2543	3133	2966, 2943	1543
[6]$^-$	$PPh_4[Fe^{III}(Tt)(CN)_3]$	[98], this work	2124	2548	3084, 3101	–	1497
[9]$^-$	$PPh_4[Fe^{III}(Ttp)(CN)_3)]$	[98]	2119	–	3108, 3129 (sh), 3139	–	1501 (coord) 1514 (free)
8	$[Fe^{III}(Tpm^*)(CN)_3]$	this work, [120]	2128	–	3143	2882, 2996	1557

[a] Not given in the literature.

XRD data indicate that iron(III)–$C_{cyanide}$ bond lengths are in average clearly longer than in their iron(II) analogues. Yet, if the iron-carbon bond lengths were function of the iron oxidation state, the opposite situation would be observed. This means that the iron(II) cyanide bonds have a greater π backbonding character than their iron(III) analogues. Enhanced electron density in the CN antibonding π* orbitals weakens the CN triple bonds, which reduces the stiffness of the associated harmonic oscillator describing the stretching vibration mode. This explains that in all presented iron(III) species, the cyanide stretches absorb at about 2120 cm^{-1}, while all iron(II) species exhibits absorption at lower frequency than 2100 cm^{-1}, thus complying with the empirical rule mentioned above. It is remarkable that, for iron(III) species, the solvation state of the samples have more influence on the cyanide stretch frequency than the nature of the scorpionate ligand. This is coherent with a limited cyanide π backbonding in these iron(III) complexes. The energy levels of the metal-centred molecular orbitals are greatly influenced by the nature of the coordinated scorpionate ligand. These energy levels in turn influence the efficiency of the π backbonding to the cyanide ligands sharing the same orbitals. The cyanide stretches in the iron(II) species are therefore much more affected by the nature of the tripodal ligand, as they range from 2016 (Tp) to 2073 cm^{-1} (Tpms).

Borohydride frequencies range from 2472 to 2548 cm^{-1}, without a clear trend regarding the iron oxidation state or the donor properties of the scorpionate ligand. However, for a given scorpionate ligand, the B–H moiety of the iron(III) species always absorbs at higher frequency than in its iron(II) analogue.

All complexes exhibit very weak CH absorptions between 2850 and 3200 cm^{-1}: the pyrazolyl C–H absorb between 3105 and 3166 cm^{-1}. Iron(II) species tend to absorb at slightly lower frequency than their iron(III) analogue, but the position of the stretches is mainly determined by its position at the ring: when the resolution is sufficient, pyrazolyl species (Tp, Ttp, Tpm, Tpe and Tpms) exhibit three small absorptions in the corresponding spectral region; $[Fe^{III}(Tp^{*})(CN)_3]^{-}$ exhibits only one C–H stretch at 3133 cm^{-1} (intermediate value) while $[Fe^{III}(Tp^{Me})(CN)_3]^{-}$ absorbs at 3121 and 3138 cm^{-1}: it is therefore quite reasonable to assume that the median wavenumber around 3126-3143 cm^{-1} corresponds to the fourth CH position in the ring, while the CH at 5- and the 3- positions absorb at slightly lower (3105–3115 cm^{-1}) and higher (3148–3166 cm^{-1}) wavenumbers respectively. The $[Fe(Tt)(CN)_3]^{2-/-}$ complexes also present two C–H absorptions in the region, but the presence of a nitrogen at the 4-position in the ring has a great impact on the resonance frequencies.

Except Tt, which contains 1,2,4-triazolyl rings but presents the same problem, all scorpionate ligands used or presented in this work are based on (eventually substituted) pyrazolyl rings. Their complexes therefore exhibit similar "fingerprint" pattern between 1500 and 600 cm^{-1}. This spectral region is usually very useful to characterise compounds by comparison with an existing spectrum but, in this case, this region is too crowded with peaks whose wavenumbers do not differ enough between two moieties to be of any use in the identification of a new, unknown species. However, five-membered rings such as imidazoles and pyrazoles exhibit a typical, isolated, weak to middle strong sharp absorption between 1500 and 1600 cm^{-1}. The position of this absorption is highly dependent on the nature of the ring substituents of the scorpionate ligand: pyrazole-based ligands (Tp, Ttp, Tpms, Tpm, Tpe) absorb between 1500 and 1520 cm^{-1}, while dimethylpyrazole-based ligands (Tp*, Tpm*) absorb above 1540 cm^{-1}. Even though its ligand bears methyl groups at the ring 3-positions, $Et_4N[Fe^{III}(Tp^{Me})(CN)_3]$ is found to absorb at 1504 cm^{-1}, that is at the same frequency as the non-methylated pyrazole-based compounds. It may be in fact the steric hindrance generated around the apical atom by the methyl group at the ring 5-position which would be responsible for the clear separation in frequency between dimethylated and non-methylated compounds. Tt complexes also provide an absorption band in this spectral region, but it is redshifted compared to the other pyrazole-based compounds, at the edge of the "fingerprint" region.

The absorption behaviour observed with natural abundance cyanide also goes for ^{13}C and ^{15}N isotope enriched cyanide compounds, except that wavenumbers of said absorption bands are all lower because ^{13}C and ^{15}N atoms are heavier than their far more natural abundant ^{12}C and ^{14}N isotopes (see Table 3.2).

Table 3.2: Isotope effect of enriched cyanides on the cyanide stretching band frequencies in some tricyanido iron(II), iron(III) and cobalt(III) complexes in FT-IR spectroscopy.

No	Name	ν_{CN} (cm^{-1})	ν_{13CN} (cm^{-1})	ν_{C15N} (cm^{-1})
–	KCN	2080	2031	2045
–	$K_2[Fe^{II}(Tp)(CN)_3]$	2056, 2037, 2016	2012, 1990, 1974	2025, 2004, 1988
[**3**]$^-$	$PPh_4[Fe^{II}(Tpm^*)(CN)_3]$	2042, 2048, 2064	2032, 2014 and 2006	2022[b]
[**1**]$^-$	$PPh_4[Fe^{III}(Tp)(CN)_3]$	2123	2079	2093
8	$[Fe^{III}(Tpm^*)(CN)_3]$	2128	2081	2096
[**7**]$^-$	$PPh_4[Fe^{III}(Tp^*)(CN)_3]$	2119	2068	2087
–	$PPh_4[Co^{III}(Tp^*)(CN)_3]$	2132	2080	2090

In the next chapters of this work, the FT-IR spectra of the presented multimetallic compounds are most of the time recorded on fresh samples, which contain a significant amount of water/solvent: either as mother liquor residue at the surface of the crystals, or in the sample as lattice molecules. The C–H stretching bands are therefore very often either only partially seen or not at all. Furthermore, their position is quite similar from one scorpionate ligand to another. It makes them a poor moiety analysis device, to the contrary of the ring stretching band above 1500 cm^{-1}. They will not be further discussed in FT-IR analyses of the polymetallic compounds in the next chapters.

---

[b] The three cyanide stretches are not always resolved and can appear, as in this case, as one absorption band.

3.1.4 Cyclic voltammetry

Recording cyclovoltammetric data should give a better insight into the electronic properties of each scorpionate-based tricyanido iron complex. Cyclic voltammograms of $PPh_4[Fe^{II}(Tpm^*)(CN)_3]$ (PPh_4[**3**]), $[Fe^{III}(Tpm^*)(CN)_3]$ (**8**), $PPh_4[Fe^{III}(Tp^*)(CN)_3]$ (PPh_4[**7**]), $PPh_4[Fe^{III}(Tt)(CN)_3]$ (PPh_4[**6**]), $PPh_4[Fe^{III}(Ttp)(CN)_3]$, $PPh_4[Fe^{III}(Tp)(CN)_3]$ (PPh_4[**1**]) and $(PPh_4)_2[Fe^{II}(Tpms)(CN)_3]$ ($(PPh_4)_2$[**5**]) were recorded under the same conditions in pure acetonitrile at room temperature. All potentials, inclusive the ones already reported in the literature,[134] are given using ferrocene/ferrocenium ($[Fc]/[Fc]^+$) as reference.[135] The cyclic voltammogram of PPh_4[**4**] and PPh_4[**2**] were not recorded, because of their insolubility in acetonitrile.

As expected for complexes of this type, all building blocks exhibit metal centred, quasi-reversible redox processes, with ratios of the anodic over the cathodic peak current close to 1 (see Figure 3.7). It corresponds to the reduction of the iron(III) to iron(II) ion.

The redox process in $PPh_4[Fe^{III}(Tp)(CN)_3]$ (PPh_4[**1**]) occurs at a half potential of $E°_{1/2}$ = -824 mV *vs* $[Fc]/[Fc]^+$ (a). It occurs at higher potential ($\Delta E°_{1/2}$ = 178 mV) than PPh_4[7] ($E°_{1/2}$ = -1002 mV, ΔE_p = 78 mV) (d) because of the weaker ligand field induced by the presence of methyl groups at the pyrazolyl ring 3- and 5-positions in Tp* and their steric and electronic inductive effect. At this concentration and scan rate (100 $mV.s^{-1}$), the potential difference between the reduction and oxidation half waves is ΔE_p = 134 mV, but smaller values can be obtained for lower concentrations.

For PPh_4[**6**], the Fe^{II}/Fe^{III} reduction wave arises at a half potential of $E°_{1/2}$ = -531 mV *vs* $[Fc]/[Fc]^+$. The potential difference between reduction and oxidation waves ΔE_p amounts to 90 mV. Compound PPh_4[**6**] is reduced at a significantly higher potential than its pyrazolyl analogue $PPh_4[Fe^{III}(Tp)(CN)_3]$ (PPh_4[**1**]) ($\Delta E°_{1/2}$ = 293 mV), which indicates that this triazolyl iron complex is electron poorer than its Tp counterpart. It is noteworthy that its redox potential is very close to that of **8** ($\Delta E°_{1/2}$ = 66 mV).

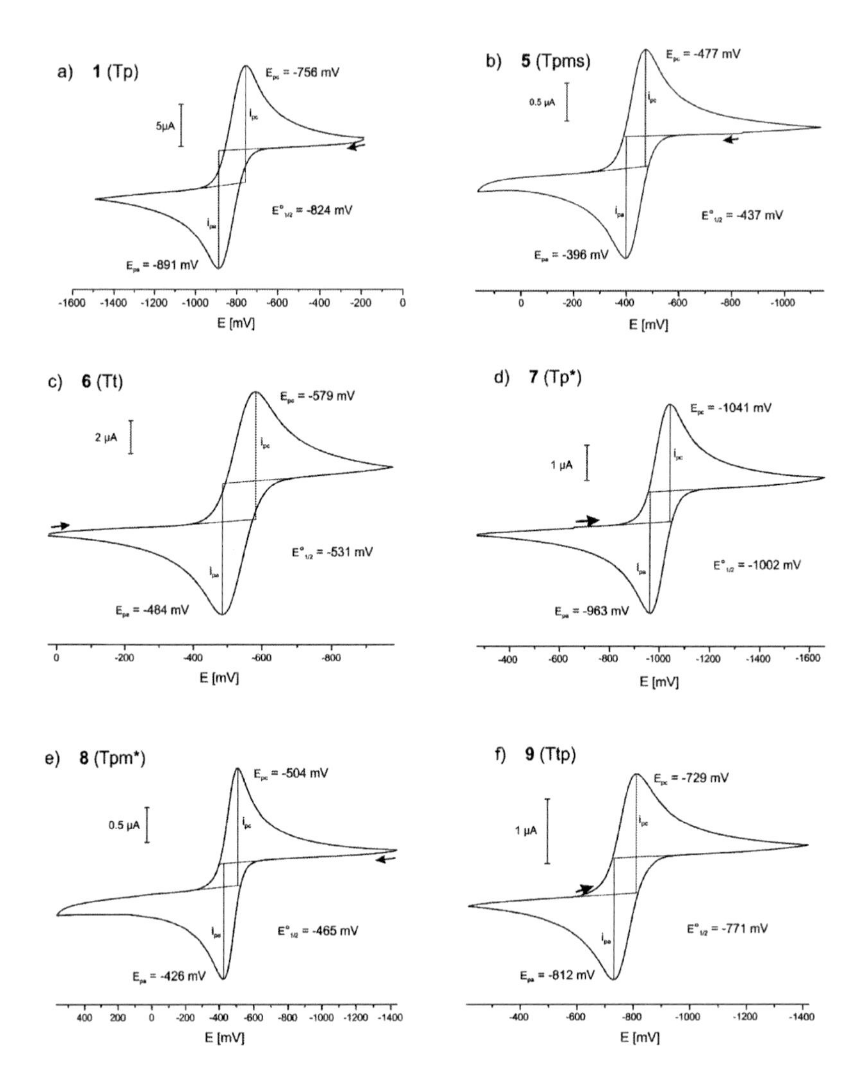

Figure 3.7: Cyclic voltammograms of a) PPh_4[**1**], b) $(PPh_4)_2$[**5**], c) PPh_4[**6**], d) PPh_4[**7**], e) PPh_4[**3**] / **8** and f) PPh_4[**9**] at room temperature in dry acetonitrile *vs* [Fc]/[Fc]$^+$. Scan rate v = 100 mV.s^{-1}, Pt/[n-Bu_4N][PF_6]/Ag. Each time, both cycles are identical, so only one is depicted here.

The reduction wave for PPh_4[**3**] in acetonitrile (e) occurs at a half potential of $E°_{1/2}$ = -465 mV *vs* [Fc]/[Fc]$^+$. The potential difference between reduction and oxidation waves ΔE_p amounts to 81 mV. The cyclic voltammogram of **8** shows the exact same values, it is therefore not depicted here. The replacement of the negatively charged Tp* ligand with its neutral Tpm* analogue, thus changing the charge of the overall iron complex, has a drastic effect on the redox potentials of the said complexes. The neutral iron(III) complex **8** is reduced at a much higher potential than its negatively charged boron analogue ($\Delta E°_{1/2}$ = 537 mV). This can be explained by a stabilisation effect of iron(III) in [**7**]$^-$ compared to its iron(II) redox partner, induced by the presence of a negative charge on the boron atom.

The reduction process in the cyclic voltammogram (f) of $PPh_4[Fe^{III}(Ttp)(CN)_3]$ (PPh_4[**9**]) occurs at a slightly higher potential ($\Delta E°_{1/2}$ = 51 mV) than for $PPh_4[Fe^{III}(Tp)(CN)_3]$ (PPh_4[**1**]). It occurs at $E°_{1/2}$ = -771 mV (ΔE_p = 83 mV).

The cyclic voltammogram of freshly dissolved $(PPh_4)_2$[**5**] (b) also features a single quasi-reversible Fe^{II}/Fe^{III} oxidation wave at $E°_{1/2}$ = -437 mV *vs* [Fc]/[Fc]$^+$. It occurs at a higher potential than all other measured tricyanido compounds. This is coherent with the stronger ligand field induced by the Tpms ligand compared to the other presented scorpionates, whose iron complexes were measured by cyclic voltammetry. The potential difference between the two half waves amounts to ΔE_p = 81 mV. This is indicative of a metal centred oxidation process without rearrangement of the coordination sphere as conceivable in case of the Tpms ligand. Finally, it is noteworthy that the half wave potential of $(PPh_4)_2$[**5**] is very close to the one of (PPh_4)[**3**] ($\Delta E°_{1/2}$ = 28 mV) and only a little higher than the one of $PPh_4[Fe^{III}(Tt)(CN)_3]$ ((PPh_4[**6**]) under similar conditions.

3.1.5 EPR spectroscopy and magnetic measurements

EPR spectroscopic measurements were carried out on the following paramagnetic iron(III) complexes at low temperature (5 K): $PPh_4[Fe^{III}(Ttp)(CN)_3]$ (PPh_4[**9**]), $PPh_4[Fe^{III}(Tt)(CN_3]$ (PPh_4[**6**]), $[Fe^{III}(Tpm^*)(CN_3]$ (**8**) and $PPh_4[Fe^{III}(Tp^*)(CN_3]$ (PPh_4[**7**]). Compound **8**, PPh_4[**9**] and PPh_4[**6**] did not give satisfying EPR data and the data are not presented here.

$PPh_4[Fe^{III}(Tp)(CN)_3]$ (PPh_4[1])

The EPR spectrum of $PPh_4[Fe^{III}(Tp)(CN_3]$ (PPh_4[**1**]) had already been recorded between 5 K and 50 K within the framework of a Parisian cooperation with Prof. M. Julve's working group in Valencia (Spain) before the beginning of this work (Figure 3.8).

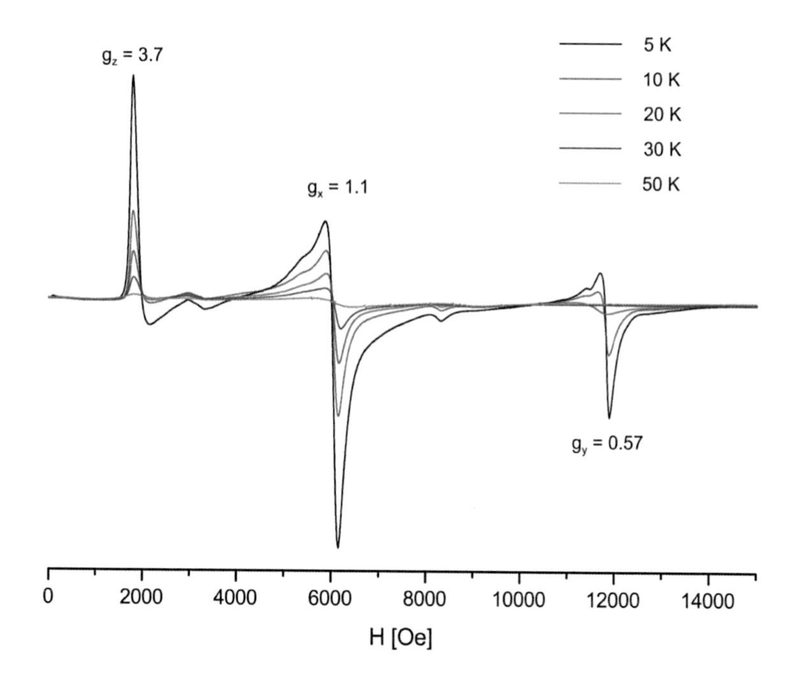

Figure 3.8: EPR spectra of ground $PPh_4[Fe^{III}(Tp)(CN)_3]$ (PPh_4[**1**]) in X-band (9.42 GHz) at 5 K, 10 K, 20 K, 30 K and 50 K.

$PPh_4[Fe^{III}(Tp)(CN_3]$ (PPh_4[**1**]) exhibits a orthorhombic g-tensor, with $g_x = 1.1$, $g_y = 0.57$ and $g_z = 3.7$. The EPR signal intensity decreases with increasing temperature. Above 50 K, the signal completely disappears in the background.

For the needs of the spin density calculations presented in the rest of this chapter, and even though the spectrum is not perfectly axial, one can define a $g_\perp$ and a $g_{//}$ as follows:

$$\begin{cases} g_\perp = g_x + g_y \\ g_\parallel = g_z \end{cases} \qquad (1)$$

In the case of $PPh_4[Fe^{III}(Tp)(CN_3]$ (PPh_4[**1**]), one obtains : $g_\perp = 0.8$ and a $g_{//} = 3.7$.

$PPh_4[Fe^{III}(Tp^*)(CN)_3]$ (PPh_4[7])

The EPR spectra of PPh_4[7] were measured at 5 K, with X and Q-band setups. The corresponding spectra are depicted in Figure 3.9.

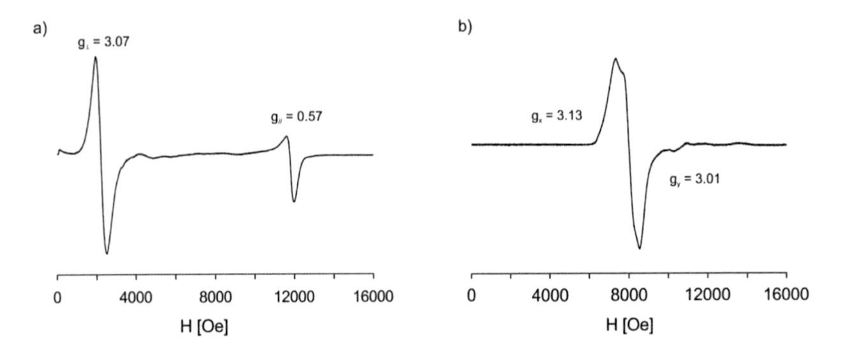

Figure 3.9: EPR spectra of ground PPh_4[**7**] at 5 K in X-band (9.42 Mhz) and in Q-band (33 Mhz).

To the contrary of the EPR X-band spectrum of $PPh_4[Fe^{III}(Tp)(CN)_3]$ (PPh_4[**1**]), the X-band spectrum of PPh_4[**7**] is axial with: $g_{\perp}$ = 3.07 and $g_{//}$ = 0.57. In the Q-band spectrum, the signal corresponding to the parallel g value is not visible. The resolution on the perpendicular g value is, in contrast, better. Two g values could be obtained: g_x = 3.13 and g_y = 3.01.

$[Fe^{III}(Tpm^*)(CN)_3]$ (8)

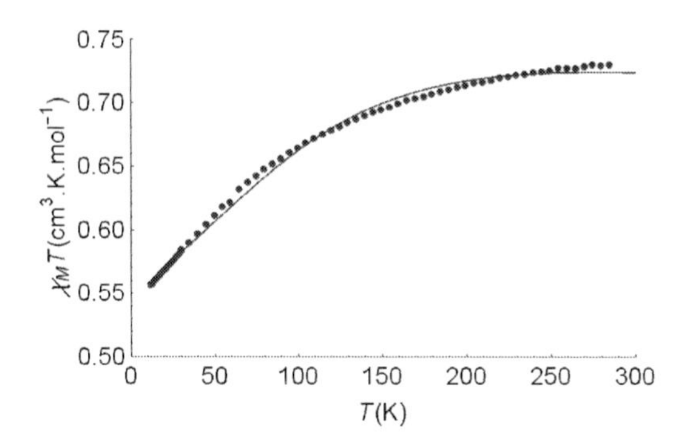

Figure 3.10: Experimental (red dots) and simulated (blue line) $\chi_M T$ product of compound **8** between 5 K and 300 K.

In order to obtain an estimate of the Landé factor g_{av} for **8**, its $\chi_M T$ product *vs* T curve was simulated using the following total Hamiltonian:

$$\mathcal{H}_{tot} = \mathcal{H}_{SO} + \mathcal{H}_{dist} + \mathcal{H}_{Ze} \tag{2}$$

The spin-orbit coupling Hamiltonian is expressed, in case of iron(III), as:

$$\mathcal{H}_{SO} = -\alpha\lambda L_{Fe} S_{Fe} \tag{3}$$

where L and S are the orbital and spin operators with L = 1 and S = ½, and α is the orbital reduction coefficient within the framework of the T–P isomorphism.[34–37] The distortion Hamiltonian is described by the following equation:

$$\mathcal{H}_{dist} = \Delta(L^2_{zFe} - \frac{1}{3}L^2_{Fe}) \tag{4}$$

where Δ is the axial distortion parameter. The Zeeman Hamiltonian is expressed as follows:

$$\mathcal{H}_{Ze} = +g_{Fe}\beta \cdot S_{\nu_{Fe}} \cdot H_\nu \tag{5}$$

with β being the Bohr magneton and H the applied magnetic field tensor along ν = x, y, z. Using Lines' model,[136] an effective Landé factor $g_{eff_{Fe}}(T)$ can be drawn out of the $\chi_M T$ data plot using equation (6) where $(\chi_M T)_{exp}$ are the experimental values of complex **8**.

$$g^2_{eff_{Fe}} = \frac{4k(\chi_M T)_{exp}}{\mathbb{N}\beta^2} \tag{6}$$

The agreement factor is calculated as the quadratic sum of the gap between the simulated and the experimental curves. The $\chi_M T$ *vs* T experimental and simulated curves of **8** are depicted in Figure 3.10. The best fit was obtained with the following parameters: $\alpha = 1.10$, which is too high since it should be smaller than 1; $\lambda = -308\ \text{cm}^{-1}$, $\Delta = -2637\ \text{cm}^{-1}$, which is quite high for this system and a temperature independent parameter (TIP) of $20 \cdot 10^{-6}\ \text{cm}^3 \cdot \text{mol}^{-1} \cdot \text{K}$. The mean $g_{eff_{Fe}}$ factor over the whole temperature range amounts to 2.59. It is 2.44 at 11 K and reaches 2.69 at room temperature, which is the value used in the rest of this chapter for **8**.

3.1.6 MAS-NMR spectroscopy

NMR spectroscopy is a widely used method in chemistry and material sciences for compound characterisation. However, it is almost exclusively used on diamagnetic compounds, whereas one often favours EPR spectroscopy for paramagnetic ones. This difference is easily explained by the problems encountered during the recording and the analysis of NMR spectroscopic data in presence of unpaired electron(s): (i) signal broadening that can make difficult or preclude the detection of the probed nuclei; (ii) very wide chemical shift ranges, far wider than "normal" diamagnetic range, (iii) truncated FIDs (if a signal is found!) which leads to analytic difficulties (e.g. baseline distortion).

Even if the first NMR spectra of paramagnetic hexacyanidometallates were published in the 1960's,[137] the design and commercialisation of new generations of spectrometers with considerably improved electronics (e.g. with improved performance of analog-to-digital converters, improved sampling frequencies) for almost two decades has been the key factor to help overcome these challenges. This brought a new, very useful tool for investigating materials with paramagnetic sources in various research fields like solid phase studies of lithium batteries and electrodes,[138–141] organometallic catalysis,[142,143] protein dynamics and structure[144–146] and magnetic molecular compounds.[82,147–150] Indeed, if an unpaired electron exhibits a non-zero probability of presence at a given magnetic active nucleus, the spin-spin and spin-lattice relaxation times of the latter are considerably shortened. However, the actual spectrometers are able to acquire reliable data sets with only small optimisations of the acquisition parameters (provided that the electron-nucleus interaction is not too strong). In other cases, the FID duration, which can be shorter than 5 ms, falls in the same order of magnitude as the electronic delay of response ("*de*" parameter). This can lead to heavily truncated FIDs, loss of signal and in some case, acoustic ringing from the probe head (severe baseline distortion). In such cases, the Hahn-echo pulse sequence can help overcoming these problems (at some cost, with a relative decrease of signal-to-noise ratio). Finally, the truncated FID can also be mathematically reconstructed, but this approach only leads to very limited spectra improvement.

Since the magnetic communication (or the "magnetic exchange interaction" in specialised terms) between metal centres in cyanide-bridged polynuclear compounds occurs through the cyanide bridges, the distribution of unpaired electrons along the cyanide ligands in the mononuclear reagents bears crucial information for the comprehension of the magnetic properties of their products. Solid-state NMR studies on hexacyanidometallates and Prussian Blue Analogues (PBAs)[82,147,148] demonstrated that NMR spectroscopy applied to paramagnetic species (improperly coined as "paramagnetic NMR") provides access to accurate local structural information and, more interestingly, to the local magnetic information.

In this perspective, $PPh_4[Fe^{III}(Tp)(CN)_3]$ (PPh_4[**1**]), $PPh_4[Fe^{III}(Tp^*)(CN)_3]$ (PPh_4[**7**]) and its carbon-based analogue $[Fe^{III}(Tpm^*)(CN)_3]$ (**8**) were synthesised using ^{13}C (99%) and ^{15}N (98%) enriched potassium cyanide at a 1.0 mmol scale and their solid-state paramagnetic NMR spectra were recorded.

The solid-state NMR spectra were acquired using magic angle spinning technique, by rotating the sample at the magic angle ($\theta \approx 54.74°$), in order to average the dipolar and anisotropic interactions that take place in condensed phases and thus increase the resolution of the spectra. This gives rise to an isotropic value, also known as δ_{iso}, and a spinning sideband pattern, regularly spaced by the MAS spin rate, which can be used to determine the chemical shift anisotropy (CSA) of the nuclei. The superposition of two or more spectra recorded at selected spin rates help discriminating the isotropic signal(s) among the side bands.

3.1.6.1 Paramagnetic NMR as a magnetic probe: theoretical background

Experimental chemical shifts (in ppm) measured for paramagnetic species can be decomposed in two main contributions:

$$\delta_T^{exp} = \delta_T^{dia} + \delta_T^{para} \tag{7}$$

δ_T^{dia} is the diamagnetic contribution to the chemical shift, and is solely due to the chemical environment of the observed nucleus; it is also the chemical shift one would measure for a structurally analogue diamagnetic species. δ_T^{para} is the paramagnetic contribution to the experimental chemical shift and is solely due to the unpaired electron. Two different mechanisms contribute to this shift:

$$\delta_T^{para} = \delta_T^{con} + \delta_T^{dip} \quad (8)$$

While δ_T^{dip}, or pseudo-contact term, is due to the purely dipolar through-space interaction between two magnetic moments and can be estimated by taking into account electronic geometric considerations, δ_T^{con} , also called Fermi contact term, is proportional to the spin density seen by the observed nucleus, that is in atomic s-orbitals:[147,148,151]

$$\rho_{is}(X) = \frac{9k_B T a_0{}^3}{\mu_0\, g_{av}{}^2\, \beta_e{}^2 (S+1)|\Psi_{is}(0)|^2}\delta_{T,iso}^{con}$$

$$= 5.077\ 10^{-5}\frac{\delta_{298,iso}^{con}}{g_{av}{}^2 (S+1)|\Psi_{is}(0)|^2} \quad (9)$$

where $\rho_{is}(X)$ is the fractional spin density on the ith s orbital given in $(au)^{-3}$. Ψ_{is} is the wave function describing a semi-occupied s orbital, and $|\Psi_{is}(0)|^2$ (given in $Å^{-3}$) is the corresponding spin density at the nucleus which value are tabulated for a given atom and a given i^{th} s orbital.[152,153] The Boltzmann constant k_B, the temperature *T*, the average *g* factor g_{av} of the compound and the electron spin quantum number S are all given in SI units, while the Bohr radius a_0 is given in angstroms, and both the magnetic constant μ_0 and Bohr magneton β_e are expressed in SI with a metre to angstrom conversion. The Fermi contact term, being a chemical shift, is given in parts per million (ppm).

It is noteworthy that both Fermi contact and pseudo-contact terms are inversely proportional to the temperature, which makes paramagnetic NMR spectroscopy data useless for quantitative analyses if the actual temperature inside the probe is unknown. Since MAS-NMR spectroscopy involves very high spinning rates of the sample (up to 67 kHz, depending on the rotor size), the sample undergoes a non-negligible increase of its internal temperature compared to the set point temperature. This can be monitored and adjusted by using an internal temperature standard. In this work we used nickelocene ($[NiCp_2]$), whose temperature dependence of its isotropic proton chemical shift is tabulated in the literature:[154]

$$T = -\frac{79477}{\delta^{exp}} - 12.89 \quad (10)$$

For each of the following measurements, some freshly ground nickelocene was thus inserted in the rotor in order to measure and adjust the temperature during the measurement. Proton NMR spectra was acquired before and after each ^{13}C and ^{15}N NMR measurement in order to check the stability of the inner temperature during the measurement. Two spectra were acquired at two different carefully selected spinning rates but at the same inner temperature in order to find out the isotropic peak.

3.1.6.2 *Solid-state NMR spectroscopy of $PPh_4[Co^{III}(Tp^*)(CN)_3]$*

Ideally, each measurement of paramagnetic chemical shift would require the use of an isostructural diamagnetic reference (see equation 7). However, as the cobalt(III) equivalents of $PPh_4[Fe^{III}(Tp)(CN)_3]$ (PPh_4[**1**]) and $[Fe^{III}(Tpm^*)(CN)_3]$ (**8**) are not known to date, the diamagnetic low-spin cobalt(III) complex $PPh_4[Co^{III}(Tp^*)(CN)_3]$ was taken as diamagnetic reference for the three species. This complex exhibits a closely related structure with the same coordination sphere and the same overall C_{3v} symmetry as the paramagnetic iron(III) complexes. Of course, it is possible that real diamagnetic contribution of each species would differ by few ppm from the reference signal shifts. The error introduced here is however small in regard to the overall large signal shifts.[155] $PPh_4[Co^{III}(Tp^*)(CN)_3]$ was synthesised[156] using ^{13}C (99%) and ^{15}N (98%) enriched potassium cyanide at a 1.0 mmol scale and their solid-state NMR spectra were recorded.

^{13}C MAS-NMR spectroscopy

Its ^{13}C spectra were recorded in a 500 MHz Bruker spectrometer equipped with a BL probe head and the chemical shift scale was referenced against TMS (adamantane as secondary reference). They were recorded with spinning rates of 10 and 6 kHz, with a relaxation delay *d*1 of 5 seconds between two scans. The superposition of the two spectra allowed the identification of three kinds of isotropic signals, as depicted in Figure 3.11.

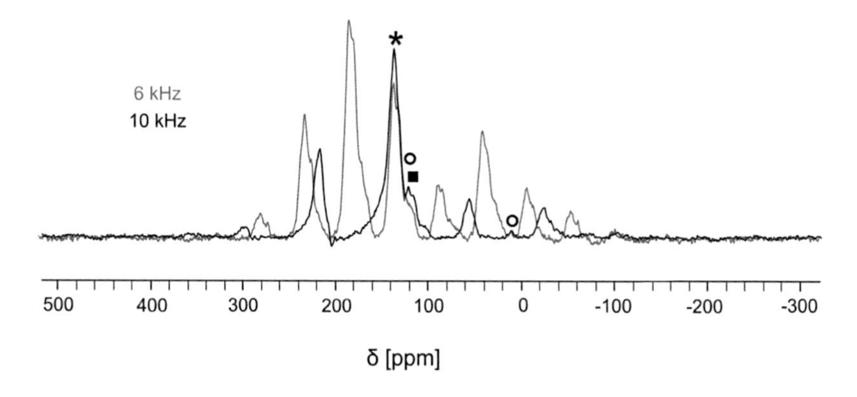

Figure 3.11: ^{13}C MAS-NMR spectra of the diamagnetic $PPh_4[Co^{III}(Tp^*)(CN)_3]$ at 6 (red) and 10 kHz (black) recorded at a ^{13}C-Larmor frequency of 125.77 MHz. The three types of isotropic signals are marked with an asterisk (isotope enriched cyanides), a black square (tetraphenylphosphonium countercation), and a circle (Tp* ligand).

Even though only the three cyanide carbons are enriched, and exhibit two isotropic shifts marked with an asterisk, the sheer amount of the tetraphenylphosphonium carbons (24 equivalents dispatched into 4 different sites) per metal complex, and its comparatively shorter relaxation rates allows its detection in the same spectrum area as a shoulder of the cyanide isotropic peak (marked as black square in Figure 3.11). These chemical shifts partly correspond to those found by NMR in solution (ν_L = 75 Mhz), where quaternary cyanide carbons are detected at about 139 ppm in acetonitrile-d_3, and tetraphenylphosphonium gives rise to four different doublets at ~118.9 ppm (4C, $^1J_{CP} \approx$ 90 Hz), 131.3 ppm (8C, $^3J_{CP}$ = 12.9 Hz), 135.6 ppm (8C, $^2J_{CP}$ = 10.5 Hz) and 136.4 ppm

(4C, $^{4}J_{CP}$ = 3.1 Hz) in the same solvent. The 4-position carbons of the Tp* ligand (3C, one chemical site) also give rise to a signal in the lower field part of the spectrum, which render unclear, at these spin rates, which carbons are exactly responsible for the shoulder, and which lie underneath the cyanide isotropic shifts. Finally, a very small isotropic peak at 12.7 ppm can be ascribed to the methyl moieties of the Tp* ligands (six carbon atoms at two different sites).

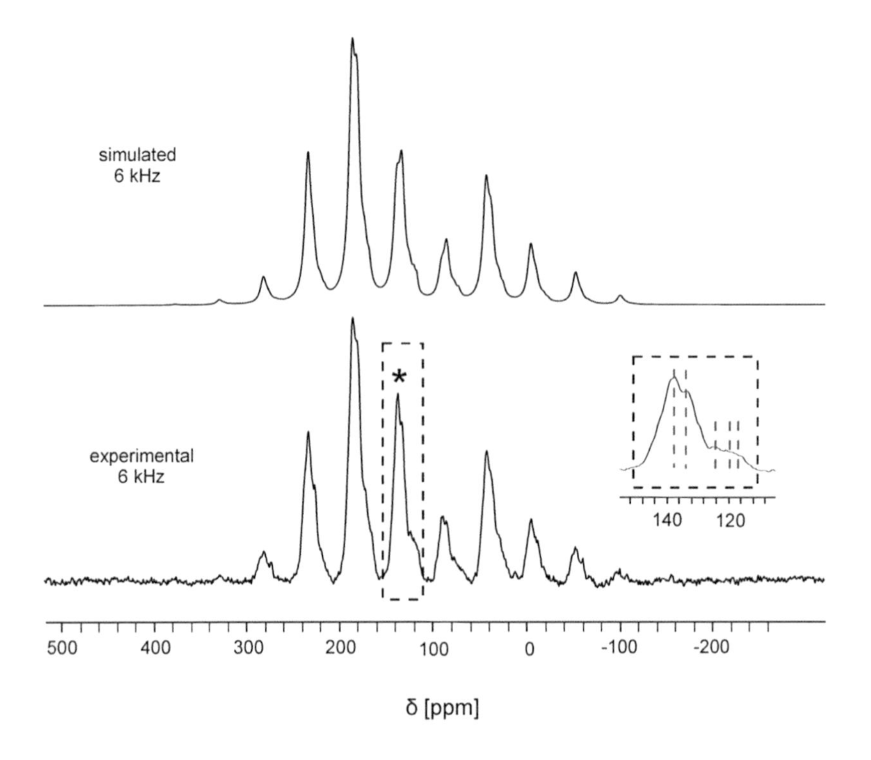

Figure 3.12: Simulated (top) and experimental (bottom) ^{13}C MAS-NMR spectra of the diamagnetic $PPh_4[Co^{III}(Tp^*)(CN)_3]$ at 6 kHz and at a ^{13}C-Larmor frequency of 125.77 MHz. The position of the isotropic peaks is marked with an asterisk, and the enlarged area is displayed in the inset.

A Herzfeld-Berger Analysis (HBA) of the (overnight) recorded 6 kHz MAS-NMR spectrum could be performed. Simulated and experimental spectra are displayed in Figure

3.12. Best results (88.4% overlap) could be obtained by considering only two different cyanide ^{13}C environments, at δ_{iso} = 138.8 and 133.2 ppm and three $[PPh_4]^+$/Tp* isotropic values (δ_{iso} = 125.1, 120.4 and 117.4 ppm). Experimental parameters obtained from the ^{13}C MAS-NMR spectra of $PPh_4[Co^{III}(Tp^*)(CN)_3]$ are summarised in the Table 3.3. Due to its small intensity, the Tp* methyl signal was not included. At these spinning rates, the broadness of the peaks does not allow sufficient resolution for the third cyanide to be resolved. The Herzfeld-Berger Analysis for the simulations are expressed using the Haeberlen convention, as recommended by IUPAC,[157] for the sake of clarity and to provide coherence with previously published work.[82,147,148]

Table 3.3: Herzfeld-Berger Analysis results for the simulation of the 6 kHz ^{13}C MAS-NMR spectrum of $PPh_4[Co^{III}(Tp^*)(CN)_3]$ using five different isotropic shifts. Best overlap: 88.35%. For specific definitions of each tensor parameters, see pages 217-218.

Compound	$PPh_4[Co^{III}(Tp^*)(CN)_3]$				
Contribution	a	b	c	d	e
$\delta_{iso}^{exp}(^{13}C)$	138.8	133.2	125.1	120.4	117.4
$\delta_{zz}^{exp}(^{13}C)$	-75.6	-20.6	-43.0	-24.4	148.3
$\delta_{yy}^{exp}(^{13}C)$	242.6	210.0	208.9	191.4	116.7
$\delta_{xx}^{exp}(^{13}C)$	249.4	210.2	209.6	194.2	87.2
$\Delta\delta^{exp}(^{13}C)$	-321.6	-230.8	-252.3	-217.2	46.3
$\eta\delta^{exp}(^{13}C)$	0.032	0.001	0.004	0.019	0.954

The two cyanide isotropic chemical shifts located at (a) δ_{iso} = 138.8 ppm and (b) 133.2 ppm are consistent with the isotropic/solution values found for other reported diamagnetic polycyanido cobalt(III) complexes, like $Cs_2K[Co^{III}(CN)_6]$ (135.1 ppm), $PPh_4[Co^{III}(bipy)(CN)_4]$ and $PPh_4[Co^{III}(phen)(CN)_4]$ (122.7 ppm/133.4 ppm and 121.5/133.1 ppm respectively in acetonitrile).[148] Both signals exhibit a CSA (Chemical Shift Anisotropy) with a strong axial symmetry, with $\eta \approx 0$, which is consistent with the MAS-NMR spectra of hexacyanidometallates reported in the literature.[82,148] The signal (a) exhibits an almost half as great anisotropy $\Delta\delta^{exp}$ than the (b) one. They are also negative, indicating that the nucleus is less shielded in two directions than it is in the third one. The cyanide bridges being linear (along the z axis for instance), the local x and y axis can be considered as intervertible, so that the anisotropy tensor contributions along these two axes are expected to be identical, but quite different from the last one.

The shoulder to the cyanide isotropic signal can be modelled as three different contributions but they cannot be ascribed to a specific carbon in the tetraphenylphosphonium salt (see inset of Figure 3.12). Their isotropic signals are slightly more shielded, as compared to the cyanide isotropic values, and their chemical shifts are 125.1 ppm, 120.4 ppm to 117.4 ppm for signal (c), (d) and (e), respectively. The first two have a strong axial symmetry and a strong negative anisotropy in the same order of magnitude as the cyanides signal sets. Signal set (e), however, exhibits a clearly lower symmetry and a very small anisotropy (46.3 ppm, to be compared with the -252.3 and -217.2 ppm for contributions (c) and (d)).

Since it is not possible to ascribe each cyanide in the structure to its NMR signal with sufficient accuracy, the following "mean" set of parameters (see Table 3.4), associated with uncertainty, will be further used in this work.

Table 3.4: Mean values and associated uncertainties for the cyanide contributions to the ^{13}C MAS-NMR spectrum of $PPh_4[Co^{III}(Tp^*)(CN)_3]$, based on the values of Table 3.3.

$PPh_4[Co^{III}(Tp^*)(CN)_3]$	Mean value [ppm]	Uncertainty [ppm]
$\delta_{T,iso}^{exp}(^{13}C)$	136.0	± 2.8
$\delta_{T,zz}^{exp}(^{13}C)$	-48.1	± 27.5
$\delta_{T,yy}^{exp}(^{13}C)$	226.3	± 16.3
$\delta_{T,xx}^{exp}(^{13}C)$	229.8	± 19.6
$\Delta\delta_{T}^{exp}(^{13}C)$	-276.2	± 45.4
$\eta\delta_{T}^{exp}(^{13}C)$	0.019	–

^{15}N MAS-NMR spectroscopy

The ^{15}N enriched sample of $PPh_4[Co^{III}(Tp*)(CN)_3]$ was prepared following the same synthetic procedure as for the ^{13}C enriched sample. Its ^{15}N NMR spectra were recorded with the same experimental setup. The ^{15}N chemical shift scale was referenced against CH_3NO_2 (with NH_4NO_3 as a secondary reference). The spectra were recorded using a Hahn echo pulse sequence, at 3, 6 and 10 kHz, with a relaxation delay *d*1 of either 60 or 120 seconds. The isotropic values were determined by superposition of the 6 and 10 kHz spectra as depicted in Figure 3.13.

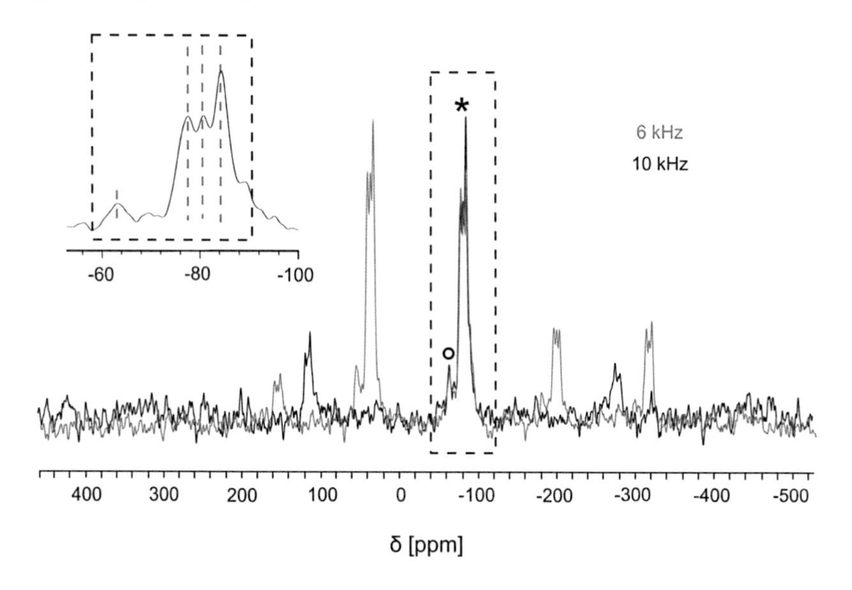

Figure 3.13: ^{15}N MAS-NMR spectra of diamagnetic $PPh_4[Co^{III}(Tp^*)(CN)_3]$ at 10 (black curve) and 6 kHz (red curve) at a Larmor frequency of 50.87 MHz. The isotropic peaks are noted with an asterisk ($PPh_4[Co^{III}(Tp^*)(CN)_3]$) and a circle (impurity). The inset shows the isotropic peaks of the 10 kHz spectrum.

At these spinning rates, the three cyanide contributions are resolved, but very few spinning bands define the overall shape of the CSA. It is also not possible to attribute precisely each isotropic value to a specific cyanide. Nonetheless, the 6 kHz spectrum was simulated and the results are reported in Table 3.5, while the experimental and simulated spectra are depicted in Figure 3.14. A fourth nitrogen environment is clearly visible at δ_{iso} = -62.4 ppm. This may correspond to a small portion of co-crystallised $PPh_4[C^{15}N]$. The three cyanide environments (a), (b) and (c) at δ_{iso} = -76.8, -80.2 and -84.0 ppm respectively, are close to the isotropic values reported in the literature for the hexacyanidocobaltate(III) (-81.9 ppm), $PPh_4[Co^{III}(bipy)(CN)_3]$ (-68.1 ppm and -81.1 ppm) and $PPh_4[Co^{III}(phen)(CN)_3]$ (-66.6 ppm and -79.9 ppm).[148] Signals (a) and (b) exhibit a strongly axial symmetry ($\eta_a \approx \eta_b \approx 0$) while the (c) environment departs slightly from axiality ($\eta_c \approx 0.216$); this is coherent with the linearity of cyanides and the previous ^{13}C results displayed in this work. Their anisotropy is bigger than for ^{13}C and ranges from -490.4 to -523.6 ppm. In comparison, the signal (d) exhibits a bigger anisotropy (-541.8 ppm) but also a lower symmetry with $\eta_d = 0.626$.

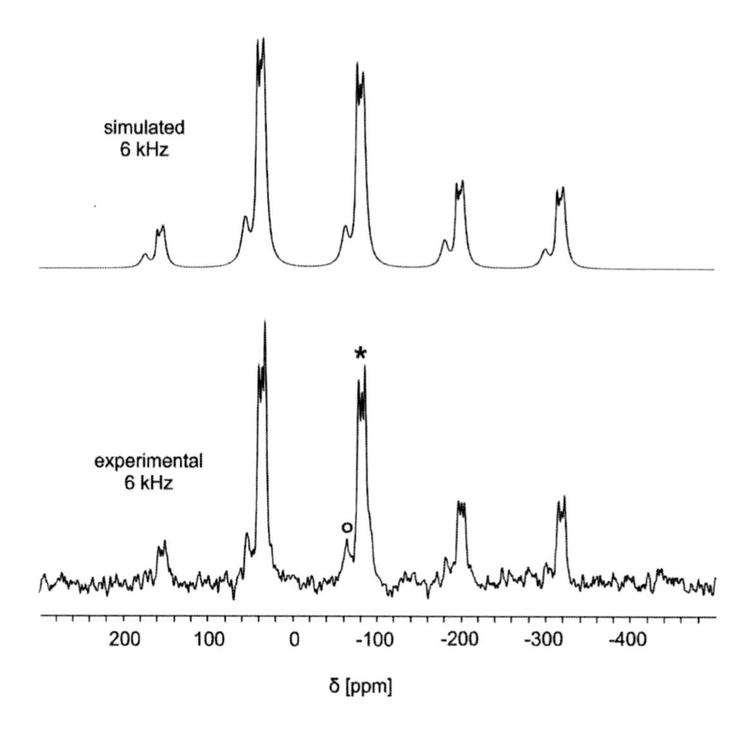

Figure 3.14: Simulated (top) and recorded (bottom) ^{15}N-MAS-NMR spectra of the diamagnetic $PPh_4[Co^{III}(Tp^*)(CN)_3]$ at 6 kHz and at a Larmor frequency of 50.87 MHz. The iron cyanide isotropic peaks are marked with an asterisk, the impurity peak is marked with a circle.

The experimental and simulated 3 kHz ^{15}N spectra are depicted in Figure 3.15, while the extracted numerical parameters are reported in Table 3.5. At 3 kHz, the dipolar interactions are not fully averaged so the three different cyanides are not resolved anymore. The spectrum can be here simulated using only two different nuclei. Despite 72 hours of acquisition, the signal-to-noise ratio is still quite low, so that the fourth environment is not clearly visible and it was therefore neglected for the simulation. Although the parameters of the signal (e) should be comparable to those of (a), (b) and (c), the signal (f) exhibits a slightly smaller anisotropy. More importantly, the overall symmetry is far less axial, with $\eta_f = 0.411$.

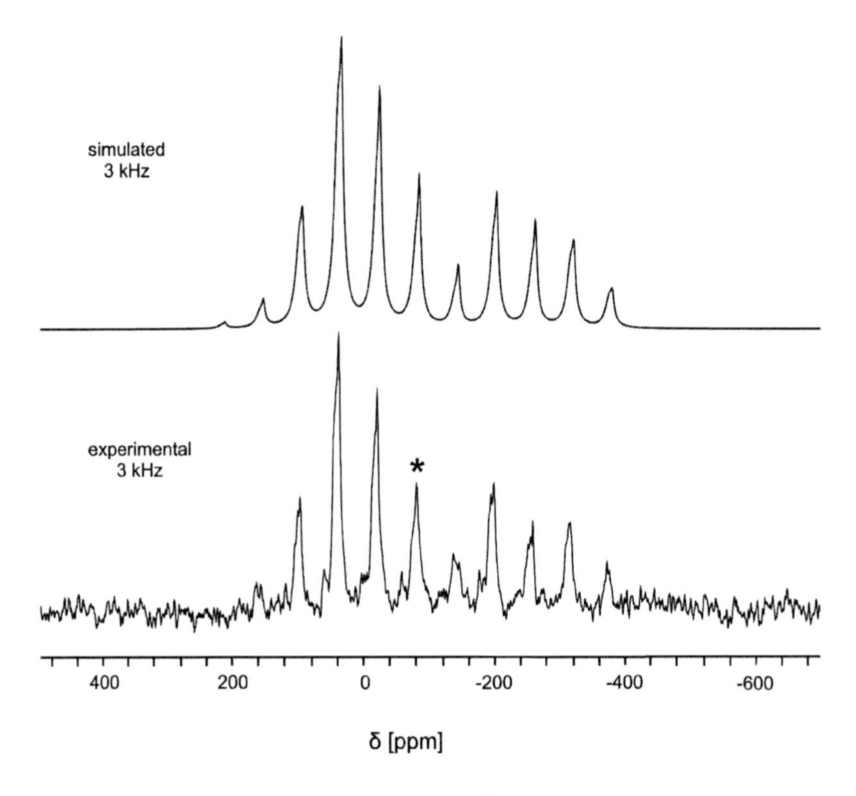

Figure 3.15: Simulated (top) and recorded (bottom) ^{15}N MAS-NMR spectra of the diamagnetic $PPh_4[Co^{III}(Tp^*)(CN)_3]$ at 3 kHz and at a Larmor frequency of 50.87 MHz. The isotropic peak is marked with an asterisk.

Table 3.5: Herzfeld-Berger Analysis results for the simulation of the 3 and 6 kHz ^{15}N MAS-NMR spectrum of $PPh_4[Co^{III}(Tp^*)(CN)_3]$ using respectively three and four different isotropic shifts. Best overlap for 3 kHz: 85.0%. For 6 kHz: 80.4%. For precise definitions of each tensor parameter, see page 226-227.

	$PPh_4[Co^{III}(Tp^*)(CN)_3]$					
Spinning rate	6 kHz				3 kHz	
Contribution	a	b	c	d	e	f
$\delta^{exp}_{T,iso}(^{15}N)$	-76.8	-80.2	-84.0	-62.4	-80.2	-85.1
$\delta^{exp}_{T,zz}(^{15}N)$	-413.8	-407.2	-433.1	-423.6	-424.4	-390.2
$\delta^{exp}_{T,yy}(^{15}N)$	85.3	83.2	52.9	5.2	91.9	4.78
$\delta^{exp}_{T,xx}(^{15}N)$	98.1	83.3	128.1	231.2	91.9	130.26
$\Delta\delta^{exp}_{T}(^{15}N)$	-505.5	-490.4	-523.6	-541.8	-516.2	-457.7
$\eta\delta^{exp}_{T}(^{15}N)$	0.038	0.0	0.216	0.626	0.0	0.411

The diamagnetic $PPh_4[Co^{III}(Tp^*)(CN)_3]$ reference will be therefore described by the following "average" set of parameters (see Table 3.6) and uncertainties further in this work.

Table 3.6: Mean values and associated uncertainty for the cyanide contribution to the ^{15}N MAS-NMR spectrum of $PPh_4[Co^{III}(Tp^*)(CN)_3]$, based on the experimental parameter sets listed in Table 3.5.

$PPh_4[Co^{III}(Tp^*)(CN)_3]$	Mean value [ppm]	Uncertainty [ppm]
$\delta^{exp}_{iso}(^{15}N)$	-80.3	± 3.5
$\delta^{exp}_{zz}(^{15}N)$	-418.0	± 15.1
$\delta^{exp}_{yy}(^{15}N)$	73.8	± 11.5
$\delta^{exp}_{xx}(^{15}N)$	103.2	± 24.9
$\Delta\delta^{exp}(^{15}N)$	-506.5	± 16.1
$\eta\delta^{exp}(^{15}N)$	0.087	–

3.1.6.3 *Paramagnetic NMR studies of $PPh_4[Fe^{III}(Tp)(CN)_3]$ (PPh_4[1])*

^{13}C MAS-NMR spectroscopy

The measurement was carry out on a 10 mg sample of dry ground crystalline ^{13}C-enriched $PPh_4[Fe^{III}(Tp)(C^*N)_3]$ in a 1.3 mm ZrO_2 rotor. The ^{13}C spectra were acquired by magic angle spinning (MAS) in a 700 MHz spectrometer equipped with a 1.3-BL probe head, and the chemical shift was referenced against TMS (with adamantane as secondary reference). The spectrum of the sample was recorded at 60 and 65.5 kHz. The temperature data for both measurements are summarised in Table 3.7.

Table 3.7: Tabulated report of ^{13}C MAS-NMR measurement of $PPh_4[Fe^{III}(Tp)(CN)_3]$ recorded with a 1.3-BL probehead and a Bruker AV-700 spectrometer

Probe	Spinning rate (kHz)	$T_{\text{set point}}$ (K)	$\delta_{NiCp_2}^{before}$ (ppm)	$\delta_{NiCp_2}^{after}$ (ppm)	T_{actual} (K)	ΔT (K)	δ_T^{iso} (ppm)
1.3-BL	60	297.5	-228.7	–	334.6	–	-3752
	65.5	264.0	-228.7	-230.3	333.4	2.4	

For each spinning rate, the temperature inside the rotor was set using a nickelocene internal reference (with δ_H = -228.7 ppm. corresponding to 334.6 K). Although the inner temperature slightly decreased overnight during the 65.5 kHz measurement, the slight shift induced by the temperature change is not noticeable on signals with such a large width at half-height ($\Delta\nu_{1/2} \approx 20000$ Hz). The spectra were acquired using a one-pulse sequence, and their baseline distortions were manually corrected. Superposition of the 60 and 65.5 kHz data allowed the determination of the isotropic shift at δ_{iso} = -3752 ppm. Another measurement at 48 kHz (not displayed) was carried out to confirm this attribution.

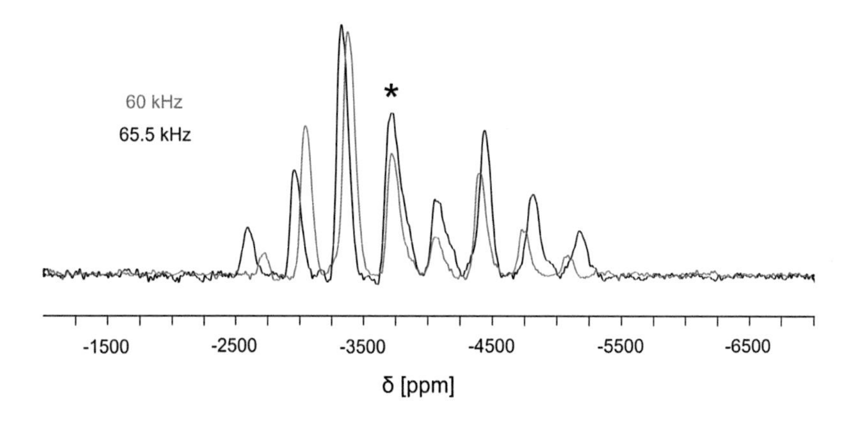

Figure 3.16: ^{13}C MAS-NMR spectra of $PPh_4[Fe^{III}(Tp)(CN)_3]$ (PPh_4[**1**]) at 60 kHz (red) and 65.5 kHz (black) at T = 334 K acquired at a ^{13}C-Larmor frequency of 175.37 MHz. The isotropic peak, which is not shifted, is marked with an asterisk.

The 60 kHz spectrum was simulated by Herzfeld-Berger analysis and the extracted parameters are reported in Table 3.14. The experimental spectrum and the simulation are depicted in Figure 3.17. Even though the different cyanide contributions are not resolved, a significantly better overlap was obtained when using two different nuclei in the simulation. Simulation and experiment only match at 84.1%, partly because of the spectral bump, which is pushed flat by the baseline correction.

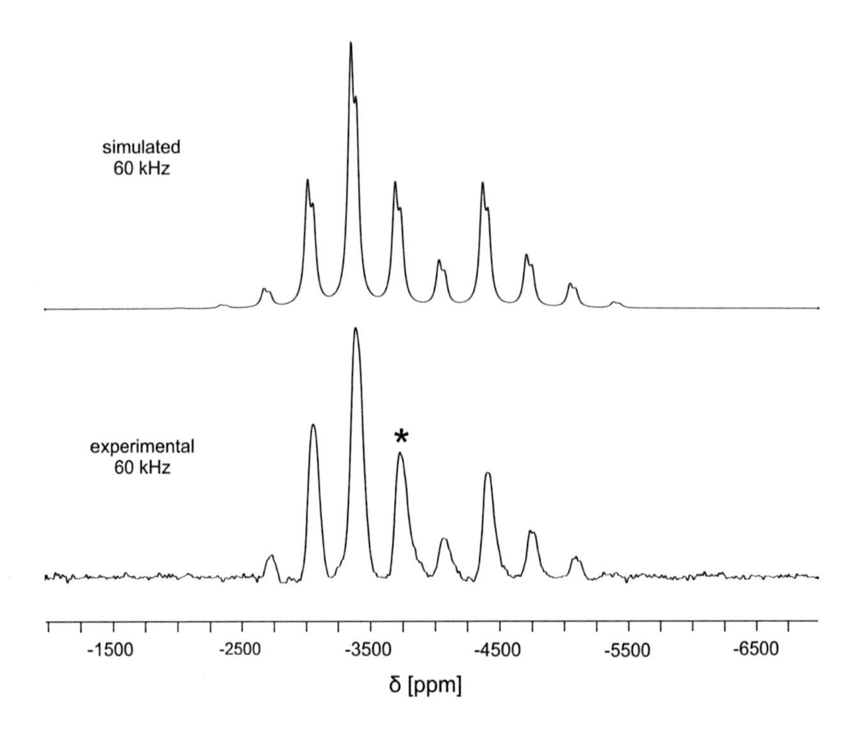

Figure 3.17: Simulated (top) and experimental (bottom) spectra of $PPh_4[Fe^{III}(Tp)(CN)_3]$ (PPh_4[**1**]) at 60 kHz at $T = 334.6$ K acquired at a Larmor frequency of 175.37 MHz.

As previously reported for the hexacyanidometallates, the isotropic chemical shifts of the cyanide carbons are strongly shifted to lower frequency compared to their reference pointing to a negative spin density on these atoms. The isotropic shifts exhibit high negative values, -3726 and -3755 ppm at 334.6 K. Both contributions are strongly axial, with found asymmetry parameters close to zero. The chemical shift anisotropy (CSA) tensor possesses two components. The equatorial component of the CSA is more deshielded than the axial one, leading to a negative anisotropy parameter ($\Delta\delta$ = -2150 and -2033 ppm for an overall chemical shift anisotropy (CSA) of about 2660 ppm). Even though the latter is in absolute value tenfold bigger than for the diamagnetic $PPh_4[Co^{III}(Tp^*)(CN)_3]$ reference, the overall symmetry of the signals remains unchanged. This leads to a negative paramagnetic contribution, which is coherent with a spin delocalisation mechanism and spin density sign for the carbon atoms observed in a previous study.[148] This is also consistent, qualitatively, with the negative sign of the spin density located on the carbon atoms in the spin density plot calculated by DFT (see Figure 3.18).

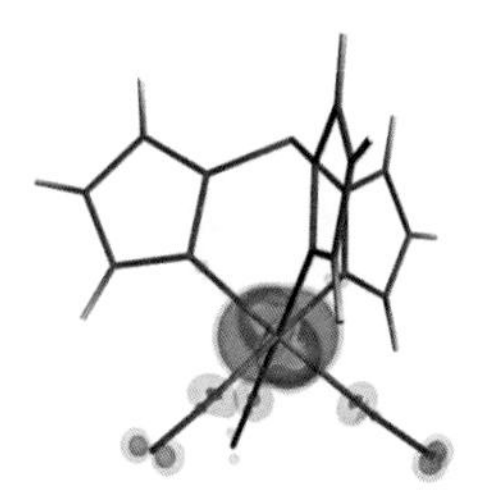

Figure 3.18: Spin density distribution of $PPh_4[Fe^{III}(Tp)(CN)_3]$ (PPh_4[**1**]) (grey atoms: carbon, blue ones: nitrogen, orange one: iron, yellow one: boron) at the [B3LYP] level of theory. Positive spin density in red, negative spin density in blue.

^{15}N MAS-NMR spectroscopy

The 4 mm rotor was prepared using 100 mg of ground ^{15}N cyanide enriched $PPh_4[Fe^{III}(Tp)(CN^*)_3]$ (PPh_4[**1**]). The ^{15}N spectra were acquired on a 400 MHz spectrometer, equipped with a 4-BL probe head. The inner temperature in the rotor was set at a nickelocene chemical shift of -245.9 ppm (6 kHz) and -245.4 ppm (10 kHz) corresponding to 310.3 and 311.0 K respectively (see Table 3.8). The chemical shifts are referenced against CH_3NO_2.

Table 3.8: Summary of experimental conditions for the ^{15}N MAS-NMR spectra of $PPh_4[Fe^{III}(Tp)(CN)_3]$ (PPh_4[**1**]) at 6 and 10 kHz at a ^{15}N-Larmor frequency of 40.55 MHz.

Probe	Spinning rate [kHz]	$T_{\text{set point}}$ [K]	$\delta^{\text{iso}}_{NiCp_2}$ [ppm]	T_{actual} [K]	δ^{iso}_{T} [ppm]
4-BL	6	308	-245.9	310.3	522, 502, 474
	10	301.8	-245.4	311.0	

The superposition of the 6 and 10 kHz spectra as displayed by Figure 3.19 allows the determination of three clearly defined different cyanide sites, though grouped in one signal bush, at δ_{iso} = 522, 502 and 474 ppm at 310 K respectively. This is 554–602 ppm shifted to higher frequency compared to the reference. Even though each isotropic signal corresponds to one of the three slightly different cyanides determined by X-Ray diffraction analysis in the iron complex, no specific attribution can be done. This positive paramagnetic contribution corresponds to a positive spin density perceived by the three nitrogen nuclei of the cyanides. This is consistent with the positive sign of the spin density found on the nitrogen atoms by DFT calculations performed on $PPh_4[Fe^{III}(Tp)(CN)_3]$ (PPh_4[**1**]) (see Figure 3.18), but also with previous paramagnetic NMR studies on hexacyanidometallates.[148]

The resolution of the ^{15}N NMR spectra is better than for the respective ^{13}C MAS-NMR spectra. It is due to a smaller dipolar coupling with the paramagnetic centre so that the relaxation is more favourable (longer), which correlates with narrower linewidths.

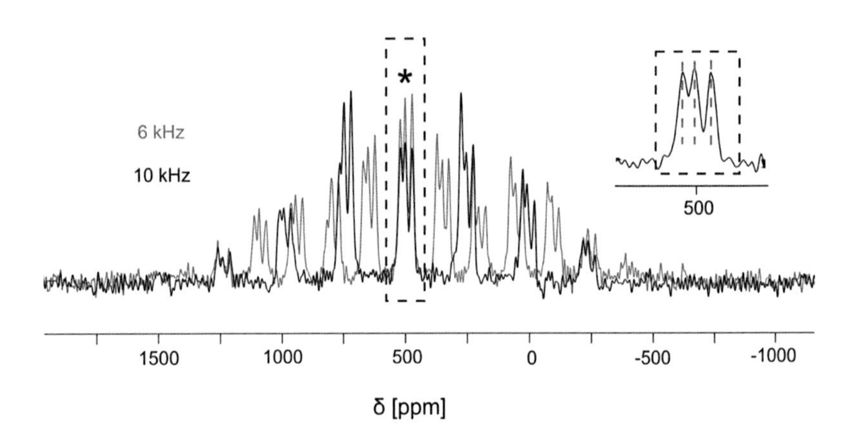

Figure 3.19: ^{15}N MAS-NMR spectra of $PPh_4[Fe^{III}(Tp)(CN)_3]$ (PPh_4[**1**]) at 6 kHz (red) and 10 kHz (black) at T = 310 K acquired at a ^{15}N-Larmor frequency of 40.55 MHz. The isotropic peak, which is not shifted, is marked with an asterisk. A zoom on the isotropic signal set of the 10 kHz spectrum is displayed right.

The 6 kHz spectrum could be simulated using HBA with 87.1% overlap. The extracted values are reported in Table 3.15 and both simulated and experimental spectra are depicted in Figure 3.20. Simulation of the spectra at 10 and 13 kHz gave similar results. More strikingly, the ^{15}N spectra of $PPh_4[Fe^{III}(Tp)(CN)_3]$ (PPh_4[**1**]) do not exhibit the same spectral symmetry as the reference compound. Indeed, while the cobalt(III) complex provides strongly axial signals (mean $\eta = 0.087$), the iron(III) species exhibits clearly non axial contributions, with asymmetry parameter ranging from 0.785 to 0.912. Simulation attempts with axial contributions failed at satisfactorily reproducing the overall shape of the signal with physically reasonable parameters. Since no change of the overall symmetry compared to the reference was found for the ^{13}C spectra, this cannot be an effect of the quadrupolar moment of the cobalt ion ($I = 7/2$, whereas ^{57}Fe has $I = 1/2$). Furthermore, this change in the spectrum symmetry between the spectrum of the diamagnetic reference $PPh_4[Co^{III}(Tp^*)(CN)_3]$ and the spectrum of $PPh_4[Fe^{III}(Tp)(CN)_3]$ (PPh_4[**1**]) is not due to a Tp structural singularity due to the lack of methyl groups compared to Tp*, because this spectral symmetry change is also observed for **8** (Tpm* ligand) and PPh_4[**7**] (Tp* ligand). Indeed, the structural symmetry around the cyanides atoms is actually axial, which is confirmed by the diamagnetic spectra of $PPh_4[Co^{III}(Tp^*)(CN)_3]$, and is expected for a linear fragment. However, the empty, antibonding cyanide molecular orbital of π^* symmetry (whose principal contribution is the p orbitals of the nitrogen atom) possesses an adequate symmetry to interact with the d orbitals containing the unpaired electron through backbonding. This results in a significant, positive spin delocalisation preferentially into one of the p orbitals of the nitrogen atoms (positive p-looking orbital contribution on the cyanide atoms in Figure 3.18). This axial, perpendicular to the CN triple bond orbital based on the nitrogen atom interacts with the nucleus centred 2s orbital of the same atom and is therefore responsible for the sweep from an axial-symmetric signal without unpaired electron to a strong non-axial one in its presence.

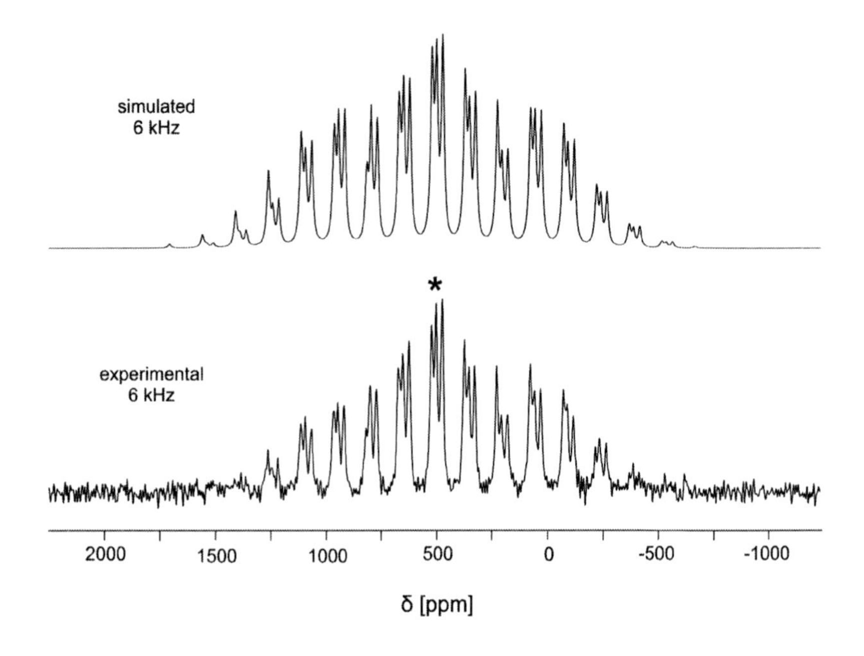

Figure 3.20: Simulated (top) and experimental (bottom) ^{15}N MAS-NMR spectrum of $PPh_4[Fe^{III}(Tp)(CN)_3]$ (PPh_4[**1**]) acquired at a Larmor frequency of 40.58 MHz, $\delta(NiCp_2)$ = -245.9 ppm (*T* = 310.3 K) and a spinning rate of 6 kHz.

Two CSA tensors of the cyanide nitrogen sites in $PPh_4[Fe^{III}(Tp)(CN)_3]$ (PPh_4[**1**]) exhibit a negative anisotropy ($\Delta\delta_{502}$ = -1277 ppm and $\Delta\delta_{474}$ = -1270 ppm), while the third one is characterised by its somewhat bigger positive anisotropy ($\Delta\delta_{522}$ = 1426 ppm). This has a major impact on the xx, yy and zz labelling, since it switches artificially the position of the xx and zz chemical shift tensors for one of the cyanide sites. The overall CSA amounts to 1960 ppm, that is 78 kHz at this Larmor frequency. This is quite high but is coherent with the CSA exhibited by $K_3[Fe(CN)_6]$ for the same nucleus.[148]

3.1.6.4 Paramagnetic NMR studies of [Fe^{III}(Tpm)(CN)$_3$] (8)*

^{13}C MAS-NMR spectroscopy

The ^{13}C MAS-NMR spectra were recorded by magic angle spinning on a 700 MHz Bruker spectrometer equipped with a 1.3-BL probe. The sample ^{13}C MAS-NMR spectrum was recorded at 55 and 60 kHz for an inner temperature of 307.1 K (δ_H([NiCp$_2$]) = -248.4 ppm) in the rotor, using a Hahn Echo pulse sequence. The temperature data for both spinning rates are summarised in Table 3.9. Since the irradiation window is too small to acquire the extremely large spectrum at once, two spectra with different irradiation windows were recorded at 60 kHz.

Table 3.9: Tabulated report of ^{13}C MAS-NMR measurement of **8** recorded with a 1.3-BL probe head and a Bruker AV-700 spectrometer.

Probe	Spinning rate (kHz)	$T_{\text{set point}}$ (K)	$\delta_{NiCp_2}^{before}$ (ppm)	$\delta_{NiCp_2}^{after}$ (ppm)	T_{actual} (K)	ΔT (K)	δ_T^{iso} (ppm)
1.3-BL	55	280	-248.5	-248.2	307.1	0.4	-4135
	60	260	-248.3	-248.5	307.1	0.3	

Only one isotropic shift was found at -4135 ppm at this temperature, as shown in Figure 3.21. The linewidth of the signal does not allow sufficient resolution to separate the different cyanide contributions of the isotropic signal (because of strong dipolar interactions with the close paramagnetic source). The paramagnetic contribution is once again strongly negative, which is consistent with the negative spin density observed by DFT calculations performed on **8** (see Figure 3.22). Even if the spectra are distorted due to the size of the CSA (span of 3200 ppm, to be compared with an experimental span of 2300 ppm for PPh$_4$[Fe^{III}(Tp)(CN)$_3$] (PPh$_4$[**1**]) on the same spectrometer), the overall symmetry of the ^{13}C MAS-NMR spectrum of **8** is clearly axial.

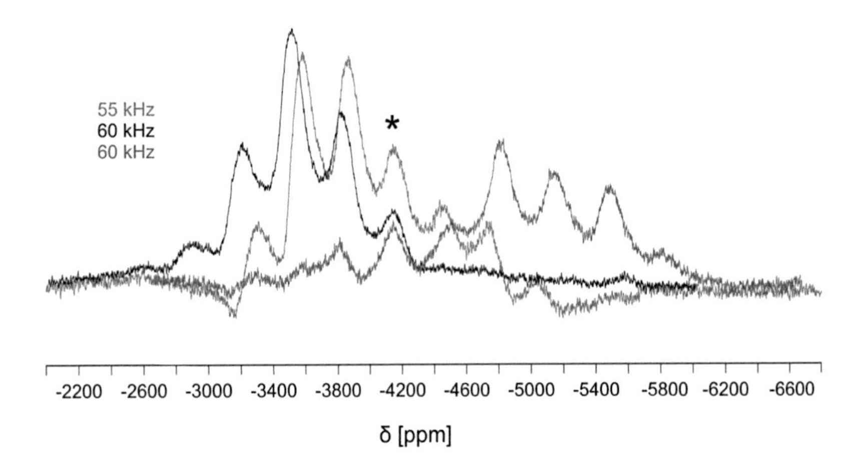

Figure 3.21: ^{13}C MAS-NMR spectra of **8** at 55 kHz (red) and 60 kHz (black and gray) at $T = 307$ K acquired at a Larmor frequency of 175.37 MHz. The isotropic peak, which is not shifted, is marked with an asterisk.

This is the same symmetry as observed for the ^{13}C MAS-NMR spectra of the diamagnetic reference $PPh_4[Co^{III}(Tp^*)(CN)_3]$ and the paramagnetic $PPh_4[Fe^{III}(Tp)(CN)_3]$ (PPh_4[**1**]) complex. It is noteworthy that the increase of the CSA for **8** compared to $PPh_4[Fe^{III}(Tp)(CN)_3]$ (PPh_4[**1**]) complex is coupled with an increase of signal width at half-height, which makes the recording of the signal even more difficult.

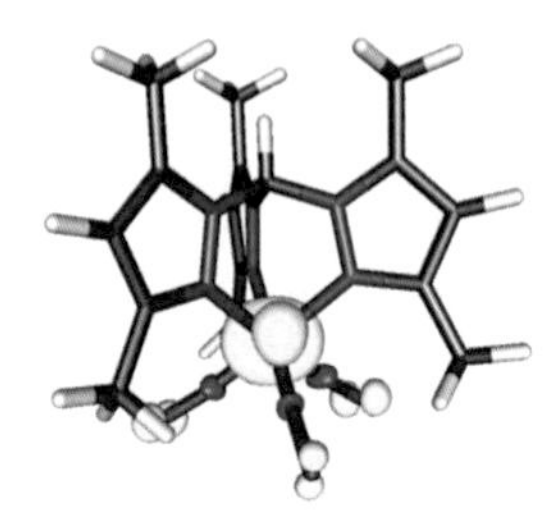

Figure 3.22: Spin density distribution of **8** at the [BP86, def2-SVP] level of theory (gray atoms: carbon, blue ones: nitrogen, red one: iron). Positive spin density in yellow, negative spin density in blue.

Since it was not possible to acquire the whole spectrum at once, no Herzfeld-Berger analysis was performed, and therefore the isotropic value at 307.1 K is the only experimental value recorded in Table 3.14.

^{15}N MAS-NMR spectroscopy

The ^{15}N MAS-NMR spectra were acquired by magic angle spinning in a 500 MHz Bruker spectrometer equipped with a 4-BL probe head. The sample was measured at 6 kHz and 13 kHz. The temperature data are summarised in Table 3.10. For each spinning rate, $\delta_H([NiCp_2])$ was set to -243.0 ppm which corresponds to an inner temperature of 314.2 K. While the temperature was stable during the whole acquisition at 13 kHz, the inner temperature of the sample increased by 1.2 K to 315.4 K for the measurement at 6 kHz. This is probably due to the fact that the 13 kHz spectra were recorded during the day (6 hours) while the acquisition of the 6 kHz counterparts took place during the night (18 hours).

Table 3.10: Tabulated report of ^{15}N MAS-NMR measurements of **8** recorded with a 4-BL probehead and a Bruker AV-500 spectrometer.

Probe	Spinning rate (kHz)	$T_{\text{set point}}$ (K)	$\delta_{NiCp_2}^{\text{before}}$ (ppm)	$\delta_{NiCp_2}^{\text{after}}$ (ppm)	T_{actual} (K)	ΔT (K)	δ_T^{iso} (ppm)
4-BL	6	323	-243.0	-242.1	314.8	1.2	727.2
	13	296	-242.9	-242.9	314.3	0[c]	

However, such a small temperature increase has little to no impact on the determination of the isotropic chemical shift(s) because of the linewidth of the signals ($\Delta\nu_{1/2}$ ~ 1900 Hz). For the same reason, only one isotropic peak was found by

[c] Since $[NiCp_2]$ is an air-sensitive paramagnetic species, its ^{1}H MAS-NMR signals are quite broad ($\Delta\nu_{1/2} \approx$ 700 Hz), which corresponds to 7.9 ppm on a 500MHz NMR spectrometer. In these conditions, the reading error on the chemical shift is quite high and can manually be evaluated to 0.2-0.3 ppm (~0.5 K).

superposition of 6 kHz and 13 kHz spectra (see Figure 3.23): since the crystal structure of **8** · 2 MeCN displays three slightly different cyanide bonds (Fe–C ranging from 1.909(3) Å to 1.920(4) Å), one would have expected three different isotropic ^{15}N signals. However, the isotropic peak exhibits a signal half-width of 1887 Hz, which is far too broad to resolve the different cyanide ligands.

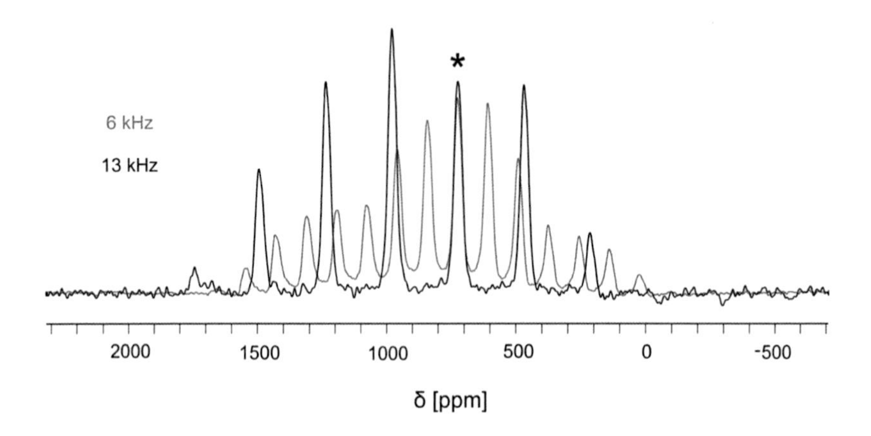

Figure 3.23: ^{15}N MAS-NMR spectra of **8** at 6 kHz (red) and 13 kHz (black) at 314 K acquired at a Larmor frequency of 50.87 MHz. The isotropic peak, which is not shifted, is marked with an asterisk.

Further investigation at both sides of the signal at 6 and 13 kHz spinning rates revealed additional spinning bands, leading to an overall CSA of 126 kHz at a Larmor frequency of 50.87 MHz, which is by far wider than the 4-BL probe head irradiation range for this nucleus. Such a large chemical shift anisotropy is indicative of an anisotropic interaction between the iron(III) metal ion and the cyanide ligand nuclei. The intramolecular dipole interaction due to the nuclear moment are negligible for ^{15}N spectra due to the low natural abundance of the adjacent ^{13}C nucleus (^{12}C nuclei having no spin), spinning side band patterns can be seen as the result of the magnetic contributions.

Since the chemical shift anisotropy increases for higher magnetic fields, the spectrum of **8** was recorded at a lower Larmor frequency (ν_L = 30.46 MHz) on a Bruker AV-300 with a

MQ probe head and the same rotor size, allowing the recording of the whole spectrum at once (see Figure 3.24). The experimental spectrum of Figure 3.24 was mathematically corrected by removing the 158 first data points of the FID in order to suppress artefacts caused by acoustic ringing probe head which leads to baseline distortion and its δ_{iso} was set at the same chemical shift as the previous 4-BL spectra (inner recorded temperature of 314.3 K). Due to the weaker magnetic field, a better resolution could be achieved with a signal width at half height for the isotropic signal of 1197 Hz, which was sufficient to show some shoulder structure ($\Delta\delta_{iso} \approx 9$ ppm), corresponding to two cyanide ligands.

This spectrum could be simulated by Herzfeld-Berger Analysis, with an overlap of 94.8%. The extracted values are reported in Table 3.15, while the simulated spectrum is depicted above the experimental one in Figure 3.24. At 314.3 K, the isotropic values are $\delta_{iso} = 713$ and 731 ppm, which are shifted by 793 and 811 ppm higher frequency from the mean isotropic value of the reference compound respectively. This corresponds to a positive spin density on the observed nuclei, in agreement with the DFT calculations for **8** (Figure 3.22). As already referred to, ^{15}N signals of **8** also undergo a change of spectral symmetry, with cyanide sites being clearly non-axial (η = 0.821 and 0.972). The anisotropy parameters are positive and in absolute value bigger than for $PPh_4[Fe^{III}(Tp)(CN)_3]$ (PPh_4[**1**]) ($\Delta\delta$ = 1630 and 1564 ppm *vs* 1426, -1277 and -1270 ppm for $PPh_4[Fe^{III}(Tp)(CN)_3]$ (PPh_4[**1**])), hinting towards a larger anisotropy in the spin density localisation of **8** than in the Tp parent compound. **8** also suffers from a bigger CSA (~2600 ppm) than its Tp counterpart, which is responsible for the recording issues.

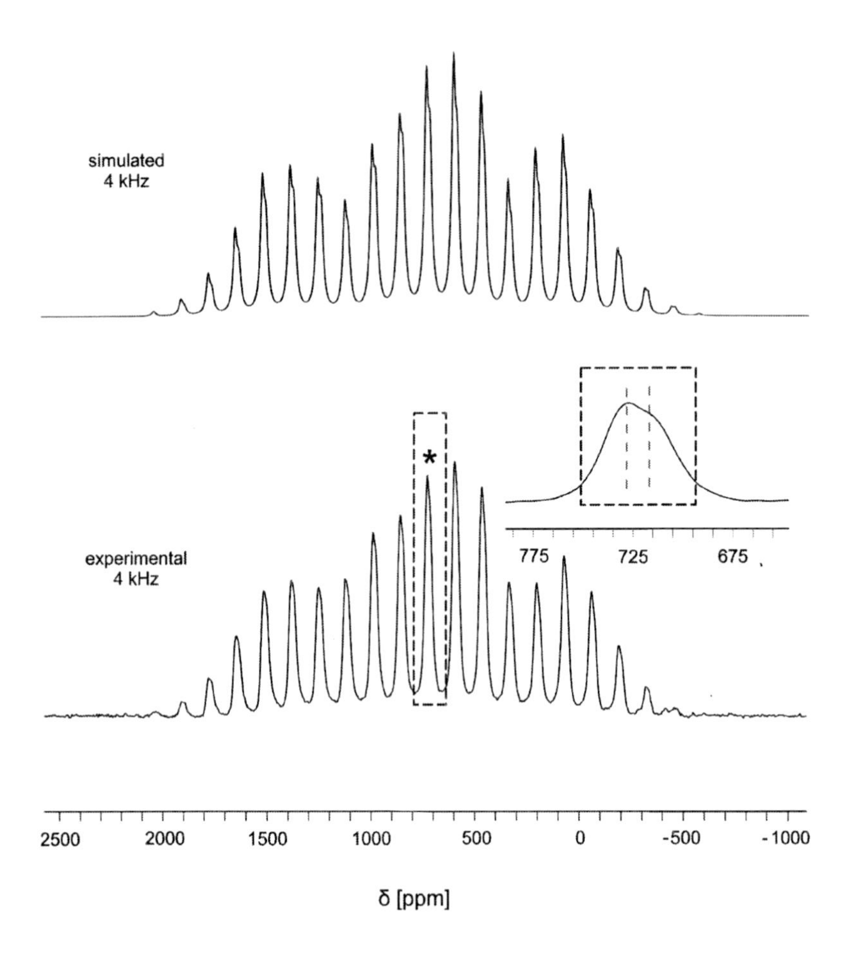

Figure 3.24: Simulated (top) and experimental (bottom) ^{15}N MAS-NMR spectra of **8** acquired at a Larmor frequency of 30.46 MHz, $\delta([NiCp_2])$ = -242.9 ppm (T = 314.3 K) and a spinning rate of 4 kHz. Overlap: 94.8%.

3.1.6.5 Paramagnetic NMR studies of $PPh_4[Fe^{III}(Tp^*)(CN)_3]$ (PPh_4[7])

^{13}C MAS-NMR spectroscopy

The measurement was carried out on 10 mg of dry ground crystalline ^{13}C-enriched PPh_4[**7**] in a 1.3 mm rotor. The ^{13}C spectra were acquired by MAS-NMR in a 700 MHz spectrometer equipped with a 1.3-BL probe head, and the chemical shift was referenced against TMS (adamantane as secondary reference). The spectrum of the sample was recorded at 50 and 60 kHz and the temperature data for both measurements are summarised in Table 3.11.

Table 3.11: Tabulated report of ^{13}C MAS-NMR measurement of PPh_4[**7**] recorded with a 1.3-BL probehead and a Bruker AV-700 spectrometer.

Probe	Spinning rate (kHz)	$T_{\text{set point}}$ (K)	$\delta_{NiCp_2}^{\text{before}}$ (ppm)	$\delta_{NiCp_2}^{\text{after}}$ (ppm)	T_{actual} (K)	ΔT (K)	δ_T^{iso} (ppm)
1.3-BL	50	280	-246.0	-245.5	310.5	0.6	-4112
	60	255	-245.7	-245.7	310.6	0^c	

The inner temperature was set by adjusting $\delta_H([NiCp_2])$ to *ca* δ = -245.7 ppm which corresponds to 310.5 K. The ^{13}C spectra were acquired using a Hahn-Echo pulse sequence. Superposition of the 50 and 60 kHz spectra (see Figure 3.25) allowed the determination of the isotropic signal at δ_{iso} = -4112 ppm at 310.5 K, which was confirmed by a third measurement at 40 kHz (not pictured).

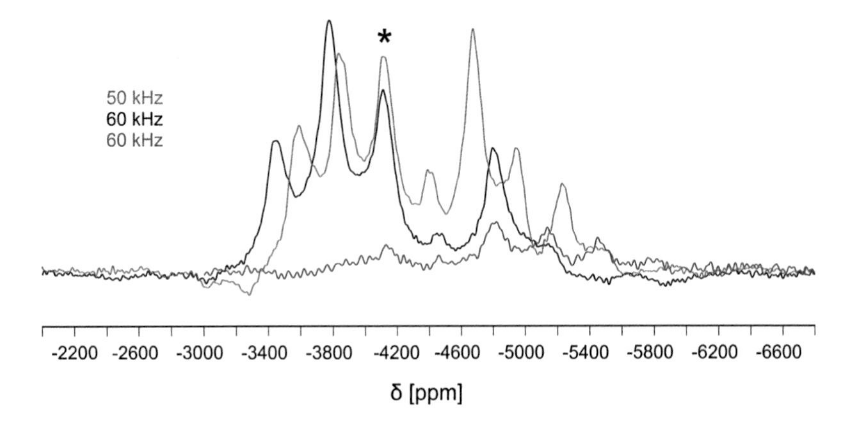

Figure 3.25: ^{13}C MAS-NMR spectra of PPh_4[**7**] at 50 kHz (red) and 60 kHz (black and gray) at $T = 311$ K acquired at a Larmor frequency of 175.37 MHz. The isotropic peak, which is not shifted, is marked with an asterisk.

The black and grey spectra of Figure 3.25, recorded at the same temperature and the same spinning rate but with different frequencies for the irradiation spectral window (O1p), illustrate that the side-band pattern is broader than the irradiation window of the 1.3-BL probe head. The 50 kHz spectrum seems to encompass the whole necessary frequency range; but spectral distortion at the edges of the signal is to be expected. It is therefore not surprising that, despite several attempts, the 50 kHz spectrum could not be simulated satisfyingly by Herzfeld-Berger analysis. Due to this only the isotropic value is reported in Table 3.14.

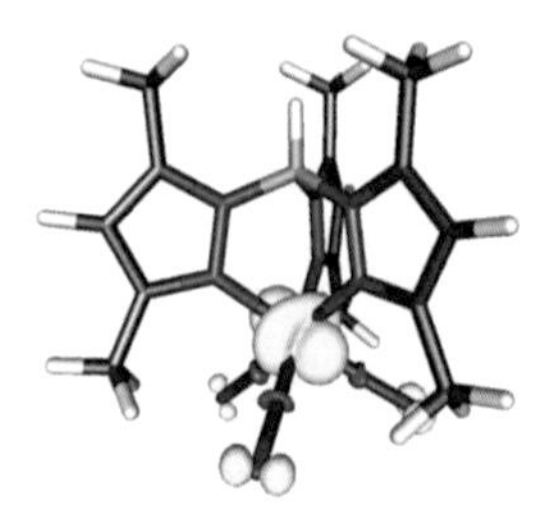

Figure 3.26: Spin density distribution of PPh_4[**7**] at the [BP86, def2-SVP] level of theory (gray atoms: carbon, blue ones: nitrogen, red one: iron, green one: boron). Positive spin density is in yellow, negative spin density in blue.

The isotropic contribution to the chemical shift is strongly negative, indicating negative spin density. This is consistent with the ^{13}C MAS-NMR for the two other tricyanido iron(III) complexes of Tp and Tpm* presented in this work, as well as DFT calculation (Figure 3.26).

The signals are also sharper compared to **8**, which leads to better spectral resolution and better accuracy on the determination of the chemical shift. However, the peaks remain broad enough to not resolve the three expected cyanide contributions. Even if the spectra depicted in Figure 3.25 are clearly strongly distorted, a closer look at the 60 kHz spectra indicates that the original, undistorted signal might be axial, like all ^{13}C MAS-NMR spectra presented so far.

^{15}N MAS-NMR spectroscopy

The ^{15}N MAS-NMR spectra were recorded on a 300 MHz spectrometer, equipped with a 4-MQ probehead, because the previously used 4-BL probehead/500 MHz spectrometer combination could not acquire the entire signal at once. The inner temperature was set to the chemical shift of nickelocene of δ = -250.1 ppm, which corresponds to an actual temperature of 304.7 K. The spectra at 8 and 10 kHz were acquired using a Hahn-echo pulse sequence, and the baselines of the spectra were mathematically corrected. Their superposition led to the identification of 3 positive isotropic shifts, at 887, 889 and 991 ppm, respectively (see Figure 3.27), which can be ascribed to each of the three cyanides ligands.

Table 3.12: Tabulated report of ^{15}N MAS-NMR measurements of PPh_4[**7**] recorded with a 4-MQ probehead and a Bruker AV-300 spectrometer.

Probe	Spinning rate (kHz)	$T_{\text{set point}}$ (K)	$\delta_{NiCp_2}^{\text{before}}$ (ppm)	$\delta_{NiCp_2}^{\text{after}}$ (ppm)	T_{actual} (K)	ΔT (K)	δ_T^{iso} (ppm)
4-MQ	8	295.5	-250.1	-250.5	304.7	0.5	887, 889 and 991
	10	293	-250.2	–	304.8	–	

Two ^{15}N cyanides present almost identical isotropic chemical shifts, at 887 and 889 ppm (T = 304.7 K) while the third signal at 991 ppm is shifted to higher frequency. The spin density in the 2s orbital of the nitrogen atoms is mainly the result of the polarisation of this orbital by the spin density delocalised from the iron(III) ion into the 2p orbitals of the said nitrogen atoms. At this stage, it is consistent with DFT, which shows positive spin density on the nitrogen atoms (see Figure 3.26).

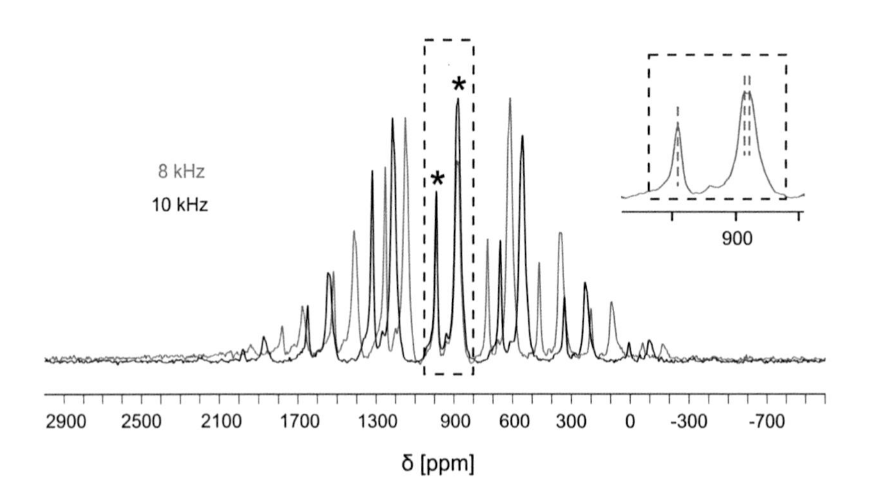

Figure 3.27: ^{15}N MAS-NMR spectra of PPh_4[**7**] at 8 kHz (red) and 10 kHz (black) at 305 K acquired at a Larmor frequency of 30.46 MHz. The isotropic peaks, which are not shifted, are marked with an asterisk.

The 8 kHz ^{15}N MAS-NMR spectrum could be simulated with Herzfeld-Berger analysis and the results are summarised in Table 3.15. Inspection of the spectrum shows the three cyanide contributions, two of them being very close to each other so that the spectrum is best simulated with only two cyanide sites, yielding an overlap of 87.6%.

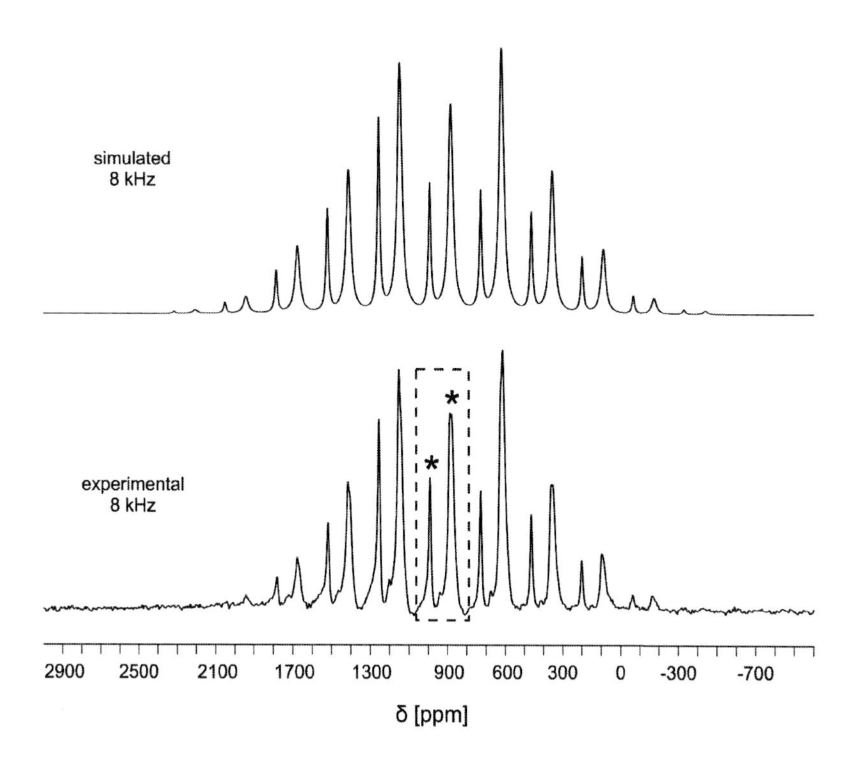

Figure 3.28: Simulated (top) and experimental (bottom) ^{15}N MAS-NMR spectrum of PPh_4[**7**] acquired at a Larmor frequency of 30.46 MHz, δ([$NiCp_2$]) = -250.3 ppm (T = 304.7 K) and a spinning rate of 8 kHz. Overlap: 87.6%.

The two CSA tensors of the nitrogen sites exhibit strongly non axial symmetry with $\eta = 0.964$ and 0.740 for the 884 ppm and 993 ppm (304.7 K) signal respectively. Because one of the signals is clearly ascribed to only one site and separated from the second site, the non-axiality of the system cannot be a calculus artefact. The anisotropy parameters of the two sites are smaller in absolute value than the anisotropy parameters found for **8** and, in average, bigger than the ones found for PPh_4[**1**]. They amount to $\Delta\delta$ = 1351 and -1519 ppm, but the sign, as already explained, is an artificial consequence of the Haeberlen convention. It was therefore treated consequently in the following spin density calculations.

3.1.6.6 Spin density calculations

Evaluation of dipolar contribution

When S = ½ (low spin iron(III) ion), the dipolar signal shift at temperature *T* can be calculated following equation (11):

$$\delta_T^{dip} = \frac{\mu_0}{4\pi}\frac{\beta_e{}^2 S(S+1)10^6}{9k_B T}\sum_j \frac{3\cos^2\theta_j - 1}{r_j{}^3}(g_\parallel{}^2 - g_\perp{}^2) \tag{11}$$

MAS-NMR experiments grant access to $\delta_{298,iso}^{para}$ by deducting the diamagnetic isotropic shift of the diamagnetic reference compound $PPh_4[Co^{III}(Tp^*)(CN)_3]$ (Table 3.4 and Table 3.6) from the experimental isotropic value $\delta_{298,iso}^{exp}$. In order to evaluate the Fermi contact term $\delta_{298,iso}^{con}$ using equation (8), one must evaluate the dipolar contribution δ_{298}^{dip} in the corresponding iron(III) complexes.

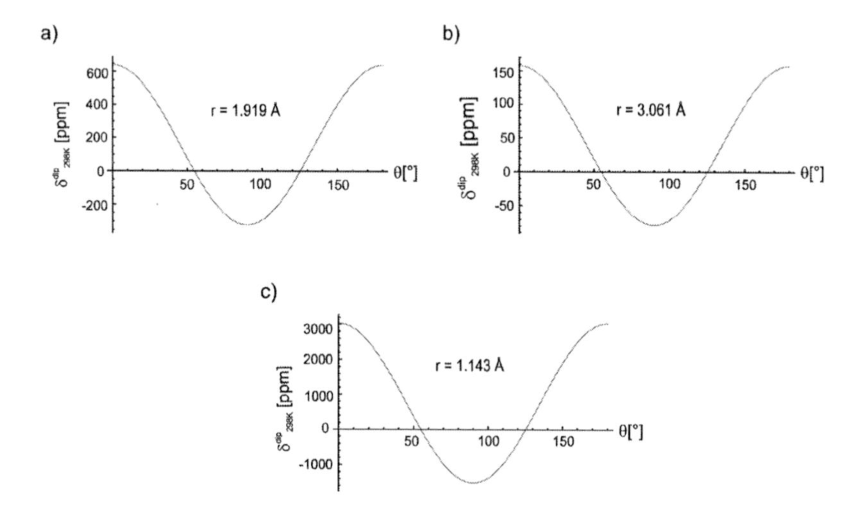

Figure 3.29: Dipolar chemical shifts (equation 11) at 298 K plotted in function of the angle θ, considering different configurations: a) effect of the spin density located on the iron (S = 1/2) in ^{13}C NMR, b) effect of the spin density located on the iron (S = 1/2) in ^{15}N NMR, c) effect of the spin density delocalised on the carbon atoms (S = 1/2, not weighted) on the dipolar chemical shifts undergone by the nitrogen atoms. The mean distances used for the plots are the mean values for $PPh_4[Fe^{III}(Tp)(CN)_3]$ reported in Table 3.13.

As shown in Figure 3.29, the value taken by this contribution is highly dependent from the angle θ between the easy magnetisation axis and the Fe···X axis, where X is the observed nucleus, and iron the location of the considered spin density. A ½-spin located on the iron ion, can have a deshielding effect up to *ca* 648 ppm or a shielding effect on the same atom up to *ca* -321 ppm at the cyanide carbons (case a)), 1.919 Å away, solely based on the geometry of the complex. Since nitrogen lies farther to the iron ion, the dipolar spin density contributions to its paramagnetic chemical shift is smaller (case b)), but can still range from +159 ppm to -79 ppm, which can have a significant impact on the calculated spin density. Due to the short triple bond distance, the electron spin density located on the carbon atom can have an impact on the neighbouring nitrogen atom, theoretically ranging from *ca* +3019 ppm to -1510 ppm for a full spin ½ (case c)). However, the spin on the carbon atoms is only a small fraction of the overall spin density, so that its effect is drastically reduced to about the same order of magnitude of the iron spin density effect on the nitrogen atoms. It is noteworthy that these contributions are cancelled for magic angle spinning ($\theta \approx 54.7°$) and its supplementary angle $\theta \approx 125.3°$. [**1**]$^-$, **8** and [**7**]$^-$ are all C_{3v} –symmetric complexes along the B/C···Fe axis. Symmetry considerations together with PND measurements carried out on [**1**]$^-$[unpublished] lead to the assumption that for all three tricyanido complexes, the easy axis of magnetisation lies along the symmetry B/C···Fe axis. This geometric configurations lead to θ angles very close to the cancelling angle: from 122.6° to 129.9° in overall for the three complexes. This ("accidental") optimal situation minimises the dipolar contributions to the chemical shifts for all sites and therefore allows the determination of the Fermi contact term with the highest possible accuracy. Table 3.13 summarises the X-Ray diffraction data and the calculated dipolar shift contributions for PPh_4[**1**], **8** and PPh_4[**7**], respectively.

It is noteworthy that for each compound, the dipolar contribution to the ^{15}N chemical shift induced by the electron spin density from the iron ion is about as high in magnitude as the dipolar contribution induced by the electron spin density located in the 2s orbital of the neighbouring carbon. As there are of opposite sign, the overall dipolar contribution can be reduced to 3 ppm to -5 ppm for PPh_4[**1**] +7 ppm to -2 ppm for PPh_4[**7**] and from +7 ppm to -2 ppm for **8**, which would be in each case negligible in regards to the involved Fermi contact shifts for these nuclei.

Table 3.13: Selected bond lengths and angles for PPh_4[**1**], **8** and PPh_4[**7**]: mean values and range calculated dipolar contributions at the observed nuclei ^{13}C and ^{15}N.

$PPh_4[Fe^{III}(Tp)(CN)_3]$ (PPh_4[**1**])	Mean value	Range
Fe–C bond length [Å]	1.919	1.911(6) – 1.930(6)
Fe···N distance [Å]	3.061	3.046 – 3.072
B···Fe–C angle [°]	126.5	124.5 – 127.6
B···Fe···N angle [°]	125.5	122.6 – 127.1
$\delta^{dip}_{298,Fe}$(C) [ppm]	+14.3	+27.3 / -8.7
$\delta^{dip}_{298,Fe}$(N) [ppm]	+0.7	+5.4 / -7.4
$\delta^{dip}_{298,C}$(N) [ppm]	-0.5	+4.9 / -3.8
$[Fe^{III}(Tpm^*)(CN)_3]$ (**8**)[d]	Mean value	Range
Fe–C bond length (Å)	1.914	1.909(3) – 1.920(4)
Fe···N distance (Å)	3.062	3.059 – 3.066
B···Fe–C angle (°)	127.0	125.1 – 129.4
B···Fe···N angle (°)	126.7	124.2 – 129.8
$\delta^{dip}_{298,Fe}$(C)	+14.5	+35.8 / -1.6
$\delta^{dip}_{298,Fe}$(N)	+2.9	+9.4 / -2.2
$\delta^{dip}_{298,C}$(N)	-2.0	1.5 / -6.3
$PPh_4[Fe^{III}(Tp^*)(CN)_3]$ (**7**)	Mean value	Range
Fe–C bond length (Å)	1.923	1.917(3) – 1.930(3)
Fe···N distance (Å)	3.075	3.072 – 3.077
B···Fe–C angle (°)	126.8	125.2 – 129.3
B···Fe···N angle (°)	126.5	124.0 – 129.9
$\delta^{dip}_{298,Fe}$(C)	+12.8	+34.0 / -0.9
$\delta^{dip}_{298,Fe}$(N)	+2.5	+9.5 / -2.5
$\delta^{dip}_{298,C}$(N)	-1.7	+1.7 / -6.3

[d] Since no g_x, g_y and g_z could be extracted from the EPR data of **8**, and since the NMR spectra of **8** have a CSA size as big as the one of PPh_4[**7**], the $g_{//}^2$ - $g_{\perp}^2$ value of PPh_4[**7**] was used for the dipolar shift contribution evaluation.

Isotropic spin densities

Since chemical shifts of paramagnetic species are strongly dependent on temperature and every spectrum was acquired at a different temperature, no direct comparison of raw data could be performed. Assuming the Curie law is valid for these species (which is very much reasonable near room temperature) and using equation (7), one can calculate the corresponding Fermi contact contribution to the chemical shift at 298 K:

$$\delta^{con}_{298,iso} = \left(\delta^{exp}_{T,iso} - \delta^{dia}_{iso}\right) T/298 - \delta^{dip}_{298,iso} \tag{12}$$

In Table 3.13, the dipolar chemical shift contributions for each compound and each nucleus are reported, along with the relevant structural information. The diamagnetic values for ^{13}C and ^{15}N are summarised in Table 3.4 and Table 3.6, respectively. The contact term gives access to the spin density felt by the nuclei i.e. that is due to unpaired electrons localised in the s orbitals (that have non zero contribution on the nuclei). This spin density can be due to direct delocalisation of the unpaired electron into these orbitals or to the polarisation of the s orbital by the p (or d in case of transition metal atom) orbital of the same atom. The extraction of the spin density on the atom itself can be deduced.

The average Landé factor was extracted from the EPR spectra for PPh_4[**1**] and PPh_4[**7**]. For **8**, they were extracted from the SQUID magnetic data.

As already mentioned, all three complexes possess negative spin densities in their carbon 2s orbitals (see Table 3.14), while the spin densities at the nitrogen atoms are positive (see Table 3.15). This is consistent with a dominating spin polarisation mechanism for the magnetic information propagation on the carbon atoms.[158,159] Such a mechanism is also strongly axial along the cyanide axis, which is consistent with the retained axial symmetry of the ^{13}C MAS-NMR spectra displayed by the three complexes.

Table 3.14: Solid-state ^{13}C NMR data[e] and spin densities[g] of the PPh_4[**1**], **8** and PPh_4[**7**].

Compound	PPh_4[**1**]		**8**	PPh_4[**7**]
$\delta^{exp}_{T,iso}(^{13}C)$	-3726	-3755	-4141	-4112
$\delta^{exp}_{T,zz}(^{13}C)$	-5159	-5110	–	–
$\delta^{exp}_{T,yy}(^{13}C)$	-3016	-3081	–	–
$\delta^{exp}_{T,xx}(^{13}C)$	-3003	-3074	–	–
$\Delta\delta^{exp}_{T}(^{13}C)$	-2150	-2033	–	–
$\eta\delta^{exp}_{T}(^{13}C)$	0.009	0.005	–	–
$\delta^{para}_{298,zz}(^{13}C)$	-5111	-5062	–	–
$\delta^{para}_{298,yy}(^{13}C)$	-3242	-3307	–	–
$\delta^{para}_{298,xx}(^{13}C)$	-3233	-3304	–	–
$\delta^{con}_{298,iso}(^{13}C)$	-4352		-4422[d]	-4440
Mean $\rho_s(^{13}C)$	-0.0441		-0.0313	-0.0356

If the previously evaluated uncertainties are taken into account, and considering that the precise attribution of the signals to the respective cyanide in the structure is not possible, the upper and lower bounds to the spin densities of $PPh_4[Fe^{III}(Tp)(CN)_3]$ (PPh_4[**1**]) on the carbon atoms are -0.0438 $(au)^{-3}$ and -0.0442 $(au)^{-3}$, respectively. The mean electron spin density located on the carbons atoms of PPh_4[**1**] is found to be -0.0441 $(au)^{-3}$ and must be compared to the mean electron spin density calculated by DFT for PPh_4[**1**] (see Figure 3.18): -0.0417 $(au)^{-3}$. Depending on the cyanide and the calculation conditions, this estimate ranges from -0.0323 $(au)^{-3}$ to -0.0517 $(au)^{-3}$, which encompasses the electron spin density value found by ^{13}C MAS-NMR.

The mean electron density located on the carbon atoms of PPh_4[**7**] (av.-0.0356 $(au)^{-3}$, ranging from -0.0358 to -0.0355 $(au)^{-3}$) was found to be lower than for PPh_4[**1**] but in the same order of magnitude, which indicates that the propagation of the magnetic information by polarisation of the orbitals located on carbon atoms might be less effective in PPh_4[**7**] than it is in PPh_4[**1**]. The calculated total spin density found by DFT

---

[e] δ in ppm. PPh_4[**1**]: T = 333.4 K, ν_{rot} = 60 kHz; **8**: T = 307.1 K, ν_{rot} = 60 kHz; PPh_4[**7**]: T = 310.5 K, ν_{rot} = 60 kHz.

calculation on the cyanide carbon atoms of PPh_4[**7**] ranges from -0.03378 to -0.03948 (au)$^{-3}$, (average: -0.0375 (au)$^{-3}$) which encompasses the NMR experimental values. This matches quite well the values found by NMR, however, the DFT calculations reveal that most of this spin density is calculated to be in the p orbitals of the carbon atoms, with only a small portion of it (in average -0.0082 (au)$^{-3}$) in the s orbital. This seems somewhat contradictory, since (i) in principle only the electron spin density located in the s orbitals can interact with the nuclei and can be reflected in the contact term (It is however worth noticing that the spin density in p orbital can be reflected to s orbital through polarisation mechanism so that the spin density present in the p orbital is indirectly reflected in the contact term); (ii) this contradicts the hypothesis of a dominant polarisation mechanism for the propagation of the magnetic information on carbon atoms; (iii) it is inconsistent with previous studies on hexacyanidometallates.[82,148] This theoretical question is still open and it is one of the issue that is currently investigated in the framework of a collaboration with theoreticians from the university of Rennes, (K. Costuas, B. Leguennic) and physicists from CEA (B. Gillon) to extract direct spin density measurement by PND.

Since EPR data are not available, SQUID magnetometry was used to obtain the $\chi_M T$ at 300 K, from which the g_{av} value can be deduced using the spin-only formula. It is worth noticing that this g effective value bears the magnetic orbital contribution which is present in low-spin iron(III) ions with $^2T_2$ electronic ground term.. The mean electron spin density found by NMR is in the same order of magnitude as for PPh_4[**7**] and PPh_4[**1**], albeit lower. It amounts to -0.0313 (au)$^{-3}$ for an average Landé factor of $g = 2.69$, but rises to -0.0355 (au)$^{-3}$ when using the g value of PPh_4[**7**] and -0.0381 (au)$^{-3}$ for the effective $g_{eff} = 2.44$ calculated at 11 K. For g_{eff} = 2.69, the lower and upper bounds are -0.0315 (au)$^{-3}$ and -0.0312 (au)$^{-3}$. This is somewhat smaller than the total electron spin densities calculated by DFT on the cyanide carbon atoms: from -0.0384 (au)$^{-3}$ to -0.0439 (au)$^{-3}$. This would correspond to average Landé factors between 2.44 and 2.27. However, and similarly to PPh_4[**7**], the calculated total spin density ascribes most of the spin density to the p orbitals of the carbon atoms, with only -0.0099 (au)$^{-3}$ in average in the s orbitals, which is insufficient to account for the observed contact shifts.

Table 3.15: Solid-state ^{15}N NMR data[f] and spin densities[g] of PPh$_4$[**1**], compound **8** and PPh$_4$[**7**].

Compound	PPh$_4$[**1**]			**8**		PPh$_4$[**7**]	
$\delta^{exp}_{T,iso}(^{15}N)$	522	502	474	731	713	884	993
$\delta^{exp}_{T,zz}(^{15}N)$	1478	-345	-380	1817	1756	1784	-20
$\delta^{exp}_{T,yy}(^{15}N)$	419	579	512	634	699	867	1125
$\delta^{exp}_{T,xx}(^{15}N)$	-332	1273	1292	-258	-314	-1	1874
$\Delta\delta^{exp}_{T}(^{15}N)$	1434	-1272	-1282	1630	1564	1351	-1519
$\eta\delta^{exp}_{T}(^{15}N)$	0.785	0.819	0.912	0.821	0.972	0.964	0.740
$\delta^{para}_{298,zz}(^{15}N)$	1432[h]	76	39	1808	1743	1719	407
$\delta^{para}_{298,yy}(^{15}N)$	360	526	456	591	659	811	1075
$\delta^{para}_{298,xx}(^{15}N)$	90[h]	1219	1238	169	110	427	1811
Mean $\delta^{con}_{298,iso}(^{15}N)$	602			847		1023	
Mean $\rho_s(^{15}N)$	0.0034			0.0034		0.0046	

In absolute values, the electron spin density located in the 2s orbitals of the nitrogen atoms of the cyanides is tenfold smaller than that found in the 2s orbitals of the carbon atoms of the cyanides. For PPh$_4$[**1**], the experimental mean spin density found by MAS-NMR averages 0.0034 (au)$^{-3}$, but ranges from 0.0032 to 0.0033 (au)$^{-3}$ for $\delta^{exp}_{T,iso}(^{15}N) = 474$ ppm and from 0.0035 (au)$^{-3}$ to 0.0036 (au)$^{-3}$ for $\delta^{exp}_{T,iso}(^{15}N) = 522$ ppm. DFT calculations performed on PPh$_4$[**1**] lead to more spread values of spin density varying from 0.0362 to 0.0078 (au)$^{-3}$, depending on the cyanide (and on the DFT solvent parameters). In every case, it is larger compared to values measured by MAS-NMR. This is not surprising, since the main propagation mechanism of the spin density to the

---

[f] δ in ppm. PPh$_4$[**1**]: T = 310.3 K, ν_{rot} = 6 kHz; Compound **8**: T = 314.3 K, ν_{rot} = 4 kHz; PPh$_4$[**7**]: T = 304.8 K, ν_{rot} = 8 kHz.

[g] ρ_s in (au)$^{-3}$.

[h] The anisotropy of this contribution is positive, which leads to an artificial change in the xx and zz Haeberlen convention compared to the reference mean parameters. In order to get meaningful results for this contribution, the chemical shift are on this occasion calculated as: $\delta^{T}_{xx} = \delta^{T}_{xx} - \delta^{ref}_{zz}$ and $\delta^{T}_{zz} = \delta^{T}_{zz} - \delta^{ref}_{xx}$ instead of the normal $\delta^{T}_{xx} = \delta^{T}_{xx} - \delta^{ref}_{xx}$ and $\delta^{T}_{zz} = \delta^{T}_{zz} - \delta^{ref}_{zz}$.

nitrogen atom should be the spin delocalisation from the iron ion t_{2g} orbitals to the antibonding π^* cyanide orbital, whose main contribution is born by the nitrogen. This electron spin density in turn polarises the s orbital of the same nucleus, which is the information obtained by the analysis of the isotropic chemical shift performed in this work. This is consistent with the symmetry change observed for the ^{15}N MAS-NMR spectra due to the unpaired electron presence in one of the p orbitals. The fraction of the spin density located in these p orbitals could be in principle evaluated by analysis of the CSA of the signal.[148] However, this requires some approximations and very few works have carry out such analysis (i.e. Baumgärtel *et al.*[148]). In all cases, they were done on systems with purely axial CSA, which is not the case here. The adaptation of axial equations to non-axial systems is definitely not trivial and therefore beyond the scope of this work.

There is significantly more spin density located in the s orbital of the cyanide nitrogen atoms of PPh_4[**7**] than in PPh_4[**1**], but it remains in the same order of magnitude. It reaches 0.0046 (au)$^{-3}$ in average, which actually encompasses two different ranges: from 0.0044 (au)$^{-3}$ to 0.0045 (au)$^{-3}$ for $\delta^{exp}_{T,iso}$(^{15}N) = 884 ppm (two cyanides) and 0.0049 (au)$^{-3}$ to 0.0050 (au)$^{-3}$ for $\delta^{exp}_{T,iso}$(^{15}N) = 993 ppm (one cyanide). The total spin densities located on the cyanide nitrogen atoms calculated by DFT performed on PPh_4[**7**] (Figure 3.26) range from 0.0269 (au)$^{-3}$ to 0.0567 (au)$^{-3}$. As expected, these values are clearly higher than the values obtained by NMR measurements, but encompass the spin density located in the p orbitals. However, and as previously noted for the ^{13}C data analysis of the same species, almost no spin density is found in the s orbitals by DFT calculations (0.0003 (au)$^{-3}$ to -0.0002 (au)$^{-3}$).

For **8**, the mean spin density measured by NMR ranges from 0.0034 (au)$^{-3}$ (g_{eff} = 2.69) to 0.0041 (au)$^{-3}$ (g_{eff} = 2.44). This is consistent with the spin densities found for the two other compounds. When the uncertainties on the various contributions and cyanide sites are taken into account, the spin density range is quite narrow. The values range from 0.0033 to 0.0034 (au)$^{-3}$ for g_{eff} = 2.69 and from 0.0040 (au)$^{-3}$ to 0.0042 (au)$^{-3}$ for g_{eff} = 2.44. It is possible to state that the data are consistent with the values of the two other compounds values, but the comparison cannot be brought further without an accurate value for the Landé factor of **8**. It is noteworthy that the DFT calculated total spin densities on the nitrogen atoms are found to be an order of magnitude bigger than the measured values. It is mainly attributed to the p orbitals of the nitrogen atoms, leaving only slightly more density in the s orbitals than for PPh_4[**7**]: 0.0006 $(au)^{-3}$ to 0.0009 $(au)^{-3}$ for **8**.

Compared to the reported $K_3[Fe^{III}(CN)_6]$,[148] the three κ^3–substituted hexacyanidometallates analysed in this work achieve a better transfer of the spin density on the carbon atoms (-0.0370 $(au)^{-3}$ in average for compounds PPh_4[**1**], PPh_4[**7**] and **8,** to be compared with -0.0275 $(au)^{-3}$ for $K_3[Fe^{III}(CN)_6]$) but a poorer transfer to the 2s orbitals of the nitrogen atoms (0.0048 $(au)^{-3}$ in average for compounds PPh_4[**1**], PPh_4[**7**] and **8**, to be compared with 0.0064 $(au)^{-3}$ for $K_3[Fe^{III}(CN)_6]$). The latter could be due to a more favourable orbital overlap between the cyanide fragments and the iron ion in the ferricyanide than in the complexes presented here.

4 Molecular squares based on the $[Fe^{III}(Tp^*)(CN)_3]^-$ ($[7]^-$) building block

As exposed in the chapter 3, the $[Fe^{III}(Tp^*)(CN)_3]^-$ (**[7]**$^-$) complex possesses a much lower redox potential than $[Fe^{III}(Tp)(CN)_3]^-$ (**[1]**$^-$). It is thus expected that ETCST phenomena in {FeCo} square would require $\{Co(L)_2(NC\text{-})\}$ subunits with lower redox potential (or a stronger ligand field). Indeed, thermo- and photo-induced charge transfer properties were reported in two $\{Fe_2Co_2\}$ molecular squares whose ligands on the cobalt side are 4,4'-di*tert*butyl-2,2'-bispyridine (dtbbpy) and bipy, respectively.[94,95,100,134] These bipy derivatives induce a stronger ligand field on the cobalt ion than the bis(*N*-alkyl)imidazolylketones (bik) do in the reported photomagnetic Tp and Ttp-based complexes by Mondal *et al.*[96–98] Thus, when replacing the dtbbpy/bipy ligands by bik in such molecular squares, no oxidation-reduction reaction should occur and the resulting $\{Fe_2Co_2\}$ molecular squares **10** and **11** (see Figure 4.1) are expected to be paramagnetic – consisting of $\{Fe^{III}_{LS}Co^{II}_{HS}\}$ pairs – over the whole temperature range.

The same applies for the $\{Fe_2Fe_2\}$ square **12** based on {Fe(Tp*)} and $\{Fe(bik)_2\}$ moieties. No charge transfer (CT) is expected. However, a switchable behaviour can be observed as the spin crossover phenomenon observed in $\{Fe^{III}_2Fe^{II}_2\}$ molecular squares is much more dependent upon the electronic environment of the iron(II) ions in the $\{Fe^{II}(L)_2(NC\text{-})_2\}$ moieties than it is on the electronic properties of the tricyanido iron(III) building block providing the *N*-bridging cyanides. Thus, the $[Fe^{II}(tpa)(NCS)_2](X)_2$ (X = anion) is a well-known example of a SCO system,[160,161] but SCO phenomena have also been detected in systems where the thiocyanate ligands have been replaced by *N*-bridging cyanide or dicyanamide ligands.[102,162]

The $\{Fe^{II}(bik)_2(NC\text{-})_2\}$ moiety has already been reported by Mondal *et al.* as a spin-state transition molecular fragment in a $\{Fe_2^{III}Fe_2^{II}\}$ mixed-valence square involving the $\{Fe^{III}(Tp)(CN)_3\}$ moiety as complex-as-ligand[107] and in a $\{Fe_2Mo_2\}$ molecular square.[98] It was also demonstrated that a spin-state transition can also be triggered at low temperature (T = 20 K) by laser light (LIESST effect). Interestingly, the most efficient wavelength at which the phenomenon is observed depends on the nature of the metalloligand (405 nm in {FeMo} systems, 735 nm in {FeFe} systems and 635 nm for the reference compound $[Fe^{II}(bik)_3]^{2+}$). Here it is expected that the analogous $\{Fe^{III}{}_2Fe^{II}{}_2\}$ molecular square **12** (see Figure 4.2), in which the Tp ancillary ligand was replaced by Tp*, retains the spin crossover behaviour of its parent compound.

4.1.1 Syntheses

$\{[Fe^{III}(Tp^*)(CN)_3]_2[Co^{II}(bik)_2]_2\}(X)_2 \cdot n\, H_2O$
(X = $[ClO_4]^-$, n = 2, 10) and (X = $[BF_4]^-$, 11)

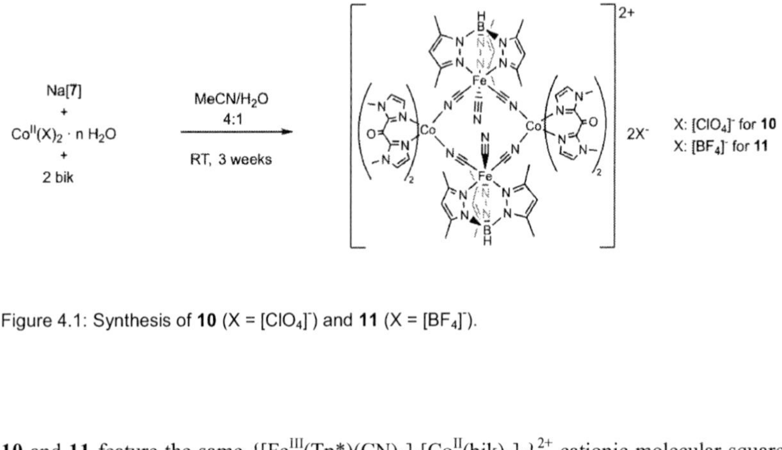

Figure 4.1: Synthesis of **10** (X = $[ClO_4]^-$) and **11** (X = $[BF_4]^-$).

10 and **11** feature the same $\{[Fe^{III}(Tp^*)(CN)_3]_2[Co^{II}(bik)_2]_2\}^{2+}$ cationic molecular square with different counteranions: **10** is a perchlorate salt while **11** is a tetrafluoroborate salt. Both compounds are synthesised by slow evaporation of a solution of the respective reagents in an acetonitrile/water (4:1) mixture over a few weeks (see Figure 4.1). In both cases, the solvent evaporation rate is crucial: if the latter is not slow enough, Na[7] tends to recrystallise before forming **10** or **11**.

$\{[Fe^{III}(Tp^*)(CN)_3]_2[Fe^{II}(bik)_2]_2\}(ClO_4)_2 \cdot 2\ H_2O$ (12)

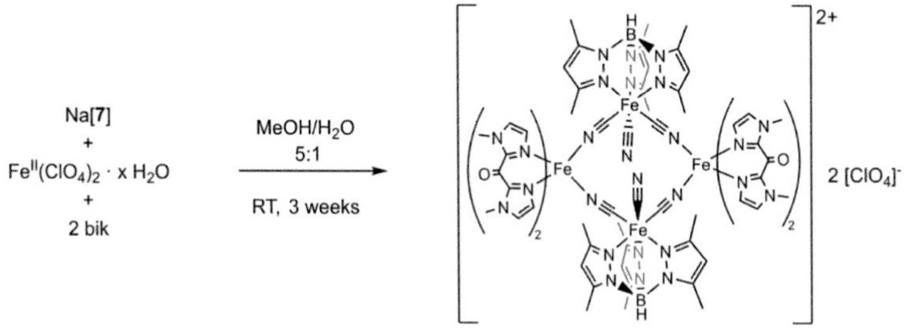

Na[**7**] + $Fe^{II}(ClO_4)_2 \cdot x\ H_2O$ + 2 bik → (MeOH/H_2O 5:1, RT, 3 weeks) → [...]$^{2+}$ 2 $[ClO_4]^-$ **12**

Figure 4.2: Synthesis of **12**.

Unlike **10** and **11**, **12** is best synthesised by slow evaporation of a methanol/water (5:1) reaction mixture at room temperature (see Figure 4.2). When the starting Na[**7**] concentration amounts to 3.33 mM, as for **10** and **11** syntheses, **12** precipitates quantitatively as microcrystals over two days in 40% isolated yield. A starting Na[**7**] concentration of 1.67 mM before slow evaporation allows the growth of crystals of **12** suitable for X-ray diffraction, provided the evaporation rate of the solvent mixture is slow enough. **12** and Na[**7**] crystallise both as red blocks; however, when the solvent evaporation rate is slow enough and **12** is formed, the mother liquor is light pink, whereas it is deep blue when Na[**7**] is recrystallised. This is due to the presence of the intensive blue complex $[Fe^{II}(bik)_3](ClO_4)_2$, obtained by dismutation of the precursor $[Fe^{II}(bik)_2(S)_2](ClO_4)_2$ in solution.

4.1.2 Structural analyses

{[Fe^{III}(Tp*)(CN)$_3$]$_2$[M^{II}(bik)$_2$]$_2$}(ClO$_4$)$_2$ · 2 H$_2$O (M = Co 10, M = Fe 12)

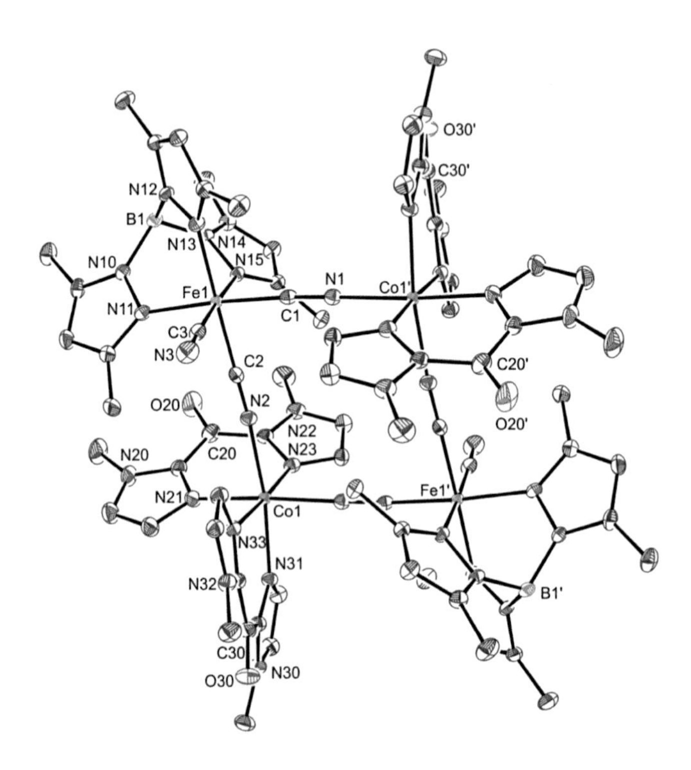

Figure 4.3: Molecular structure of the cationic unit of molecular square **10** at 200 K. Atoms are displayed as 30% probability ellipsoids. Hydrogen atoms, water solvent molecules and perchlorate counteranions are omitted for clarity. Equivalent atoms (noted with an apostrophe) within the molecular square are generated with the following symmetry operations: -x, 1-y, 1-z. Selected bond lengths (Å) and angles (°) for **10**: Fe1–C1 1.917(4), Fe1–C2 1.911(4), Fe1–C3 1.920(4), Fe1–N11 2.004(3), Fe1–N13 2.006(3), Fe1–N15 1.981(3), Co1'–N1 2.108(3), Co1-N2 2.109(3), Co1–N21 2.116(3), Co1–N23 2.153(3), Co1–N31 2.129(3), Co1–N33 2.148(3), C20–O20 1.223(5), C30–O30 1.222(5), C1-Fe1-C2 84.74(15), C1-Fe1-C3 87.14(15), C2-Fe1-C3 85.78(16), N11-Fe1-N13 89.62(12), N11-Fe1-N15 89.84(12), N13-Fe1-N15 89.60(12), C1-Fe1-N13 91.63(14), C1-Fe1-N15 90.95(14), C2-Fe1-N11 93.99(14), C2-Fe1-N15 92.69(13), C3-Fe1-N11 92.03(14), C3-Fe1-N13 91.81(14), N1'-Co1-N2 96.15(12), N1'-Co1-N23 89.26(12), N1'-Co1-N31 88.61(12), N1'-Co1-N33 87.80(12), N2-Co1-N21 90.16(12), N2-Co1-N23 90.01(12), N2-Co1-N33 91.62(12), N21- Co1-N23 85.14(12), N21-Co1-N31 85.56(12), N21-Co1-N33 97.64(12), N23-Co1-N31 94.50(12), N31-Co1-N33 84.10(12), Fe1-C1-N1 177.8(3), Fe1-C2-N2 176.3(3), Fe1-C3-N3 177.3(3), Co1-N2-C2 175.5(3), Co1-N1'-C1' 178.7(3), Fe1-N11-N10-B1 4.1, Fe1-N13-N12-B1 6.0, Fe1-N15-N14-B1 4.3.

Selected bond lengths (Å) and angles (°) for **12**: Fe1–C1 1.910(9), Fe1–C2 1.930(8), Fe1–C3 1.920(9), Fe1–N11 2.013(6), Fe1–N13 2.024(6), Fe1–N15 1.985(6), Fe2–N1' 1.976(7), Fe2–N2 1.961(6), Fe2–N31 1.995(6), Fe2–N33 1.993(6), Fe2–N21 1.973(7), Fe2–N23 1.990(6), C1-Fe1-C2 84.7(3), C1-Fe1-C3 85.8(3), C2-Fe1-C3 87.3(3), C3-Fe1-N11 91.6(3), C2-Fe1-N11 91.5(3), C1-Fe1-N15 93.0(3), N15-Fe1-C2 91.4(3), N15-Fe1-N11 89.6(2), N13-Fe1-C1 94.4(3), N13-Fe1-C3 91.4(3), N11-Fe1-N13 89.3(2), N13-Fe1-N15 89.8(2), Fe1-C1-N1 174.4(6), Fe1-C2-N2 177.1(6), Fe1-C3-N3 177.6(7), N1'-Fe2-N2 94.4(2), N1'-Fe2-N31 90.1(2), N1'-Fe2-N23 91.0(3), N31-Fe2-N23 93.9(2), N23-Fe2-N2 86.9(2), N31-Fe2-N21 86.7(2), N23-Fe2-N21 88.5(3), N2-Fe2-N21 88.8(3), N1'-Fe2-N33 88.3(3), N31-Fe2-N33 89.4(3), N2-Fe2-N33 89.9(2), N21-Fe2-N33 92.3(3), C1'-N1'-Fe2 179.1(6), C2-N2-Fe2 174.3(6), B1-N10-N11-Fe1 9.7, B1-N12-N13-Fe1 5.5, B1-N14-N15-Fe1 6.2, Fe2···C20-O20 170.3, Fe2···C30-O30 175.2.

At 200 K, **12** was found to be isostructural with **10**, so only the cationic unit of **10** is depicted in Figure 4.3. Selected bond lengths and angles for **10** and **12** are listed in the caption. Both compounds crystallise in the triclinic space group $P\bar{1}$ with one chemical formula per unit cell but only half of one in the asymmetric unit. Their structure consists of a centrosymmetric dicationic cyanide-bridged tetranuclear heterobimetallic $\{Fe_2M_2\}$ complex, two perchlorate anions and two water lattice molecules. The tetranuclear unit is made of two $\{Fe^{III}(Tp^*)(CN)_3\}$ complex units acting as metalloligands (though *cis*-coordinated cyanides) toward two divalent metal ions (M = Co^{II} for **10**, M = Fe^{II} for **12**) whose coordination sphere is completed by two bik ligands, thus providing a [2+2]-type diamond-like distorted centrosymmetric molecular square (Fe1···M···Fe1' angle = 96.6° and 96.0°; M···Fe1···M' angle = 83.4° and 84.0° for **10** and **12**, respectively). Indeed, the Fe···Co edges in **10** are 5.166 Å and 5.167 Å long, and the Fe1···Fe2 edges of **12** are also of identical length (5.038 Å).

The two iron atoms denoted as Fe1 are in a slightly distorted octahedral C_3N_3 environment formed by the three imine moieties of the pyrazolyl rings of a *fac*-coordinating Tp* ligand and the carbon atoms of three cyanides. Two of the three cyanides act as bridging ligands between the two iron and the two cobalt/iron ions. Due to crystallographic symmetry, the remaining terminal cyanide ligands are orientated in *trans* position in respect to the plane containing all four metal atoms. The Fe-$C_{cyanide}$ bond lengths range from 1.911(4) – 1.920(4) Å (for **10**) and 1.910(9) – 1.930(8) Å (for **12**). For both complexes, this value is above 1.900 Å and clearly corresponds to low-spin iron(III) ions (Fe1). The Fe–N_{pz} bond lengths are longer than their Fe–C counterparts with a mean value of 1.997 Å and 2.007 Å for **10** and **12**, respectively. All three cyanides bind their iron centre in an almost linear way, with Fe1-C-N angles equal to or wider than

174.4(6) Å. The octahedral distortion, defined as the sum of the deviations to 90° of the twelve angles around the metal atom, for the iron(III) environment amounts to 26.4° in **10** and 26.8° for **10**, which is far more distorted than in PPh_4[**7**] (17.9°).

The octahedral coordination sphere of each cobalt(II) and iron(II) ion is completed by two bidentate *cis*-coordinating bik ligands, leading to an octahedral N_6 environment, for which the octahedral distortion amounts to 39.6° (**10**) and 23.2° (**12**).

In **10**, the two cyanide nitrogens exhibit equally long bonds (2.108 Å) to the cobalt ion. The Co-N_{im} bond lengths are slightly longer, and range from 2.116(3) to 2.153(3) Å, in agreement with a high-spin state for the cobalt(II) ions. The C=O moieties of both bik ligands are notably bent with respect to the Co-C_{ketone} vector and show bent angles of 173.02° and 167.52° for Co1···C20-O20 and Co1···C30-O30, respectively. The bite angle of the bik ligand are 85.14(12)° and 84.10(12)°.

In **12**, by contrast, the average Fe2–N bond lengths amount 1.981 Å. This compares well with the distances observed in previously reported $\{Fe^{III}_{LS}Fe^{II}_{HS}\}$ analogous molecular squares.[107,109] The C=O moieties of the bik ligands are only slightly bent (170.3° and 175.2°). The bite angles of the bik ligands are 88.5(3)° and 89.4(3)°.

In both complexes, one of the cyanide bridges is slightly bent on the N–M metal side (C2-N2-Co1 = 175.5(3)° and C2-N2-Fe2 = 174.3(6)°) while the second one is almost linear (C1'-N1'-Co1 = 178.7(3)°, C1'-N1'-Fe2 = 179.1(6)°).

Each square unit is well separated from the others by perchlorate anions and lattice water molecules, the shortest metal-metal distance being 9.60 Å in **10**, 9.46 Å in **12**. In both cases, the water molecules are hydrogen-bonded to the non-bridging cyanides and to an oxygen atom of the perchlorate counter ion.

4.1.3 Fourier Transform InfraRed spectroscopy

$\{[Fe^{III}(Tp^*)(CN)_3]_2[Co^{II}(bik)_2]_2\}(X)_2 \cdot n\ H_2O$ ($X = [ClO_4]^-$, n = 2, **10**) and ($X = [BF_4]^-$, **11**)

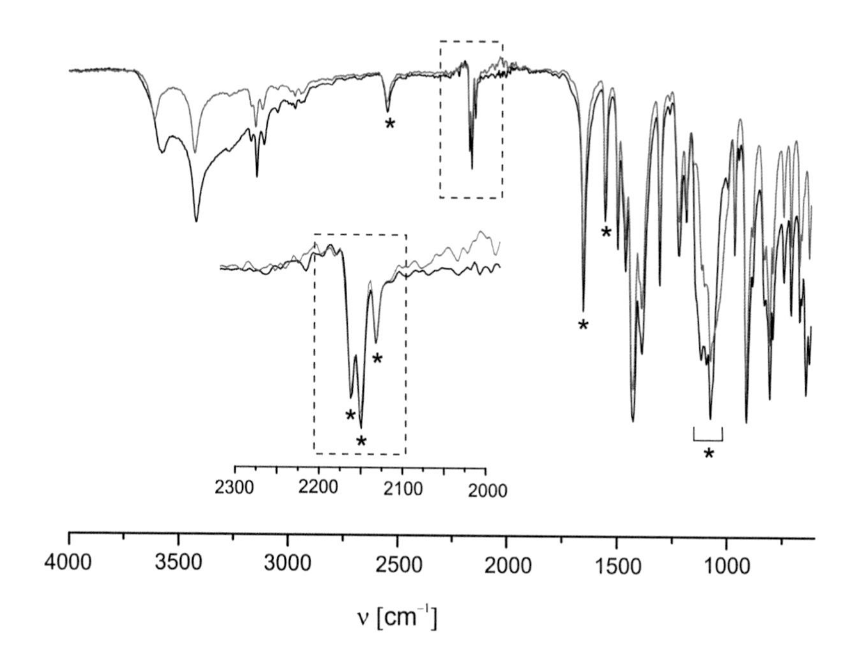

Figure 4.4: FT-IR (ATR) transmission spectrum of freshly filtered **10** (black curve) and **11** (red curve) between 4000 and 600 cm^{-1} with a 4 cm^{-1} resolution. Selected IR vibration bands in cm^{-1} and their intensities are marked with an asterisk: 1059 (vs), 1078 (s), 1102 (s), 1541 (w), 1639 (s), 2133 (vw), 2149 (w), 2159 (w), 2539 (vw) for **10**; 1050 (br, vs), 1059 (vs), 1088 (s), 1542 (w), 1639 (s), 2133 (vw), 2150 (w), 2160 (vw), 2539 (vw) for **11**.

FT-IR absorption spectra of freshly filtered samples of **10** and **11** (see Figure 4.4) were recorded at room temperature using an ATR module. The spectra of the two compounds are almost identical except for the vibrations corresponding to the counteranions, which strongly supports the occurrence of similar square motifs. The presence of $\{Fe^{III}(Tp^*)(CN)_3\}$ units is detectable by its sharp B–H stretching band at 2539 cm^{-1} and its dimethylated pyrazolyl ring stretch vibration at 1541 cm^{-1} characteristic of Tp* ligands. Three different cyanide stretching vibrations account for two types of bridging cyanides at 2149 and 2159 cm^{-1}, and a terminal *C*-bound third one at 2133 cm^{-1} (hydrogen-bonded). Those values are typical for $\{Fe^{III}_{LS}Co^{II}_{HS}\}$ pairs, which is consistent with the X-ray diffraction bond length analysis. The ketone moieties of the cobalt-bound bik ligands give rise to a characteristic absorption at 1639 cm^{-1} for both compounds. The signature of perchlorate ions can be found as a broad, strong absorption at about 1059 cm^{-1}, while the tetrafluoroborate anions absorb at about the same frequency: 1050 cm^{-1}.

$\{[Fe^{III}(Tp^*)(CN)_3]_2[Fe^{II}(bik)_2]_2\}(ClO_4)_2 \cdot 2\ H_2O$ (12)

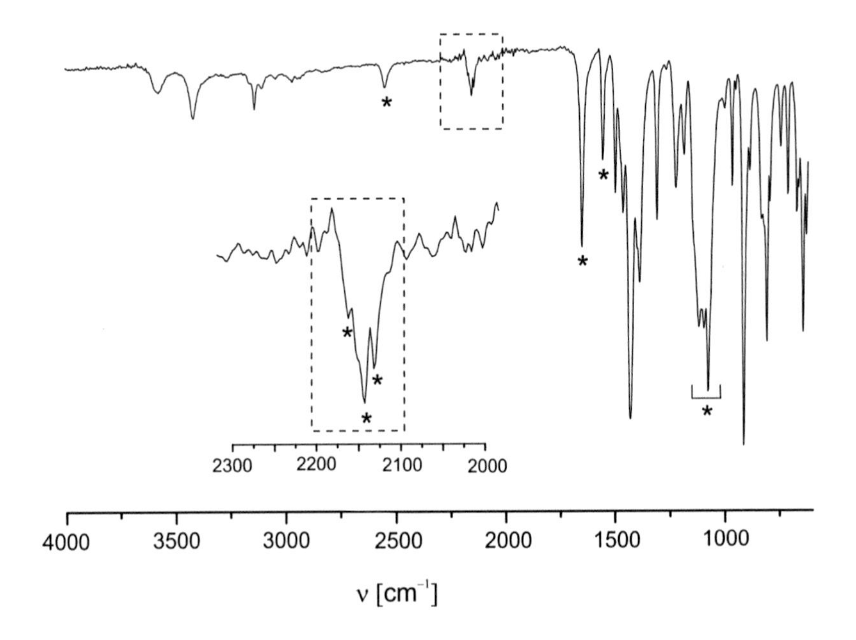

Figure 4.5: FT-IR (ATR) transmission spectrum of **12** between 4000 and 600 cm^{-1} with a 4 cm^{-1} resolution. Selected IR vibration bands in cm^{-1} and their intensities are marked with an asterisk: 1054 (sh, s), 1064 (vs), 1093 (vs, br), 1524 (vw), 1541 (w), 1634 (m), 2132 (vw), 2147 (vw), 2160 (vw), 2538 (vw).

The IR spectrum of **12** is almost identical to those of **10** and **11**. This was to be expected, considering that the only difference between **10** and **12** is the nature of the divalent metal bound to the bik ligands (cobalt(II) for **10**, iron(II) for **12**). This change of metal redshifts the ketone vibration band by 5 cm^{-1}, while the vibrations attributed to the Tp* ligand (Tp* C–H vibration pattern above 2800 cm^{-1}, pyrazolyl ring stretch at 1541 cm^{-1}) do not experience any frequency shift. The only notable change between **12** and **10** concerns the cyanide vibration bands: even if their position remains unchanged in both compounds (2132, 2147 and 2160 cm^{-1} for **12**, to be compared with 2133, 2149 and 2159 cm^{-1} for **10**), their intensity is much lower in the case of **12**, and, especially in case of freshly filtered samples, can almost disappear in the background noise if the IR spectrum is acquired with too few scans.

4.1.4 SQUID magnetometry

$\{[Fe^{III}(Tp^*)(CN)_3]_2[Co^{II}(bik)_2]_2\}(BF_4)_2 \cdot$ n Solvent (11)

The magnetic properties of **11** have been investigated by SQUID magnetometry and the $\chi_M T$ *vs T* curve, along with its simulation, is depicted in Figure 4.6.a and b.

For the fresh sample, the measured $\chi_M T$ product is 8.24 $cm^3 \cdot mol^{-1} \cdot K$ at 300 K. This value is coherent with the occurrence of four non-interacting metal ions: two low-spin iron(III) ion (*ca* $\chi_M T$ = 0.75 $cm^3 \cdot mol^{-1} \cdot K$) and two high-spin cobalt(II) ions (*ca* $\chi_M T$ = 2.7-3.6 $cm^3 \cdot mol^{-1} \cdot K$), which all exhibit a first order orbital magnetic moment (so that the spin only formula does not apply). Upon cooling, the $\chi_M T$ product first smoothly decreases down to 39.8 K (with $\chi_M T$ = 7.41 $cm^3 \cdot mol^{-1} \cdot K$), then increases down to 10 K, reaching a maxima at 10 K with $\chi_M T$ = 9.51 $cm^3 \cdot mol^{-1} \cdot K$. The smooth decrease between 300 K and 39.8 K is likely due to the effect of the spin-orbit coupling for both iron(III) and cobalt(II) ions, whereas the further increase of $\chi_M T$ at lower temperature likely accounts for the occurrence of ferromagnetic interactions between the paramagnetic ions through the cyanide bridges. The decrease at low temperature could be due to antiferromagnetic interactions. Such behaviour has been already observed in other

{FeCo} square molecules.[88] In order to rationalise the magnetic behaviour and to give support to this assumption, the $\chi_M T$ curve has been simulated as described here below.

The $\{Fe^{III}_2Co^{II}_2\}$ molecular square can be described by the following total Hamiltonian $\mathcal{H}_{tot}$:

$$\mathcal{H}_{tot} = \mathcal{H}_{int} + \mathcal{H}_{SO} + \mathcal{H}_{dist} + \mathcal{H}_{Ze} \tag{13}$$

with $\mathcal{H}_{int}$ being the contribution due to the magnetic exchange interactions between the low-spin iron(III) and high-spin cobalt(II) metallic centres. Since the compound is a distorted [2+2] molecular square and the magnetic interaction is structurally dependent, two different coupling constants J_1 and J_2 are expected. The adequate spin interaction Hamiltonian is therefore:

$$\mathcal{H}_{int} = -J_1(S_{Co1} \cdot S_{Fe1} + S_{Co2} \cdot S_{Fe2}) - J_2(S_{Co1} \cdot S_{Fe2} + S_{Co2} \cdot S_{Fe1}) \tag{14}$$

L and S are respectively the orbital and spin operators with L = 1 and S = 3/2 in the T-P isomorphism approach;[136,163–165] $\mathcal{H}_{SO}$ is the Hamiltonian describing the spin-orbit coupling in the cobalt ions. Both cobalt ions are identical and therefore simulated with identical spin-orbit coupling constant (λ) and orbital reduction factor (α). This interaction is orientation dependent along the ν = x, y and z axes:

$$\mathcal{H}_{SO} = \sum_{i=1}^{2} -\frac{3}{2}\alpha\lambda\, L_{Co_i} S_{Co_i} \tag{15}$$

The same applies for the $\mathcal{H}_{dist}$ Hamiltonian, or "distortion Hamiltonian", which is, with the spin-orbit coupling, responsible for the anisotropy of the system:

$$\mathcal{H}_{dist} = \sum_{i=1}^{2} \Delta\,(L^2_{z\,Co_i} - \frac{1}{3} L^2_{Co_i}) \tag{16}$$

where Δ is the axial distortion parameter.

The last Hamiltonian in equation (13), $\mathcal{H}_{Ze}$, describes the Zeeman interactions between the iron and cobalt ions and the applied magnetic field. In our case, it is expressed as follows:

$$\mathcal{H}_{Ze} = \sum_{i=1}^{2}\left(-\frac{3\alpha}{2}L_{\nu_{Co_i}} + g_e S_{\nu_{Co_i}}\right)\beta \cdot H_\nu + \sum_{i=1}^{2} + g_{eff_{Fe}}\beta \cdot S_{\nu_{Fe_i}} \cdot H_\nu \tag{17}$$

with ν = x, y and z axes.

In order to take into account the spin-orbit coupling on the iron ions, which cannot be directly calculated because of the huge amount of time necessary for such a calculation, an approach close to the Lines' model with a fictive temperature-dependent Landé factor $g_{eff_{Fe}} = f(T)$ was selected.

The $g_{eff_{Fe}}(T)$ function was calculated as:

$$g_{eff_{Fe}}^2 = \frac{4k(\chi_M T)_{exp}}{\mathbb{N}\beta^2} \tag{18}$$

where $(\chi_M T)_{exp}$ are the experimental values of the $[Fe^{III}(Tp)(CN)_3]^-$ ($[\mathbf{1}]^-$) complex. The obtained Landé factor function is plotted *vs* temperature in Figure 4.6.c.

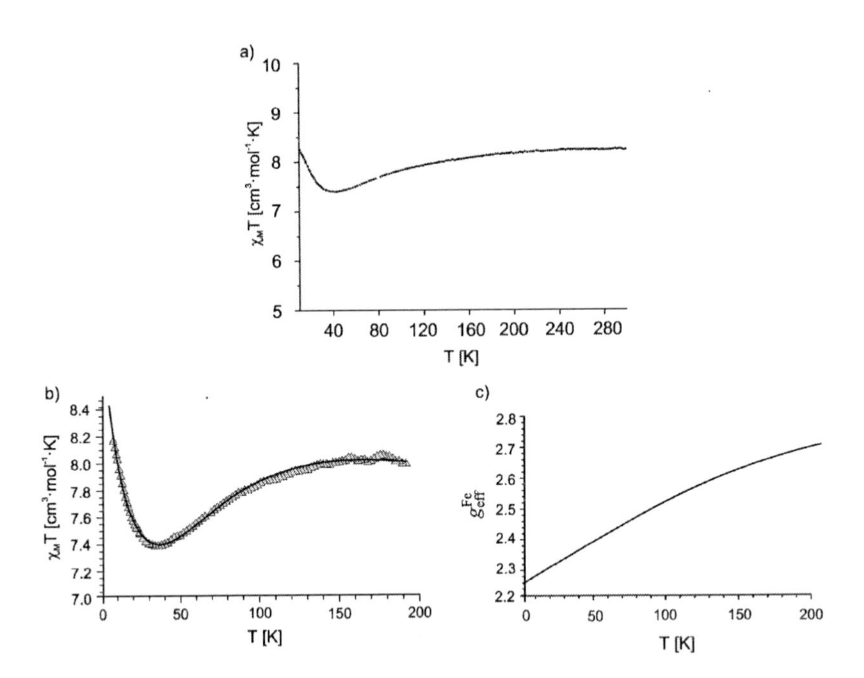

Figure 4.6: a) $\chi_M T$ *vs* T curves of **11** at H = 5000 Oe between 300 K and 10 K.
b) Experimental (blue triangles) and simulated (black curve) $\chi_M T$ *vs* T curves of fresh **11** between 200 K and 2 K.
c) Temperature-dependent Landé factor $g_{\text{eff}_{\text{Fe}}} = \text{f}(T)$ *vs* T. The sample was prepared as follows: m_{sample} = 6.3 mg, m_{film} = 8.5 mg, m_{paratone} = 4.0 mg.

The best fit curve shown in Figure 4.6.b is obtained for J_1 = 6.85 cm^{-1}, J_2 = 23.35 cm^{-1}, λ = -122.90 cm^{-1}, α = 0.79, Δ = -83.04 cm^{-1}. The significant difference between the two coupling constants could appear surprising, however one has to consider that two exchange interaction pathways coexist in this molecule (which is not a real square) as (i) the cyanide bridges do not have exactly the same geometry, (ii) more importantly, the relative orientation of the magnetic orbitals can be different through the two different pathways. Indeed, the agreement factor is significantly improved compared to models with about the same values for J_1 and J_2, and, in the case of this fit, is excellent: 2.59×10^{-5}. Some temperature-independent paramagnetism was also introduced in the fit parameters and found to amount to $258.40\cdot10^{-6}$ cm^3·mol^{-1}·K. This positive value is consistent with the slight positive slope exhibited by the compound at higher temperature.

The $\chi_M T$ *vs* T curve of two parent $\{Fe_2Co_2\}$ molecular squares were simulated with approximately the same model by Pardo *et al.* using two different J values.[88] However, the gap between the two coupling constants (J_1 = 5.4 cm^{-1} and J_2 = 11.1 cm^{-1} for the first square, J_1 = 8.1 cm^{-1} and J_2 = 11.0 cm^{-1} for the second one) is not as high as for the parameter set obtained for the presented fit. If the found J_1 = 6.85 cm^{-1} is of the same order of magnitude as the J values from the literature, the value of J_2 is clearly higher.

$\{[Fe^{III}(Tp^*)(CN)_3]_2[Co^{II}(bik)_2]_2\}(ClO_4)_2 \cdot 2\ H_2O$ (10)

The magnetic properties of **10** have been investigated by SQUID magnetometry and the resulting $\chi_M T$ *vs* T curve (see Figure 4.7.a.) was simulated with the same model as for **11** without the temperature-dependant g factor, for which no improvement was found.

The general aspect of the $\chi_M T$ *vs* T curve resembles that of **10**. Indeed, the connectivity of the molecular square is identical, while the tetrafluoroborate anions of **11** are replaced by perchlorates in **10**. At high temperature, the $\chi_M T$ product is almost constant and amounts 8.1 cm^3·mol^{-1}·K. The minimum of the ferromagnetic curves appears for 40.6 K and amounts to 7.62 cm^3·mol^{-1}·K. After a short increase (9.10 cm^3·mol^{-1}·K at 8.8 K), the $\chi_M T$ product rapidly decreases, presumably due to long range intermolecular interactions between the molecular square units.

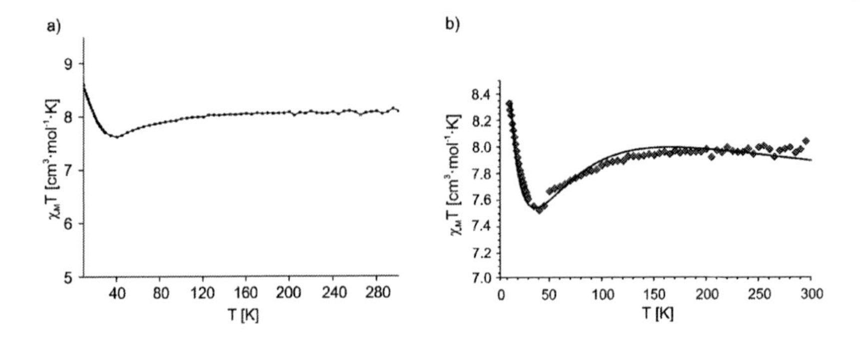

Figure 4.7 – Magnetic measurements for freshly filtered **10**:
a) $\chi_M T$ *vs* T curve between 2 and 360 K, H = 5000 Oe.
b) Zoom on the experimental (red diamond) and simulated (black curve) $\chi_M T$ *vs* T curve between 2 K and 300 K. The sample was prepared as follows: m_{sample} = 7.8 mg, m_{capsule} = 29.3 mg.

At this stage the moderate quality of the magnetic data leads only to approximate values of the electronic parameters. The best fit of the $\chi_M T$ *vs* T curve (Figure 4.7.a and b) is obtained here for J_1 = 14.00 cm^{-1}, J_2 = 1.43 cm^{-1}, λ = -128.87 cm^{-1}, α = -0.913 and Δ = -511.18 cm^{-1}. Again, there is a difference between the coupling constants J_1 and J_2 and this time, J_2 is in the expected range for similar iron-cobalt molecular square[88] but J_1 is rather small. However, the calculated Landé factor for the irons ions amounts to $g_{\text{eff}_{\text{Fe}}}$ = 2.79, which is slightly too high (expected $g_{\text{eff}_{\text{Fe}}}$ = 2.5–2.7). New measurements will be carried out to extract better estimate but it seems clear that the squares need to be fitted with different J values. This could be supported by DFT calculations to extract theoretical estimates of the magnetic coupling.

$\{[Fe^{III}(Tp^*)(CN)_3]_2[Fe^{II}(bik)_2]_2\}(ClO_4)_2 \cdot 2\ H_2O$ (12)

The magnetic properties of **12** have been investigated between 2 K and 365 K. Fresh crystals (m_{sample} = 3.8 mg) were removed from their mother liquor directly before the measurement and the sample was introduced into the SQUID magnetometer at 200 K to avoid solvent loss. The sample was measured between 2 and 365 K upon heating and cooling. The thermal dependence of the $\chi_M T$ product for fresh (black curve) and *in situ* desolvated **12** (red curve) is depicted in Figure 4.8.

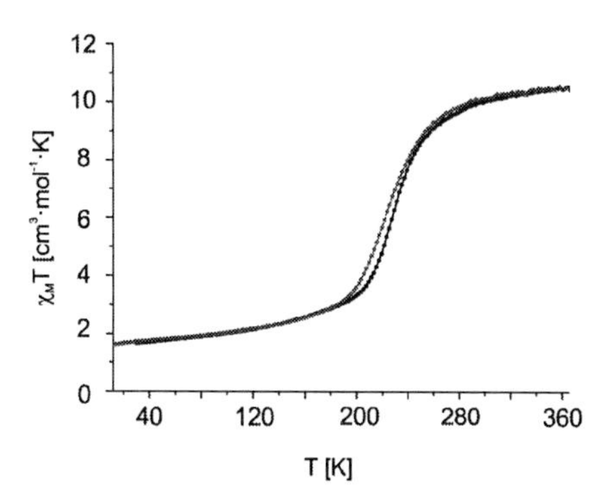

Figure 4.8: $\chi_M T$ *vs* T curves of **12** (m_{sample} = 3.8 mg, $m_{capsule}$ = 35.8 mg) at H = 5000 Oe (black: fresh sample – from 35 K to 365 K; red: after desolvation inside the SQUID magnetometer, from 365 K to 10 K).

Both solvated and desolvated samples of **12** exhibit a sigmoidal shape, with a strong increase of the $\chi_M T$ product between 200 K and 275 K. This accounts for the occurrence of a spin transition with $T_{1/2}$ = 227 K for the solvated sample of **12**. The increase of the $\chi_M T$ value from 2.34 cm^3·mol^{-1}·K. (T = 140 K) to 10.54 cm^3·mol^{-1}·K (at 365 K) is coherent with a spin crossover on both iron(II) ions. Actually, the saturation value at high temperature reaches the expected value for two non-interacting low-spin iron(III) ions (0.7 cm^3·mol^{-1}·K for compounds with high-orbital contribution) and two non-interacting high-spin iron(II) ions ($\chi_M T$ = 3.6 cm^3·mol^{-1}·K per iron, with g = 2.2). At 12 K, the value of the $\chi_M T$ product (1.40 cm^3·mol^{-1}·K) is somewhat higher than that expected for the two low spin iron(III) ions, and may account for the presence of residual high-spin iron(II) ions. The *in-situ* desolvated **12** exhibits almost the same transition. While $\chi_M T$ product values at high and low temperatures remain identical to those obtained for the fresh sample, $T_{1/2}$ is slightly shifted toward lower temperature, i.e. 222 K. This behaviour is perfectly reversible and no hysteresis effect is observed.

LIESST (Light Induced Excited Spin-State Trapping) effect was probed on a fresh sample at 20 K by measuring the magnetisation *vs* time upon laser light irradiation. Similar conditions as those used for the previously reported {FeMo}[166] and {FeFe}[98,107] square complexes (which included the $\{Fe(bik)_2(NC\text{-})_2\}$ subunits) were used (see Figure 4.9.a). The sample is photosensitive to all six wavelengths it was exposed to (with power *ca* 5-10 $mW \cdot cm^2$), the 635, 808 and 900 nm laser sources being the most efficient, the increase of $\chi_M T$ being remarkably abrupt and the saturation being reached quickly.

Note: the magnetisation increase upon switching off the laser light is due to a thermal effect. Indeed, when irradiated by powerful laser sources, the temperature of the sample locally increases, which reduces the recorded magnetisation, and is not compensated by the SQUID temperature control unit. Directly after they are turned off, the temperature of the sample decreases back to 20 K, which enhances the recorded magnetisation.

The magnetisation at saturation (and after the laser source was turned off – so the starting and end points are recorded at the same 20 K temperature) reached with those three wavelengths amounts 12.5 $cm^3 \cdot mol^{-1} \cdot K$, which is higher than the saturation value obtained at 365 K for the bulk measurements. Although the small amount of sample (m_{sample} = 0.3 mg) used for photomagnetic measurements notably increases the uncertainty on the absolute value of the $\chi_M T$ product, this may account for the presence of intramolecular ferromagnetic interactions. It is reasonable to assume that under those conditions, the conversion from low-spin iron(II) ions to high-spin iron(II) ions is complete.

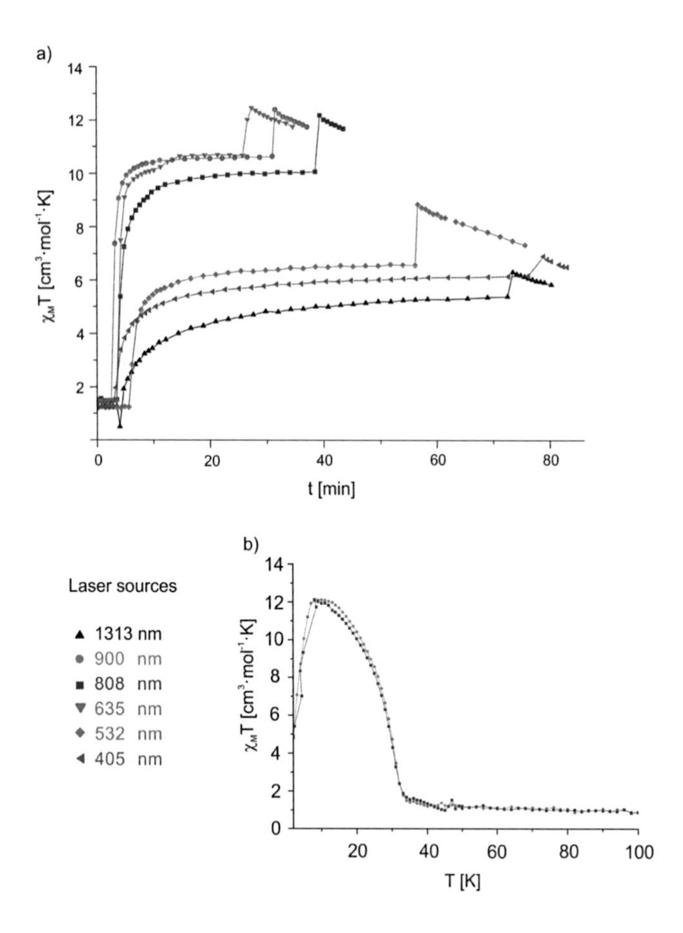

Figure 4.9: a) $\chi_M T$ *vs* time (min) curves of a freshly filtered m_{sample} = 0.3 mg sample of **12** irradiated by 405, 532, 635, 808, 900 and 1313 nm laser lights at 20 K and H = 10000 Oe. The laser source was switched on at t = 2.5–5.6 min, depending on the wavelength. The laser source were switched off at t = 73 (1313 nm), 31 (900 nm), 38.5 (808 nm), 26 (635 nm), 56 (532 nm) and 76 min (405 nm). The sample photo-induced magnetisation was reset between two photomagnetic experiments by heating the sample to 200 K *in-situ*.
b) $\chi_M T$ vs T curves: The same sample was irradiated at 808 nm (wine red) and 900 nm (grey) at 2 K and the temperature was gradually increased to 100 K at 0.5 K·min^{-1} (H = 10000 Oe).

The photo-induced high-spin metastable state is stable up to T_{LIESST} = 35 K (after irradiation at 808 nm and 900 nm, and heating the sample at 0.5 K·min^{-1}). The maximum $\chi_M T$ value reached at T = 10 K (approximately 12.1 cm^3·mol^{-1}·K for the 808 nm laser source and the 900 nm laser source) is quite high and may point to the occurrence of unexpected ferromagnetic interactions between the iron(II) and iron(III) ions. Compared to the two other $\{Fe_2Mo_2\}$[166] and $\{Fe_2Fe_2\}$[107] photomagnetic molecular squares reported in the literature and based on $\{Fe(bik)_2(NC\text{-})_2\}$ subunits, **12** possesses a slightly lower T_{LIESST} (45–48 K for both literature-known compounds). The $\{Fe_2Mo_2\}$ compound[166] undergoes a maximum effect under the 405 nm laser light. **12** and the literature-known $\{Fe_2Fe_2\}$[107] compound, however, undergo a maximum effect for the same 700–900 nm laser range. They also reach their respective saturation after 20 minutes if irradiated with their most efficient wavelength, while the $\{Fe_2Mo_2\}$ compound[166] needs 40 minutes.

5 Polymetallic cyanide-bridged transition metal complexes using the [Fe(Tpm*)(CN)$_3$] (8) building block

As shown in the chapter 3, **8** has a redox potential higher than PPh$_4$[Fe^{III}(Tp)(CN)$_3$] (PPh$_4$[**1**], $E°_{1/2}$ = -824 mV) and close to that of PPh$_4$[Fe^{III}(Tt)(CN)$_3$] (PPh$_4$[**6**], $E°_{1/2}$ = -531 mV). A major change also lies in the charge of the iron(III) species: while **8** is neutral, [Fe^{III}(L)(CN)$_3$]$^-$ (L = Tp*, Tp, Tt) are monoanionic complexes. This tends to affect the formation/crystallisation of polynuclear assemblies, as crystals of **8** are very often recovered after slow evaporation of the solution. The reactivity of this building block was investigated towards the metal ions Co^{II} and Mn^{II} to produce {FeCo} and {FeMn} molecular chains (section 5.1), and towards partially blocked cobalt subunits [Co^{II}(L)$_2$(S)$_2$]$^{2+}$ (L = bik, bim; S = acetonitrile, water) to produce {Fe$_2$Co$_2$} molecular squares (section 5.2). The reaction of **8** with [Mn^{II}(bik)$_2$(S)$_2$]$^{2+}$ led to separate crystallisation of **8** and [Mn^{II}(bik)$_3$](ClO$_4$)$_2$.

5.1 Cyanide-bridged coordination polymers

To the best of our knowledge, all literature-known double-zigzag molecular chains involving blocked iron(III) building blocks and metal ions have a planar topology,[87,103,115,167–169] as depicted in Figure 5.1.a, where the coordination sphere of the transition metal is completed by two solvent molecules in *trans* position from each other. The only notable exception is the 2,4-ribbon-like one-dimensional chain {[Fe^{III}(Tp)(CN)$_3$]$_4$[Fe^{II}(H$_2$O)$_2$][Fe^{II}]}$_\infty$ from Zhang *et al.*[170] in which two coordinated water molecules are arranged in *cis*-position (Figure 5.1b).

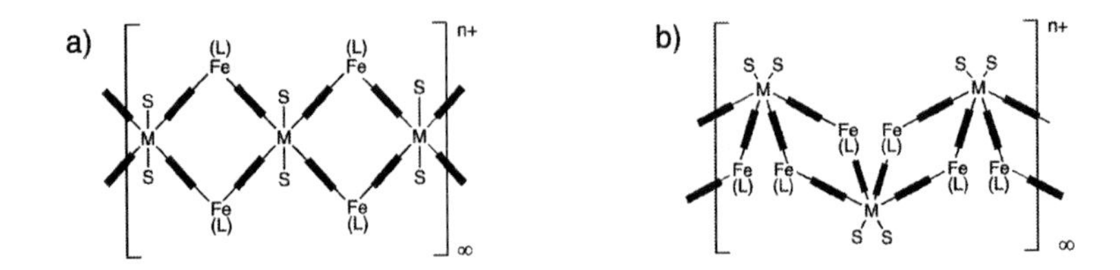

Figure 5.1: Schematic representation of two types of cyanide-bridged double-zigzag molecular chains. Bridging cyanides are represented as black rectangles. L = tripodal capping ligand, S = coordinated solvent molecule and M = transition metal ion. Non-bridging cyanides are omitted for clarity.
a) planar double-zigzag chain, where the cyanides bridges are in one plane and the two remaining positions are in *trans*.
b) cranked double-zigzag chain, where the two "links" are connected in a *cis* way, and where solvent molecules coordinate the metal ion in *cis* fashion.

5.1.1 Syntheses

$\{\{[Fe^{III}(Tpm^*)(CN)_3]_2[Co^{II}(H_2O)_2]\}(ClO_4)_2 \cdot 2\ H_2O\}_\infty$ (13)

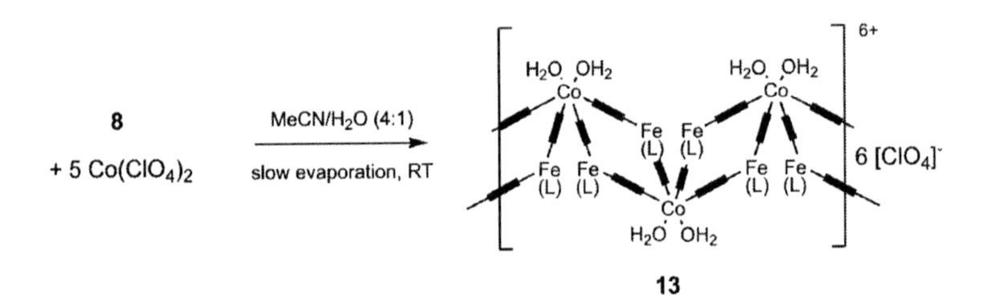

Figure 5.2: Synthesis of the cyanide bridged coordination polymer **13**. The blocking ligand L is a Tpm* ligand. Bridging cyanides are represented as black rectangles. Non bridging cyanides are omitted for clarity.

Slow evaporation of acetonitrile/water mixtures of the $[Fe^{III}(Tpm^*)(CN)_3]$ building block (**8**) and $Co^{II}(ClO_4)_2 \cdot x\ H_2O$ results in the formation of red (micro)crystals of **13**. Equimolar solutions tend to crystallise when little crystallisation solvent is left and produce a mixture of crystals of **13** and **8**. Better results were obtained using an acetonitrile/water (4:1) solvent mixture and five equivalents of $Co^{II}(ClO_4)_2 \cdot x\ H_2O$ (see Figure 5.2). Under these conditions, **13** crystallises from much more solvent, which allows storing of "fresh" substance. Since yellow **8** is neutral and only sparingly soluble in water, it tends to crystallise first as acetonitrile evaporates. Crystallisation of **13** therefore mostly occurs on the surface of these reagent crystals, slowly "consuming" them as they grow, until they disappear after two months. This behaviour can be attributed to the presence of an equilibrium in solution, in which the cobalt(II) concentration plays a key role.

While the redissolution of **13** in water or water/acetonitrile mixtures led to dissociation into **8** and $Co^{II}(ClO_4)_2$ (ultimately reforming **13** after slow evaporation of the resulting solution), redissolution of **13** in pure acetonitrile results in the formation of another, unknown {FeCo} species, in which, according to the infrared spectrum, all cyanide ligands are bridging.

{{[Fe^III^(Tpm*)(CN)~3~]~2~[Mn^II^(MeCN)~2~]}(ClO~4~)~2~ · 2 MeCN}~∞~ (14)

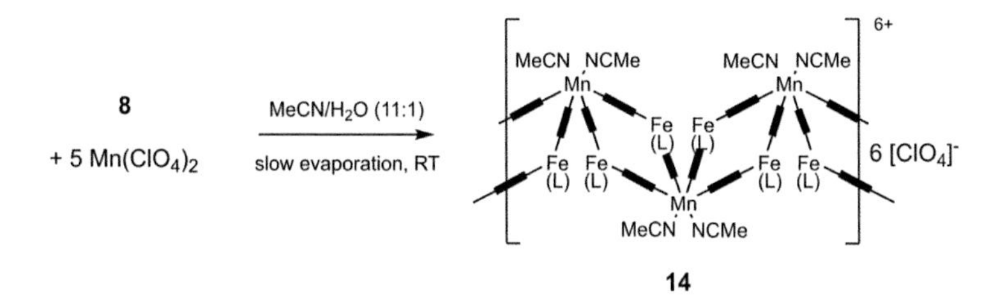

Figure 5.3: Synthesis of the cyanide-bridged coordination polymer **14**. L blocking ligand is a Tpm*. Bridging cyanides are represented as black rectangles. Non bridging cyanides are omitted for clarity

The synthesis of **14** is very similar to that of **13** (Figure 5.3), and encounters the same manganese(II) concentration problems. It is therefore best crystallised with five equivalents manganese(II) perchlorate in acetonitrile/water mixtures within a few weeks.

The redissolution in pure acetonitrile also led to micro-crystallisation of an insoluble "all-cyanide-bridged" species. **14** was produced from acetonitrile/water (11:1) mixture as red rod-like crystals, while solvent mixtures containing more water (5:1) provided more ill-defined crystals of another species, whose IR resembles that of **13**, and presumably is a coordination polymer in which the coordinated acetonitrile molecules are replaced by water.

5.1.2 Structural analyses

$\{\{[Fe^{III}(Tpm^*)(CN)_3]_2[Co^{II}(H_2O)_2]\}(ClO_4)_2 \cdot 2\ H_2O\}_\infty$ (13)

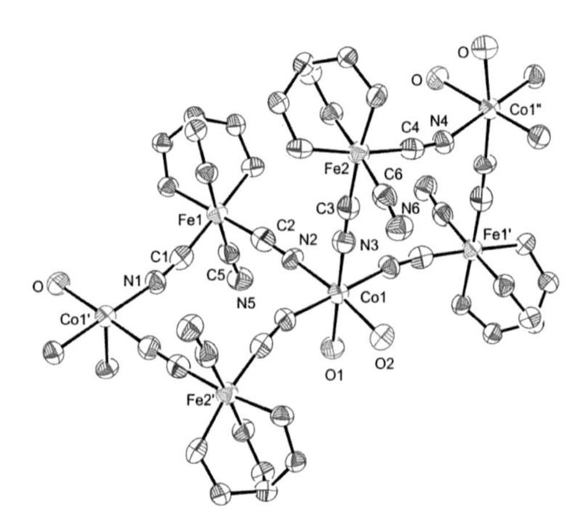

Figure 5.4: View of a fragment of the cationic chain in **13**. Atoms are displayed as 30% probability ellipsoids. Hydrogen atoms, dimethylpyrazolyl carbon atoms of the Tpm* ligands, solvent lattice molecules and perchlorate counteranions are omitted for clarity. Equivalent atoms are generated with a combination of the following symmetry operations: +x, -y, -1/2+z. The colour code for atoms is the following: grey = carbon, blue = nitrogen, orange = iron, green = cobalt, red = oxygen.

13 crystallises in the monoclinic space group *Cc*, with the following cell parameters at 200 K:

a = 25.898(5) Å, β = 16.657(3) Å, c = 13.278(3) Å, β = 115.00(3)°

Due to the reaction kinetics, **13** crystallises as badly intergrown and/or very small crystals, and the limited quality of the XRD data does not allow to get accurate structural data but it is sufficient to identify the nature of the compound (see Figure 5.4).

13 consists of a double-zigzag $\{Fe_2Co\}_\infty$ polycationic cranked one-dimensional coordination polymer, with perchlorate anions, as schematically depicted in Figure 5.1.b: The cobalt(II) ions are connected to each other by four $\{Fe^{III}(Tpm^*)(CN)_3\}$ moieties which act as bridging ligands through two out of the three cyanides, the third one remaining non bridging. The cobalt(II) coordination sphere is completed by two *cis*-coordinating water molecules. The remaining two non-bridging cyanides in each $\{Fe_2Co_2\}$ square unit (see Figure 5.4) point in opposite directions (*trans*) in respect to the plane defined by the four metal atoms. At least four lattice water molecules are present per formula unit. Because of the limited quality of the structural data, bond lengths and angles will not be further discussed.

$\{\{[Fe^{III}(Tpm^*)(CN)_3]_2[Mn^{II}(MeCN)_2]\}(ClO_4)_2 \cdot 2\ MeCN\}_\infty$ (14)

14 crystallises in the monoclinic space group *P2/c*. Its structure consists of a cranked double-zigzag cationic $\{Fe_2Mn\}_\infty$ chain (see Figure 5.1.b) running along the *c* axis and two acetonitrile lattice molecules. A perspective view of two square-shaped links of the molecular chain **14** is represented in Figure 5.5. Selected angles (°) and bond lengths (Å) are listed in the caption. As for **13**, each manganese ion are connected to four $\{Fe(Tpm^*)(CN)_3\}$ complex units acting as bridging metalloligands through two *cis* cyanide groups. The third cyanide ligand of the $\{Fe(Tpm^*)(CN)_3\}$ units is non-bridging. The manganese(II) coordination sphere is completed by two *cis*-coordinating acetonitrile molecules. The remaining two non-bridging cyanides in each $\{Fe_2Mn_2\}$ square unit point in opposite directions (*trans*) in respect to the plane defined by the four metal atoms.

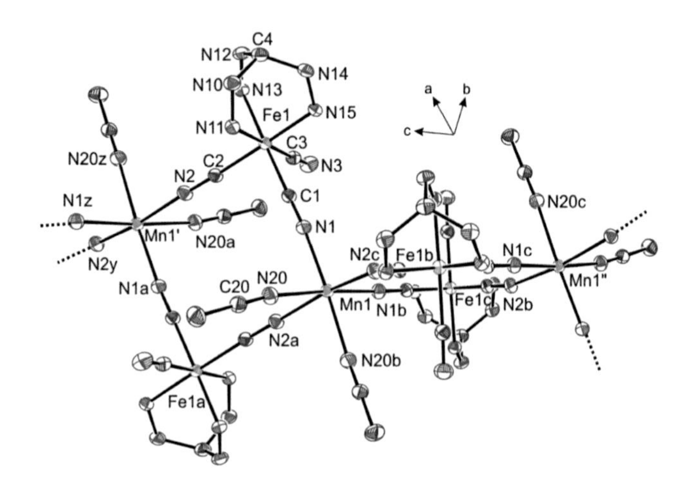

Figure 5.5: View of a fragment of the cationic chain in **14**. Atoms are displayed as 30% probability ellipsoids. Hydrogen atoms, dimethylpyrazolyl carbons of the Tpm* ligands, lattice solvent molecules and perchlorate counteranions are omitted for clarity. Equivalent atoms generated through the 2-fold screw axis along the *b* unique axis are noted with either letters or apostrophe and are generated with a combination of the following symmetry operations: -x,+y,1/2-z; +x,1-y,-1/2+z; +x,1-y,1/2+z.
Selected bonds lengths (Å) and angles (°) for **14**: Fe1–C1 1.914(2), Fe1–C2 1.913(2), Fe1–C3 1.912(2), Fe1–N11 2.0005(19), Fe1–N13 2.0008(19), Fe1–N15 1.9707(19), Mn1–N1 2.186(2), Mn1–N2a 2.192(2), Mn1–N20 2.322(2), Fe1-C1-N1 179.5(2), Fe1-C2-N2 177.5(2), Fe1-C3-N3 177.2(2), C1-N1-Mn1 173.84(19), C2-N2-Mn1' 164.43(19), Mn1-N20-C20 168.7(2), C1-Fe1-C2 87.55(9), C1-Fe1-C3 88.41(10), C2-Fe1-C3 87.82(10), N11-Fe1-N13 88.95(8), N11-Fe1-N15 87.39(8), N13-Fe1-N15 86.75(8), C1-Fe1-N11 93.00(9), C1-Fe1-N15 93.16(9), C2-Fe1-N11 91.10(9), C2-Fe1-N13 92.59(9), C3-Fe1-N13 90.23(9), C3-Fe1-N15 93.10(9), N1-Mn1-N2a 91.70(8), N1-Mn1-N20 89.26(8), N20-Mn1-N2a 87.67(8), N2a-Mn1-N20b 82.52(8), N20-Mn1-N20b 86.72(12), N1-Mn1-N2c 97.43(8), N1-Mn1-N1b 94.77(11), C4-N10-N11-Fe1 -6.2, C4-N12-N13-Fe1 2.7, C4-N14-N15-Fe1 -3.8.

Two out of three cyanide ligands *N*-coordinate two manganese ions to form a series of cyanide bridged $\{Fe_2Mn_2\}$ squares. The manganese-iron edge distances are 5.205 Å and 5.235Å, while the square angles are close to orthogonality (88.56° and 91.44° at the iron and manganese ions respectively). While the C1-N1-Mn1 angles depart only slightly from linearity (173.84(19)°), C2≡N2 binds Mn1' in a cranked way (164.43(19)°). Each manganese ion is involved in two such squares, placed in *cis* in its coordination sphere, so that the two remaining positions are occupied by acetonitrile molecules in *cis* position to each other (Mn1-N20-C20 angle = 168.7(2)°). This leads to a significantly distorted N_6 manganese coordination sphere with an octahedral distortion of 47.3° (defined as the sum of the deviation to 90° of the twelve angles of the octahedron). This is, however, comparable with values found for other {FeMn} compounds in this work (**18** and **20**) and in the literature.[88,171] The Mn–$N_{cyanide}$ bond lengths are equally long (average: 2.189 Å), while the Mn–N_{MeCN} bonds are longer: 2.322 Å. These values are comparable with Mn–$N_{cyanide}$ bond lengths found for **18** and **20**, and consistent with the manganese(II) spin and oxidation states. The iron ions lie in a C_3N_3 environment, with three quasi-linearly bonding cyanides and an octahedral distortion of 26.3°. The iron-carbon bonds are very similar (average 1.913 Å) and consistent with a low-spin iron(III) ion. The Fe–N_{pz} bonds also exhibit similar lengths (average: 1.991 Å). They are in average smaller than the Fe–N_{pz} bonds found in PPh_4[**3**], but match the values of **8** quite well. The torsion angles of the pyrazolyl rings range from 2.6° to 6.2°, while the iron-bridgehead carbon distance amounts to 3.003 Å. Finally, the chains are quite well "isolated" along the *a* and *b* axis with the smallest metal-metal distance being 9.26 Å, partly because of inserted perchlorate piles between the chains along the *a* axis.

5.1.3 Fourier Transform InfraRed spectroscopy

$\{\{[Fe^{III}(Tpm^*)(CN)_3]_2[Co^{II}(H_2O)_2]\}(ClO_4)_2 \cdot 2\ H_2O\}_\infty$ (13)

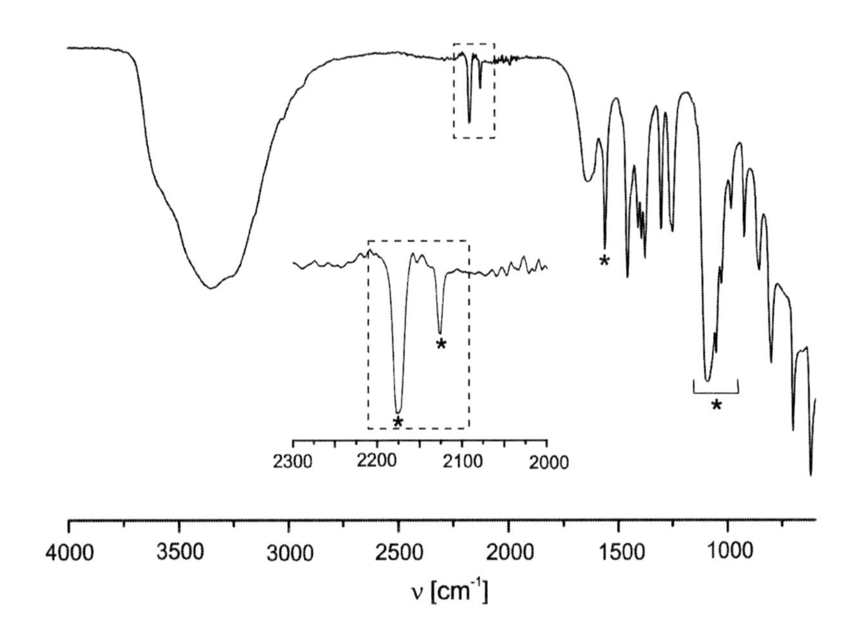

Figure 5.6: FT-IR (ATR) transmission spectrum of freshly filtered **13** between 4000 and 600 cm^{-1} with a 4 cm^{-1} resolution. Selected IR vibration bands in cm^{-1} and their intensities are marked with an asterisk: 986 (w), 1029 (m), 1053 (s), 1090 (br, s), 1562 (m), 2126 (vw), 2177 (vw).

A FT-IR spectrum of freshly filtered **13** was recorded at room temperature. Its spectrum is depicted in Figure 5.6, and selected IR vibration frequencies are listed in its caption. The four vibrations at 986, 1029, 1053 and 1090 cm^{-1} are typical of uncoordinated perchlorate anions. The Tpm* ligands of the $\{Fe(Tpm^*)(CN)_3\}$ moieties exhibit a sharp pyrazolyl ring stretch at 1562cm^{-1}. Two cyanide stretches are visible at higher frequencies than 2100 cm^{-1}, accounting for non-reduced iron(III) ions. The 2126 cm^{-1} vibration can be assigned to the non-bridging cyanides, while the unresolved 2177 cm^{-1} vibration can be attributed to the two bridging cyanide ligands. It is noteworthy that, when **13** is dried, the

2126 cm^{-1} completely disappears to leave only a slightly shifted 2170 cm^{-1} bridging cyanide stretch.

$\{\{[Fe^{III}(Tpm^*)(CN)_3]_2[Mn^{II}(MeCN)_2]\}(ClO_4)_2 \cdot 2\ MeCN\}_\infty$ (14)

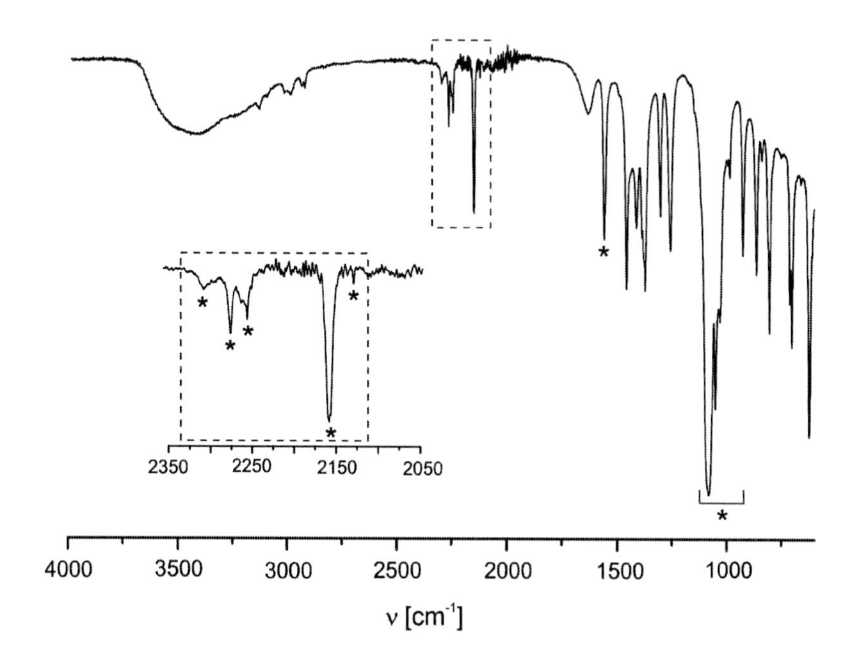

Figure 5.7: FT-IR (ATR) transmission spectrum of freshly filtered **14** between 4000 and 600 cm^{-1} with a 1 cm^{-1} resolution. Selected IR vibration bands in cm^{-1} and their intensities are marked with an asterisk: 985 (w), 1031 (s), 1052 (vs), 1084 (br, vs), 1564 (m), 2129 (vw), 2158 (w), 2253 (vw), 2272 (vw), 2304 (vw).

FT-IR spectrum of freshly filtered **14** resembles **13**. It is depicted in Figure 5.3, with a higher resolution than **13** in order to better resolve the 2250-2300 cm^{-1} vibrations. Selected IR vibration band positions and their intensities are listed in the caption. As for **13**, the spectrum displays the characteristic four strong absorptions of perchlorate anions.

The Tpm* pyrazolyl ring stretch is also visible at 1564 cm^{-1}, while cyanide stretches account for a +III oxidation state for iron ions. The bridging cyanides absorb at about 20 cm^{-1} lower frequency than **13**; this blueshift along the chemical period from Mn^{2+} to Ni^{2+} is literature-known and is related to the M–N bond strength that follows the Irving-Williams series.[133] Even though the IR spectrum of **14** depicted in Figure 5.7 corresponds to the solvated compound, the non-bridging cyanide stretch almost disappears in the background noise, as already observed in case of dry **13** samples.

5.1.4 SQUID magnetometry

$\{\{[Fe^{III}(Tpm^*)(CN)_3]_2[Co^{II}(H_2O)_2]\}(ClO_4)_2 \cdot 2\ H_2O\}_\infty$ (13)

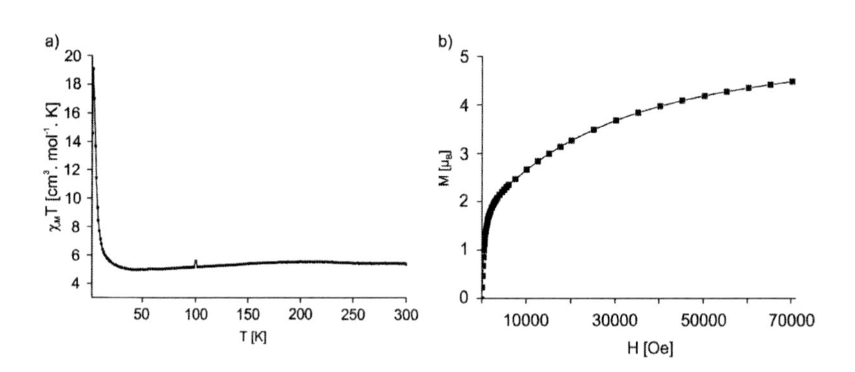

Figure 5.8: Magnetic properties of freshly filtered **13** (m_{sample} = 15.8 mg, m_{film} = 8.5 mg):
a) $\chi_M T$ *vs* T curve between 2 K and 300 K, H = 1500 Oe.
b) M *vs* H curve at 2 K, for magnetic fields from 0 to 70000 Oe.

The magnetic properties of freshly filtered **13** were investigated by SQUID magnetometry. The $\chi_M T$ product *vs* T curve between 2 K and 300 K is depicted in Figure 5.8.a, while the M *vs* H curve measured at 2 K for magnetic fields from 0 to 70000 Oe is depicted in Figure 5.8.b. **13** exhibits a typical ferromagnetic behaviour, as previously observed for other {Fe_2Co} cyanide-bridged double-zigzag molecular chains.[103] The $\chi_M T$ product reaches 5.41 $cm^3 \cdot mol^{-1} \cdot K$ at 300 K. For two independent low-spin iron(III) ions at 0.7 $cm^3 \cdot mol^{-1} \cdot K$ each, and a high-spin cobalt(II) ion (2.8 – 3.6 $cm^3 \cdot mol^{-1} \cdot K$), the expected $\chi_M T$ product value ranges from 4.2 $cm^3 \cdot mol^{-1} \cdot K$ to 5.0 $cm^3 \cdot mol^{-1} \cdot K$. This is slightly lower than the experimental value, but the $\chi_M T$ product of the precursor **8** at 300 K is nearly 0.8 $cm^3 \cdot mol^{-1} \cdot K$, which makes higher values more plausible. Between 300 K and 50 K, the $\chi_M T$ curve decreases slightly, because of the spin-orbit coupling of the iron(III) and cobalt(II) ions. From 40 K to 3 K, it increases drastically as the temperature decreases to reach 19.09 $cm^3 \cdot mol^{-1} \cdot K$ at 3 K, accounting for a typical long range ferromagnetic behaviour. The slight decrease of the $\chi_M T$ product at 2 K (14.59 $cm^3 \cdot mol^{-1} \cdot K$) can be either ascribed to antiferromagnetic interactions between the adjacent chains or to saturation effect. (if M saturates, the $\chi_M T$ product decreases).

The steep increase experienced by **13** in the *M vs H* curve at 2 K (see Figure 5.8) for low field values is consistent with a ferromagnetic behaviour. At 70000 Oe, the magnetisation has not reached a plateau yet but, from the curve inflexion, the value reached at 70000 Oe must not be far away from its plateau magnetisation value: indeed, it only amounts to 4.48 μ_B, instead of the expected 5 μ_B. No hysteresis effect is shown by **13** when the magnetic field intensity is lowered and both curves *M vs H* curves are superposable. In contrast with a number of previously reported {Fe_2Co} chains,[87,103] no out-of-phase signal was detected in *ac* measurement at this temperature and under zero DC field, which excludes slow relaxation of the magnetisation (single chain magnet (SCM) behaviour).

$\{\{[Fe^{III}(Tpm^*)(CN)_3]_2[Mn^{II}(MeCN)_2]\}(ClO_4)_2 \cdot 2\ MeCN\}_\infty$ (14)

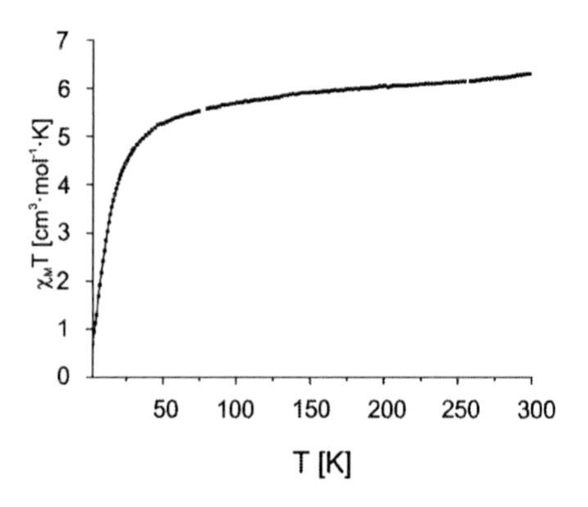

Figure 5.9: Magnetic properties of freshly filtered **14** (m_{sample} = 5.1 mg, $m_{capsule}$ = 41.1 mg): $\chi_M T$ *vs* T curve between 2 K and 300 K, H = 2000 Oe.

The magnetic properties of freshly filtered **14** were acquired by a SQUID magnetometer between 2 and 300 K and are displayed in Figure 5.9. At room temperature, the $\chi_M T$ product is 6.31 $cm^3 \cdot mol^{-1} \cdot K$. This value is close to that expected (6.2 $cm^3 \cdot mol^{-1} \cdot K$) for three non-interacting ions: one manganese(II) ion (4.5 $cm^3 \cdot mol^{-1} \cdot K$) and two low-spin $\{Fe^{III}(Tpm^*)(CN)_3\}$ units, (*ca* 0.8 $cm^3 \cdot mol^{-1} \cdot K$ each). Upon cooling, the $\chi_M T$ product first decreases slowly and then more abruptly below 50 K. The first smooth decrease is likely due to the effect of the spin-orbit coupling of the low-spin iron(III) ions (that exhibit a T ground term). The pronounced decrease at lower temperatures is due to the occurrence of intramolecular Fe^{III}-Mn^{II} antiferromagnetic interactions, but it does not bounce back to higher values at lower temperatures as a ferrimagnetic compound is expected to. Instead, the $\chi_M T$ product decreases rapidly, to reach 0.7 $cm^3 \cdot mol^{-1} \cdot K$ at 2 K. This is far lower than the minimum expected value 3.1 $cm^3 \cdot mol^{-1} \cdot K$ that can be obtained for such ferrimagnetic system. Jiang *et al.* observed a similar magnetic behaviour for a flat double-zigzag $\{Fe_2Mn\}$ chain based on Tp ligand[171] instead of Tpm*. This is to be ascribed to additional intermolecular antiferromagnetic interaction between adjacent molecular chains.

5.2 Cyanide-bridged tetranuclear molecular complexes

5.2.1 Syntheses

$\{[Fe^{II}(Tpm^*)(CN)_3]_2[Co^{III}(bik)_2]_2\}(BF_4)_2 \cdot 7\ H_2O$ (15)

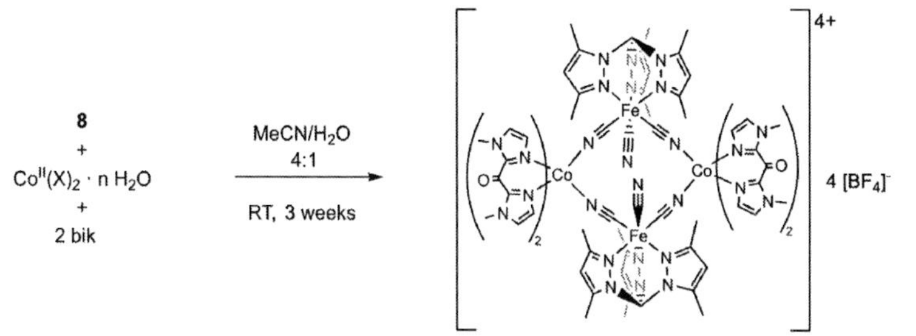

8 + $Co^{II}(X)_2 \cdot n\ H_2O$ + 2 bik → (MeCN/H_2O 4:1, RT, 3 weeks) → $[\ldots]^{4+}$ 4 $[BF_4]^-$ **15**

Figure 5.10: Synthesis of **15**.

15 was synthesised in a mixture of acetonitrile/water 4:1. $Co^{II}(BF_4)_2 \cdot 6\ H_2O$ is *in situ* precoordinated to two equivalents of bik ligand to form the yellowish-pink $[Co^{II}(bik)_2(S)_2](BF_4)_2$ complex. This solution was added dropwise into the orange solution of **8**. The solution turned immediately dark green, which indicates that the redox process between the iron and cobalt ions readily occurs upon addition. This is not surprising, considering that the electron transfer is also observed upon addition in the analogue square $\{[Fe^{II}(Ttp)(CN)_3]_2[Co^{II}(bik)_2]_2\}(BF_4)$[96–98] which is also diamagnetic at room temperature, and that the tricyanido iron(III) reagent involved is less easily reducible than **8**. Slow evaporation of the reaction mixture provided deep green diamond shaped crystals.

$\{[Fe^{II}(Tpm^*)(CN)_3]_2[Co^{III}(bim)_2]_2\}(BF_4)_2 \cdot 12\ H_2O$ (16)

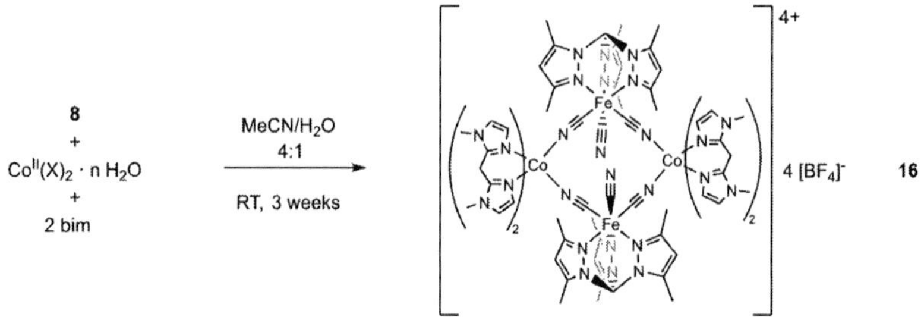

8 + $Co^{II}(X)_2 \cdot n\ H_2O$ + 2 bim $\xrightarrow[\text{RT, 3 weeks}]{\text{MeCN/H}_2\text{O 4:1}}$ [complex]$^{4+}$ 4 $[BF_4]^-$ **16**

Figure 5.11: Synthesis of **16**.

16 was synthesised using the same method as for **15**, and by replacing bik by bim. However, the reaction mixture did not turn green but instead darkened into a pink blackish solution. Dark brown block-like crystals were obtained by slow evaporation of the reaction mixture. Analogue reactions using $Co^{II}(ClO_4)_2 \cdot 6\ H_2O$ or $Fe^{II}(X)_2 \cdot x\ H_2O$ ($X = [ClO_4]^-$, $[BF_4]^-$ and $[NO_3]^-$) all underwent rapid oxidation of bim into bik, which was revealed by either infrared spectroscopy (appearance of the typical ~1670 cm^{-1} ketone vibration), X-ray diffraction and, for iron complexes, appearance of the characteristic deep blue tinge of the $[Fe^{II}(bik)_3]^{2+}$ cation.

5.2.2 Structural analyses

{[Fe^{II}(Tpm*)$(CN)_3$]$_2$[Co^{III}(bik)$_2$]$_2$}$(BF_4)_2$ · 7 H_2O (15)

Compound **15** crystallises in the monoclinic space group $P2_1$ ($Z = 4$). Its crystal structure consists of a tetracationic cyanide-bridged {Fe_2Co_2} molecular square, two tetrafluoroborate anions and seven lattice water molecules. The tetranuclear unit is made of two {Fe^{II}(Tpm*)$(CN)_3$} complex units acting as metalloligands (though *cis*-coordinated cyanides) towards two cobalt ions whose coordination sphere is completed by two bik ligands. A perspective view of the cationic unit of compound **15** is depicted in Figure 5.12, and selected bond lengths and angles are listed in the caption. It is noteworthy that the cationic {Fe_2Co_2} unit is not centrosymmetric as it is usually observed.[98] The Fe···Co edges lengths are almost identical and average 4.91 Å. This value is smaller than 5 Å, which is usually associated with a diamagnetic {$Fe_2^{II}Co_2^{III}$} spin and oxidation state. Even though the C1-Fe-C2, C3-Fe2-C4, N1-Co1-N3 and N2-Co2-N4 angles only vary slightly from orthogonality (88.4°-92.6°), the Fe···Co···Fe and Co···Fe···Co angles are farther away from the ideal 90° and measure on average 85.9° and 94.1°, respectively.

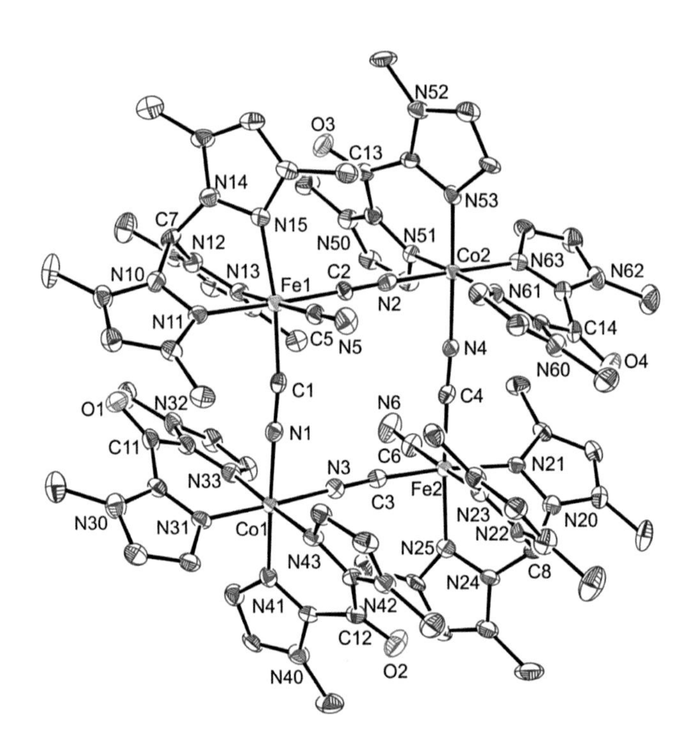

Figure 5.12: Perspective view of the cationic unit in **15**. Atoms are displayed as 30% probability ellipsoids. Hydrogen atoms, lattice solvent molecules and tetrafluoroborate counteranions are omitted for clarity.

Selected bonds lengths (Å) and angles (°) for **15**: Fe1–C1 1.875(9), Fe1–C2 1.877(8), Fe1–C5 1.879(9), Fe1–N11 2.032(7), Fe1–N13 2.029(7), Fe1–N15 2.041(7), Fe2–C3 1.864(8), Fe2–C4 1.876(9), Fe2–C6 1.885(9), Fe2–N21 2.023(7), Fe2–N23 2.025(7), Fe2–N25 2.034(7), Co1–N1 1.880(7), Co1–N3 1.892(7), Co1–N31 1.939(7), Co1–N33 1.915(7), Co1–N41 1.920(7), Co1–N43 1.933(7), Co2–N2 1.890(7), Co2–N4 1.904(7), Co2–N51 1.914(7), Co2–N53 1.898(7), Co2–N61 1.936(6), Co2–N63 1.938(7), C11–O1 1.229(11), C12–O2 1.209(11), C13–O3 1.213(11), C14–O4 1.219(11), Fe1-C1-N1 172.5(7), Fe1-C2-N2 177.9(7), Fe1-C5-N5 173.5(7), Fe2-C3-N3 175.5(7), Fe2-C4-N4 176.2(7), Fe2-C6-N6 173.5(9), Co1-N1-C1 167.2(6), Co1-N3-C3 171.6(6), Co2-N2-N2 174.6(6), Co2-N4-C4 172.3(6), C1-Fe1-C2 92.6(3), C1-Fe1-C5 90.0(3), C2-Fe1-C5 85.7(3), N11-Fe1-N13 88.4(3), N11-Fe1-N15 84.7(3), N13-Fe1-N15 85.8(3), C1-Fe1-N11 88.9(3), C1-Fe1-N13 89.6(3), C5-Fe1-N11 95.4(3), C5-Fe1-N15 95.0(3), C2-Fe1-N13 90.5(3), C2-Fe1-N15 93.8(3), C3-Fe2-C4 92.5(3), C3-Fe2-C6 89.9(4), C4-Fe2-C6 86.6(4), N21-Fe2-N23 87.9(3), N21-Fe2-N25 84.7(3), N23-Fe2-N25 88.2(3), C3-Fe2-N23 89.3(3), C3-Fe2-N25 91.1(3), C6-Fe2-N21 93.1(3), C6-Fe2-N25 94.2(4), C4-Fe2-N23 91.0(3), C4-Fe2-N11 91.8(3), N1-Co1-N3 88.5(3), N1-Co1-N33 88.5(3), N3-Co1-N33 89.6(3), N31-Co1-N41 92.0(3), N31-Co1-N43 92.4(3), N43-Co1-N41 89.9(3), N1-Co1-N31 90.1(3), N1-Co1-N43 90.6(3), N3-Co1-N41 89.5(3), N3-Co1-N43 88.7(3), N33-Co1-N31 89.4(3), N33-Co1-N41 90.9(3), N2-Co2-N4 89.8(3), N2-Co2-N51 89.3(3), N4-Co2-N51 90.0(3), N53-Co2-N63 91.2(3), N53-Co2-N61 91.6(3), N61-Co2-N63 88.9(3), N2-Co2-N53 88.8(3), N2-Co2-N61 90.2(3), N4-Co2-N61 88.5(3), N4-Co2-N63 90.1(3), N51-Co2-N53 90.0(3), N51-Co2-N63 91.6(3), C7-N10-N11-Fe1 -8.6, C7-N12-N13-Fe1 2.1, C7-N14-N15-Fe1 -0.4, C8-N20-N21-Fe2 5.1, C8-N22-N23-Fe2 2.2, C8-N24-N25-Fe2 -3.6, Co1···C11-O1 157.4, Co1···C12-O2 161.6, Co2···C13-O3 159.4, Co2···C14-O4 161.0.

In **15**, the iron ions lie in a slightly distorted C_3N_3 octahedral environment (similar to that of **8**) while the cobalt ions are in a distorted N_6 octahedral coordination sphere formed by two *N*-coordinated cyanides in *cis* position and two pairs of *cis* coordinated bidentate bis(*N*-methylimidazolyl)ketone (bik) ligands, each featuring two imine-like *N*-donors. Even though their coordination environment is identical, the iron ions exhibit quite different octahedral distortion: 34.2° and 27.1° for Fe1 and Fe2, respectively. This is slightly more distorted than in PPh_4[**3**] (23.3°–27.1°) but compares well with the octahedral distortion (35.4°) exhibited by the octanuclear BF_4@{$[Fe^{II}(Tpm^*)(CN)_3]_4[Fe^{II}(H_2O)_3]_4$} complex reported by Shi *et al.* in 2008.[120] The mean Fe–C bond lengths range from 1.864(8) to 1.885(9) Å, with mean values of 1.875 Å and 1.877 Å for Fe2 and Fe1 respectively. These values (inferior to 1.9 Å) are typical for low-spin iron(II) ions, which is coherent with the observed {Fe-CN-Co} edge lengths and the literature.[88,93,95–98,134,172,173] The two iron ions exhibit comparable $Fe–N_{pz}$ bond lengths, with mean values of 2.034 Å for Fe1 and 2.027 Å for Fe2. The $Fe–N_{pz}$ bond lengths in the three structures of PPh_4[**3**] are on average 2.035 Å, 2.034 Å and 2.038 Å. One of the bridging cyanides of Fe1 binds it almost linearly (Fe1-C2-N2 = 177.9(7)°) while the second bridging cyanide is slightly bent on the iron side (Fe1-C1-N1 = 172.5(7)°). On the other side, the bridging cyanides of Fe2 deviate slightly from linearity but experience similar bending angles (176.2(7)° and 175.5(7)°). The two terminal cyanides are orientated in *trans* in respect to the plane of the square, and exhibit the same binding angle of 173.5(7)° towards their respective iron ion. The two Tpm* exhibit similar, quite small pyrazolyl torsion angles (0.4–8.6° for Fe1, 2.2–5.1° for Fe2) as well as comparable metal-bridgehead atom distances (3.029 Å and 3.027 Å).

The octahedral distortion of the cobalt coordination sphere is moderate (11.9° for Co1 and 9.4 for Co2) as expected for low-spin cobalt(III) ions. The cyanides on the cobalt side are more bent around Co1 (167.2(6)°–171.6(6)°) than around Co2 (172.3(6)°–174.6(6)°). The cyanide nitrogen atoms are all equidistant from their respective cobalt ions, with an average bond length of 1.892 Å. The $Co–N_{im}$ bond lengths are slightly more elongated than their $Co–N_{cyanide}$ counterparts and average 1.927 Å for Co1 and 1.922 Å for Co2, and each bik ligand exhibits a bite angle close to the ideal 90° (88.9(3)°–90.0(3)°). Those bond lengths are typical values for low-spin cobalt(III) in comparable environment.[96–98] The C=O moieties of the bik ligands are notably bent with respect to the $Co-C_{ketone}$ vector and exhibit similar angles for Co1 and Co2: from 157.4° to 161.6°. This is far more bent than in the $\{Fe^{III}_2Co^{II}_2\}$ square **13** (*ca.* 170°) but compares well with similar diamagnetic $\{Fe^{II}_2Co^{III}_2\}$ squares.[96–98]

The smallest intermolecular distance between two metal ions of adjacent molecular square units reaches 7.77 Å (Fe1···Fe2) along the *c* axis. Despite this moderate intermolecular distance, the molecular squares are quite spatially isolated from each other, as the smallest distance between the pyrazolyl rings of the Tpm* ligands coordinating said iron ions is 4.0 Å, that is too long for π interactions to take place between the moieties. However, weak interactions take place between the CH and CH_3 moieties of the pyrazolyl rings of two neighbouring molecules and a fluorine of the $[BF_4]^-$ anions.

$\{[Fe^{II}(Tpm^*)(CN)_3]_2[Co^{III}(bim)_2]_2\}(BF_4)_2 \cdot 12\ H_2O$ (16)

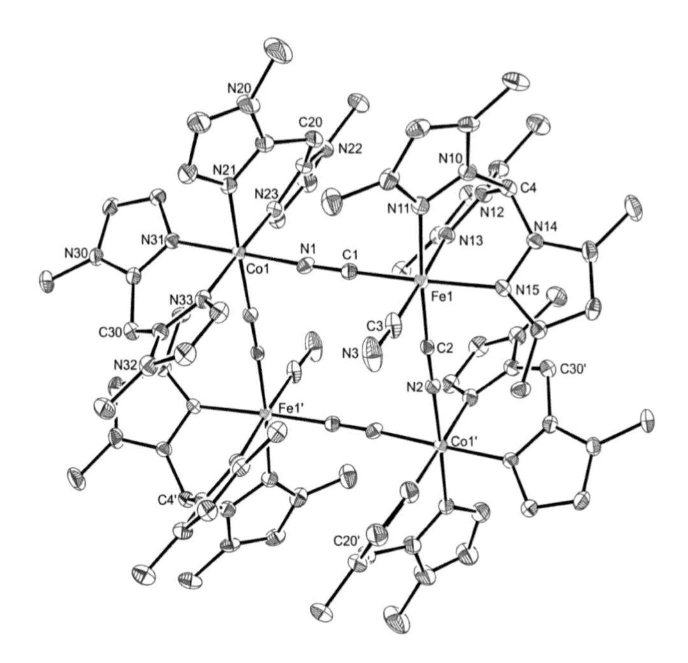

Figure 5.13: Perspective view of the cationic unit of **16**. Atoms are displayed as 30% probability ellipsoids. Hydrogen atoms, lattice water molecules and tetrafluoroborate counteranions are omitted for clarity. The atoms noted with an apostrophe are generated by the following symmetry operations: -x, 1-y, -z.

Selected bonds lengths (Å) and angles (°) for **16**: Fe1–C1 1.865(5), Fe1–C2 1.871(5), Fe1–C3 1.895(6), Fe1–N11 2.029(5), Fe1–N13 2.002(4), Fe1–N15 2.042(4), Co1–N1 1.880(5), Co1–N2' 1.884(4), Co1–N21 1.918(4), Co1–N23 1.916(5), Co1–N31 1.919(4), Co1–N33 1.933(5), Fe1-C1-N1 179.0(5), Fe1-C2-N2 177.4(4), Fe1-C3-N3 175.2(5), C1-Fe1-C2 90.2(2), C1-Fe1-C3 88.3(2), C2-Fe1-C3 88.9(2), C1-Fe1-N13 89.6(2), C2-Fe1-N13 91.06(19), C1-Fe1-N11 93.3(2), C3-Fe1-N11 93.0(2), N13-Fe1-N11 87.17(18), C2-Fe1-N15 90.00(19), C3-Fe1-N15 94.9(2), N13-Fe1-N15 87.21(17), N1-Fe1-N15 86.42(18), C1-N1-Co1 174.9(5), C2'-N2'-Co1 177.4(4), N1-Co1-N2' 89.84(18), N1-Co1-N23 89.68(19), N2'-Co1-N23 90.12(19), N1-Co1-N21 89.45(19), N21-Co1-N23 88.3(2), N2'-Co1-N31 90.04(18), N23-Co1-N31 91.69(19), N21-Co1-N31 90.70(19), N1-Co1-N33 89.50(19), N2'-Co1-N33 89.25(18), N21-Co1-N33 92.3(2), N31-Co1-N33 89.13(18), C4-N10-N11-Fe1 -9.98, C4-N12-N13-Fe1 -7.65, C4-N14-N15-Fe1 -11.36.

At 200 K, **16** crystallises in the monoclinic group *P21/n*. Like **15**, it consists of a {Fe_2Co_2} cyanide-bridged molecular square cationic unit, four tetrafluoroborate counteranions and twelve water lattice molecules. A perspective view of the tetranuclear unit of **16** is depicted in Figure 5.13, and selected bond lengths and angles are listed in its caption. Contrary to **15**, but much more classically, the molecular square unit of **16** is centrosymmetric. The intramolecular connectivity remains, however, unchanged, and each metal ion retains the same coordination sphere: N_6 for the cobalt ions and C_3N_3 for the iron ions. While the octahedral distortion around the cobalt atoms is identical in both molecular squares (9.6° in **16**, to be compared with 9.4° and 11.9° in **15**), the iron octahedral distortion amounts to 25.5° in **16** (27.1° and 37.2° in **15**). The iron-cobalt distances are about the same length as in **15** and average 4.907 Å, which is consistent with the {$Fe^{II}_{LS}Co^{III}_{LS}$} diamagnetic state. The angles between the metals and the angles between the cyanides, are very close to orthogonality, closer than they are in **15**. The Fe–C and Fe–N bond distances of **16** are sensibly the same as in **15** and are consistent with low-spin iron(II) ions. However, in **16**, the cyanide ligands *C*-bind the iron ion in an almost linear way, while it is not the case in **15**. The non-bridging cyanide diverts the most from linearity with a Fe1-C3-N3 angle of 175.2(5)°.

The bridging cyanide ligands *N*-bind the cobalt atoms in an almost linear way, with biting angles of 174.9(5)° and 177.4(4)° at N1 and N2' respectively. This leads to a far lesser distortion of the cyanide-bridged square motive in **16** than it is in **15**. Co–$N_{cyanide}$ bond lengths average 1.882 Å, that is the same distance as that found in **16**. The Co–N_{im} bonds also average 1.922 Å, but the gap in length amounts to 0.017 Å in **15**, while in **16**, it amounts to 0.024 and 0.040 Å for Co1 and Co2 respectively. The two bim ligands have bite angles close to orthogonality (90.04(18)° and 89.19(18)°).

The non-bridging cyanides are involved in a hydrogen bond network with the water molecules along the *b* axis. In spite of a small intermolecular Fe···Fe distance of 7.81 Å, the molecular squares are well "isolated" from each other: the shortest distances between the centroids of nearby heterocycles amount to 4.04 Å, which is too long for π-interactions to take place but weak interactions take place between the CH_3- moieties of the pyrazolyl rings of two neighbouring molecular squares and one of the fluorines of a $[BF_4]^-$ counteranion.

5.2.3 Fourier Transform InfraRed spectroscopy

$\{[Fe^{II}(Tpm^*)(CN)_3]_2[Co^{III}(bik)_2]_2\}(BF_4)_2 \cdot 7\,H_2O$ (15)

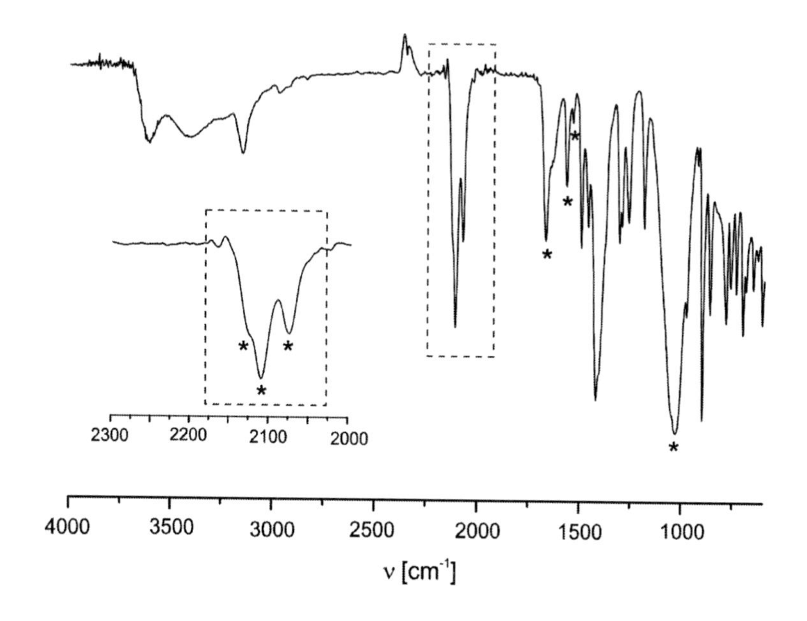

Figure 5.14: FT-IR (ATR) transmission spectrum of fresh **15** between 4000 cm^{-1} and 600 cm^{-1} with a 4 cm^{-1} resolution. Selected IR vibration bands in cm^{-1} and their intensities are marked with an asterisk: 1043 (br, vs), 1542 (vw), 1568 (w), 1672 (m), 2075 (m), 2114 (s), 2128 (sh, m).

The cyanide stretching vibration in **15** at 2075, 2114 and 2128 cm^{-1} are typical of iron(II) oxidation state, which is consistent to the X-ray diffraction data (*vide supra*). The Tpm* pyrazole rings stretch at 1568 cm^{-1}. The ketone moiety of the bik ligands bound to the cobalt(III) ion have a characteristic stretch at 1672 cm^{-1}, that is about 42 cm^{-1} at higher frequency than the free ligand. The imidazolyl stretch is found at 1542 cm^{-1}, that is 20 cm^{-1} blueshifted compared to free bik. The $[BF_4]^-$ anions also display a characteristic set of stretches in the form of a very strong broad absorption at 1043 cm^{-1}.

$\{[Fe^{II}(Tpm^*)(CN)_3]_2[Co^{III}(bim)_2]_2\}(BF_4)_2 \cdot 12\ H_2O$ (16)

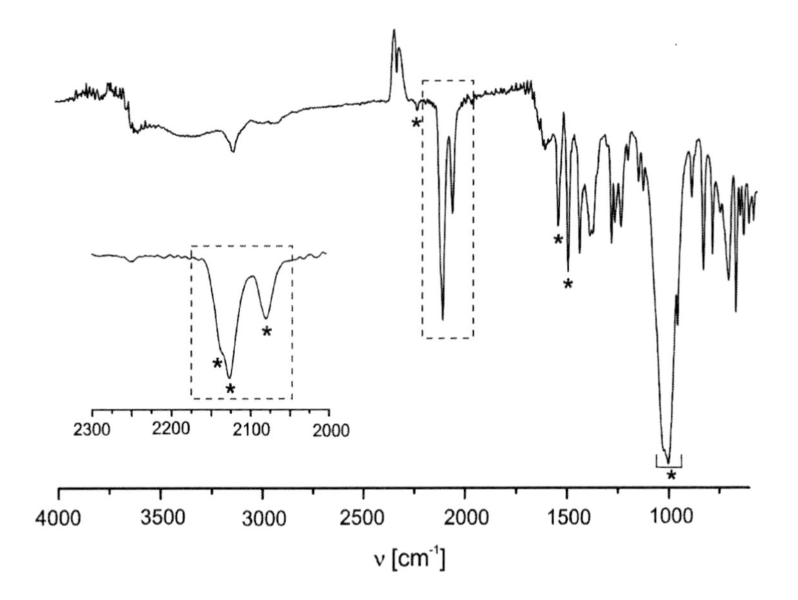

Figure 5.15: FT-IR (ATR) transmission spectrum of fresh filtered **16** between 4000 and 600 cm^{-1} with a 4 cm^{-1} resolution. Selected IR vibration bands in cm^{-1} and their intensities are marked with an asterisk: 1032 (br, vs), 1519 (m), 1567 (m), 2077 (w), 2124 (s), 2134 (m), 2249 (vw).

The FT-IR spectrum of fresh **16**, recorded at room temperature, also indicates that the iron-cobalt molecular square exhibits a $\{Fe^{II}_{LS}Co^{III}_{LS}\}$ ground state, with a cyanide vibration band pattern being very close to that of **15** (at 2077, 2124 and 2134 cm^{-1} respectively). The complete absence of a ketonic vibration at about 1670 cm^{-1} is a strong indication that the blocking ligands of the cobalt ions are actually bim and did not oxidise into bik (unlike it is usually observed under aerobic conditions). Two vibrations can be attributed to five-membered ring stretches: the tripodal Tpm* is responsible for the vibration at 1567 cm^{-1}, while the imidazolyl heterocycles of the bim ligands come at lower frequency: 1519 cm^{-1}. This is redshifted compared to free bim, where it comes at 1528 cm^{-1}. The characteristic ill-defined broad absorption at 1032 cm^{-1} is again due to the four $[BF_4]^-$ anions. The small absorption at 2249 cm^{-1} corresponds to free (not bound) acetonitrile in the sample.

5.2.4 SQUID magnetometry

Solvated phases and desolvated phases of **15** and **16** were analysed by SQUID magnetometry between 2 and 400 K. Both samples gave small, negative $\chi_M T$ products over the full temperature range, which is typical of diamagnetic compounds. These results were not modified upon desolvating the samples at 400 K (under helium reduced pressure) *in situ* in the SQUID magnetometer. This is consistent with the structural and infrared spectra analyses that account for a diamagnetic $\{Fe^{II}{}_{LS}Co^{III}{}_{LS}\}$ ground state in both compounds.

Solvated (fresh) and dehydrated samples of **15** and **16** were also tested for photomagnetism at 20 K, but no magnetic reaction was observed when irradiated with laser light at 808 and 532 nm. This is coherent with the fact that no thermo-induced ETCST phenomenon is observed by SQUID magnetometry below 400 K. A compound exhibits a thermo-induced spin transition only if the energy gap between the low-spin and high-spin potential curves is not too high. If this gap is too high, the compound remains low-spin over the whole temperature range. Because of the parallelism observed for the Jablonski curve of the spin transition and the ETCST effect, it is reasonable to think that if the high-spin potential curve corresponding to the paramagnetic state is high in energy compared to the low-spin potential curve, the ETCST is energetically unfavourable.

6 Cyanide-bridged molecular multimetallic complexes using the $[Fe(Tp)(CN)_3]^-$ ($[\mathbf{1}]^-$) building block

The $[Fe^{III}(Tp)(CN)_3]^-$ ($[\mathbf{1}]^-$) complex is the longest-known complex based on cyanide and scorpionate ligands reported in the literature. Since the complex was first reported by Lescouëzec *et al.* in 2002,[114] it was used as "complex-as-ligand" (metalloligand) to produce a wide range of cyanide and scorpionate ligands $\{Fe_xM_y\}$ clusters, and coordination polymers with M = Mn, Fe, Co, Ni and Cu. [87,88,94,98,134,174–178]

Depending on the nature of the second metal, various properties can be obtained for the corresponding clusters: for example, iron-nickel and iron-copper compounds based on $[\mathbf{1}]^-$ units are more prone to show single molecule magnet or single chain magnet (SMM or SCM) behaviour,[174] while switchable magnetic systems are rather observed for $\{Fe_xFe_y\}$ and $\{Fe_xCo_y\}$ compounds.[98,107]

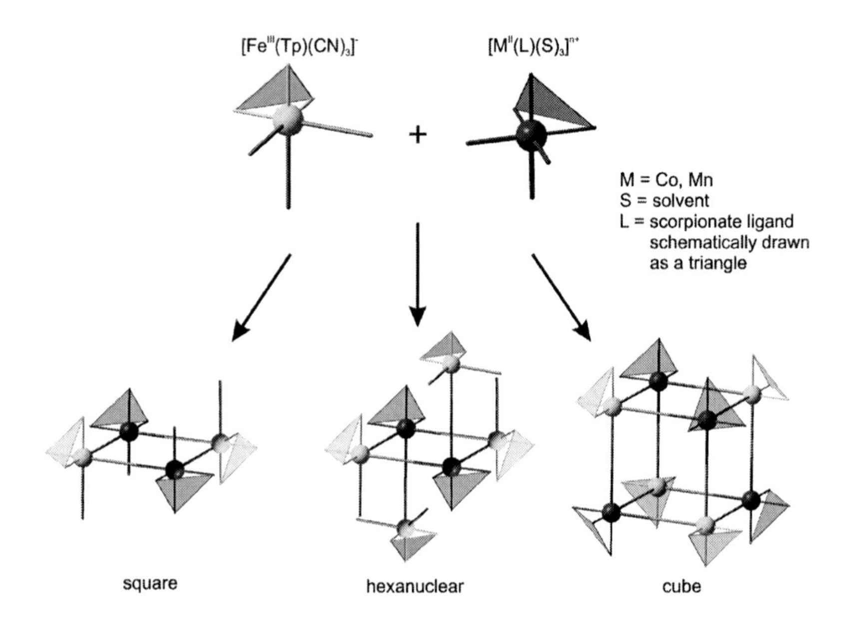

Figure 6.1: Some conceivable cyanide-bridged clusters using $[Fe^{III}(Tp)(CN)_3]^-$ (**[1]**$^-$) as complex-as-ligand in respect to partially blocked *fac*-$[M^{II}(L)(S)_3]^{n+}$ cationic units (M = Co, Mn, L = scorpionate ligand, S = solvent, n = 1, 2).

As already mentioned in the introduction of this work, the topology of the products obtained by self-assembly of substituted cyanidometallates and partially blocked cationic units is dependent on the topology of precursors. However, other parameters like the nature of e.g. metal ions, solvents, blocking ligands (in this chapter, Tp for the iron metalloligand), counterions, stoichiometric ratios, relative solubilities of all possible species and type of crystallisation (slow-evaporation, layering with a non-solvent or slow-diffusion of reagents) are expected to play a crucial role in the nature of the obtained products. Though it is not possible to draw conclusions about the effects of each parameter, this chapter illustrates this diversity (see Figure 6.1) by presenting some of the products obtained from reactions involving **[1]**$^-$ and partially blocked *fac*-$[M^{II}(L)(Solvent)_3]^{n+}$ cationic units, where M is either a cobalt(II) (**17**, **19**, **22** and **21**) or manganese(II) ion (**18** and **20**), L = the scorpionate ligand Tpm* (**17**, **18**), Tpe (**19**, **20** and **21**) or Ttp (**22**).

6.1 Molecular squares

6.1.1 Syntheses

$\{[Fe^{III}(Tp)(CN)_3]_2[Co^{II}(Tpm^*)(MeOH)]_2\}(ClO_4)_2 \cdot 2MeOH$ (17)

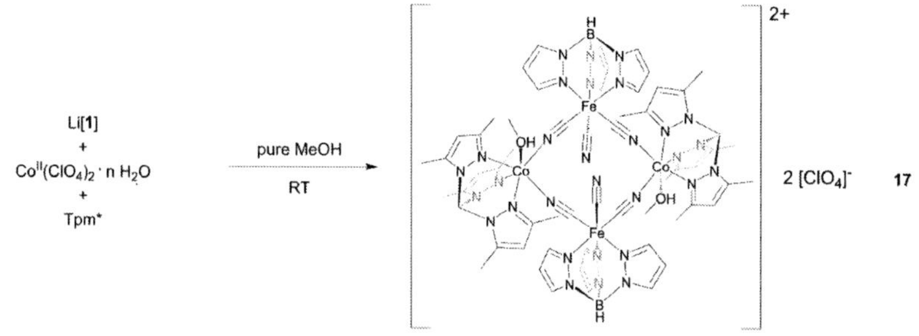

Li[1] + $Co^{II}(ClO_4)_2 \cdot n\ H_2O$ + Tpm* → (pure MeOH, RT) → [ ]$^{2+}$ 2 $[ClO_4]^-$ **17**

Figure 6.2: Synthesis of **17**.

17 was synthesised in pure methanol at room temperature (see Figure 6.2). It crystallises as big red blocks. The obtained yields for **17** are quite low (36%), but crystals were collected quite early in the crystallisation process to avoid the crystallisation of either side-products and/or reagents.

$\{[Fe^{III}(Tp)(CN)_3]_2[Mn^{II}(Tpm^*)(DMF)]_2\}(ClO_4)_2 \cdot 3\ DMF \cdot 2\ H_2O$ (18)

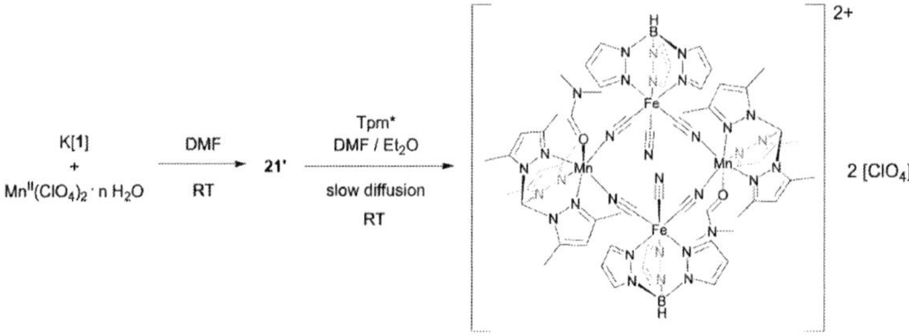

Figure 6.3: Synthesis for **18**.

18 was obtained in a synthesis inspired from those known to lead to $\{Fe_4M_4\}$ cubes in the literature,[93,176,179,180] using Tpm* instead of Tpe derivatives. When solid manganese(II) perchlorate salt was added to a DMF yellow solution of K[**1**], it slowly dissolved to produce a blood red solution (see Figure 6.3). The resulting product **18'** was precipitated as a red oily product through addition of diethyl ether. After being washed with DMF/Et_2O, **18'** was obtained as a deep red solid. The IR spectrum of **18'** shows very strong perchlorate peaks at 1083 cm^{-1} (broad) and 1047 cm^{-1}, and three cyanide peaks that can be ascribed to terminal cyanides (2125 cm^{-1} and 2132 cm^{-1}) and bridging cyanides (2150 cm^{-1}). This hints towards the synthesis of a square-like structure for **18'**. This is supported by the existence of the parent molecular square $\{[Fe(Tp^*)(CN)_3]_2[Mn^{II}(DMF)_4]_2\}(ClO_4)_2$ obtained under similar conditions by Li *et al.* in 2005 (DMF layered with diethyl ether).[172] Addition of solid Tpm* to a solution of **18'** in DMF afforded an orange solution, whose layering with diethyl ether afforded orange crystals of **18** after a few weeks. **18** was obtained in low yield (10%) due to the high solubility of the compound in DMF-diethyl ether mixtures.

6.1.2 Structural analyses

$\{[Fe^{III}(Tp)(CN)_3]_2[Co^{II}(Tpm^*)(MeOH)]_2\}(ClO_4)_2 \cdot 2MeOH$ (17)

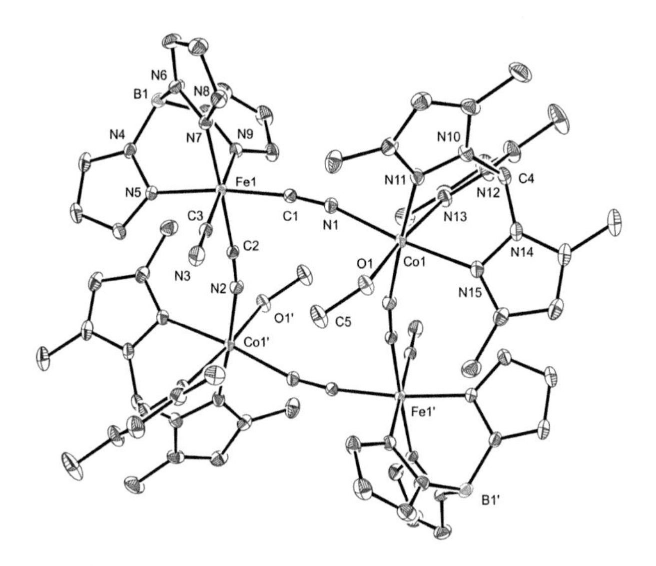

Figure 6.4: Perspective view of the cationic unit of molecular square **17**. Atoms are displayed as 30% probability ellipsoids. Hydrogen atoms, lattice solvent molecules and perchlorate counteranions are omitted for clarity. Equivalent atoms (noted with apostrophe) within the molecular square are generated with the following symmetry operations: 2-x, 1-y, 1-z. Selected bond lengths (Å) and angles (°) for **17**: Fe1–C1 1.9265(12), Fe1–C2 1.9188(13), Fe1–C3 1.9243(13), Fe1–N5 1.9701(10), Fe1–N7 1.9764(11), Fe1–N9 1.9733(11), Co1–N1 2.1123(11), Co1–N2' 2.0780(11), Co1–O1 2.0701(10), Co1–N11 2.1125(11), Co1–N13 2.1395(11), Co1–N15 2.1771(11), C1-Fe1-C2 86.20(5), C1-Fe1-C3 84.91(5), C2-Fe1-C3 90.69(5), N7-Fe1-N9 87.75(5), N5-Fe1-N7 89.17(4), N5-Fe1-N9 88.70(4), C1-Fe1-N9 94.96(5), C1-Fe1-N7 96.66(5), C2-Fe1-N9 89.79(5), C2-Fe1-N5 88.12(5), C3-Fe1-N5 91.48(5), C3-Fe1-N7 91.78(5), N1-Co1-O1 93.84(4), N1-Co1-N2' 90.36(4), O1-Co1-N2' 90.11(4), N11-Co1-N15 82.19(4), N11-Co1-N13 87.00(5), N13-Co1-N15 83.12(4), O1-Co1-N11 89.71(4), O1-Co1-N15 89.04(4), N1-Co1-N11 93.05(4), N1-Co1-N13 93.78(4), N2'-Co1-N13 92.74(5), N2'-Co1-N15 94.40(4), Fe1-C1-N1 171.78(11), Fe1-C2-N2 173.83(11), Fe1-C3-N3 179.48(12), Co1-N1-C1 167.52(10), Co1-N2'-C2' 168.90(10), Co1-O1-C5 129.36(10).

Compound **17** crystallises in the triclinic space group $P\bar{1}$ ($Z = 1$). The structure consists of a centrosymmetric dicationic cyanide-bridged tetranuclear heterobimetallic $\{Fe_2Co_2\}$ complex, two perchlorate ions and two lattice methanol molecules. Although no electron density could be attributed to lattice water molecules, elemental analysis calculations including four water molecules per molecular square account well for the found values. A perspective view of compound **17** cationic unit is depicted in Figure 6.4 and selected bond lengths and angles are listed in its caption. In the tetranuclear entity, two $\{Fe^{III}(Tp)(CN)_3\}$ complexes coordinate two cobalt(II) ions through *cis* cyanide ligands, thus building a slightly distorted centrosymmetric molecular square (C1-Fe1-C2 angle = 86.20(5)° and N1-Co1-N2' angle = 90.36(4)°). Each corner is alternately occupied by an iron or a cobalt ion. At 200 K, the Fe···Co edges are almost identical (5.122 Å and 5.101 Å) and their angles at the corners differ very slightly from orthogonality (Fe1-Co1-Fe1' angle = 92.5° and Co1-Fe1-Co1' angle = 87.5°). Those values are close to those found for similar $\{Fe_{LS}^{III}Co_{HS}^{II}\}$ cyanide-bridged molecular squares, while $\{Fe_{LS}^{II}Co_{LS}^{III}\}$ molecular squares usually display smaller edge lengths under 5.0 Å.[94–98,134]

The two iron atoms lie in a slightly distorted octahedral C_3N_3 environment formed by three imine moieties from the pyrazolyl rings of a *fac*-coordinating Tp ligand and the carbon atoms of three cyanides. Two out of the three cyanides act as bridging ligands between the two iron and the two cobalt ions. The remaining terminal cyanide ligands are orientated in *trans* position in respect to the plane containing all four metal atoms. The Fe-$C_{cyanide}$ bonds are relatively similar; their lengths range from 1.9265(12) to 1.9188(13) Å, which are typical values for low-spin cyanido iron(III) complexes.[87] The Fe–N_{pz} bond lengths are also of similar length but are a little longer than their Fe–C counterparts with a mean value of 1.973 Å. The bridging cyanides are slightly bent on the iron side (Fe1-C1-N1 angle = 171.78(11)° and Fe1-C2-N2 angle = 173.83(11)°); the non-bridging cyanide is connected linearly (Fe1-C3-N3 angle = 179.48(12)°). The octahedral distortion for the iron(III) environment is 30.93°, which is a little more distorted than in the tricyanido iron building block (25.77° in the $[\mathbf{1}]^-$).[114]

The octahedral coordination sphere of each cobalt ion is completed by a tridentate *fac*-coordinating *N*-donor Tpm* ligand and a coordinated methanol molecule, leading to an octahedral N_5O environment, for which the octahedral distortion amounts to 37.2°. The cyanide bridges are notably bent on the cobalt side (C1-N1-Co1 angle = 167.5(1)° and C2'-N2'-Co1 angle = 168.9(1)°), while the Co–$N_{cyanide}$ bond lengths amount to 2.112(1) Å and 2.078(1) Å respectively. Co–N_{pz} bond lengths range from 2.139 Å to

2.177 Å, that is a little longer than the two Co–$N_{cyanide}$ bonds, but in adequation with distances reported for Co^{II}–N_{Tpm*} values.[41] More importantly, these reported Co–N bond lengths are consistent with cobalt(II) high-spin states values found in the literature.[10,41,88,103]

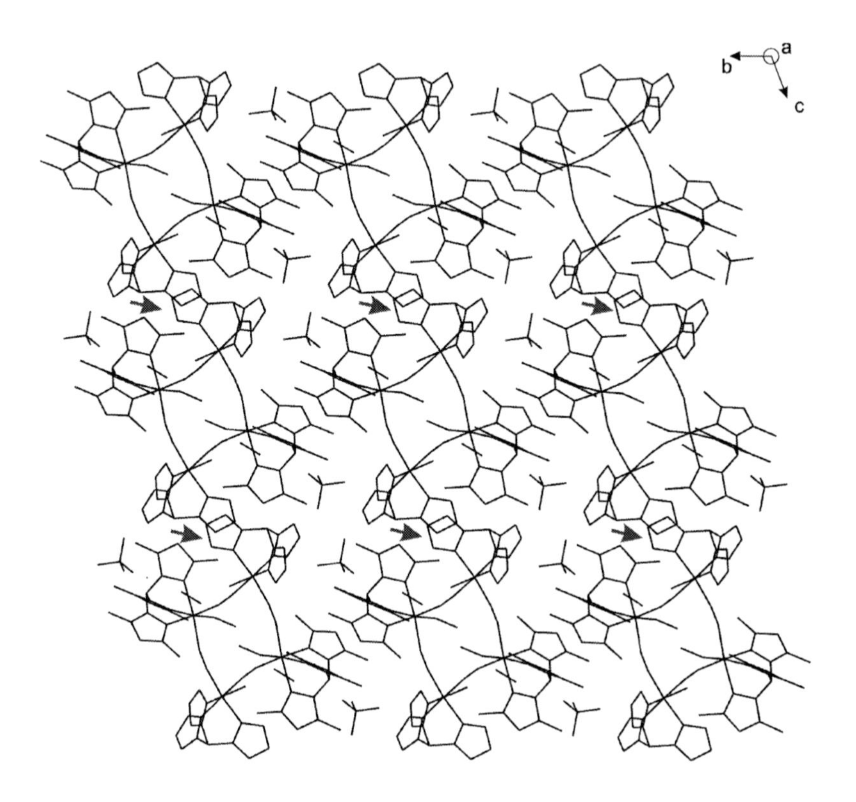

Figure 6.5: Crystal packing of **17** along the *a* axis. Piles of cations and anions are eclipsed. π-interactions between the pyrazoles of adjacents piles along the *b* axis are indicated by the red arrows.

Molecular squares of **17** and perchlorate anions are piled up in a segregate manner along the *a* axis, and alternate cation and anion piles along the *c* axis (see Figure 6.5). In each *a* axis-along pile of cations, molecular squares are ordered in a stairway manner, so that one edge of the square is at a right angle to the opposite edge (alternate iron-cobalt configuration) of the next square. Communication between different squares in the same pile is facilitated through two symmetric hydrogen bonding patterns depicted in Figure 6.6. The cobalt methanolic ligand is involved in a strong hydrogen bond with the free methanol lattice molecule, which in turn interacts with the non-bonding cyanide ligand of the next molecular square. The smallest intermolecular distance between two metal ions along the *a* axis is 9.617 Å (Fe···Co distance). Interactions between cation piles occur along the *b* axis through π-π interactions: the N4-N5 pyrazolyl moiety of the Tp ligand (iron side) and the equivalent pyrazolyl Tp ring in the molecular square of the next pile overlap partially ($Centroid_{square1} \cdots Centroid_{square2}$ distance = 3.52 Å). The smallest intermolecular distance between two metals ions from two adjacent piles is 7.570 Å, which is smaller than the smallest intermolecular metal-metal distance inside a pile.

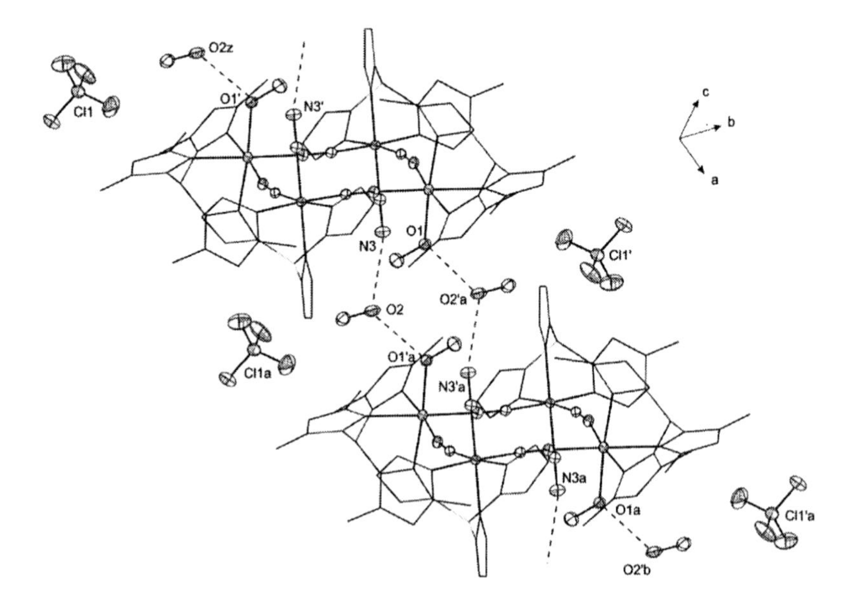

Figure 6.6 - Perspective view of **17** with its network of hydrogen bonds (dotted lines) within a cationic molecular pile. Letters in atom labels refer to consecutive cationic units in the crystal packing.

$\{[Fe^{III}(Tp)(CN)_3]_2[Mn^{II}(Tpm^*)(DMF)]_2\}(ClO_4)_2 \cdot 3\ DMF \cdot 2\ H_2O$ (18)

Compound **18** crystallises in the triclinic space group $P\bar{1}$, (Z = 1). The crystal structure consists of dicationic cyanide-bridged $\{Fe_2Mn_2\}$ complex, perchlorates anions, and lattice molecules (three DMF and one water molecule per square). A perspective view of the cationic unit is depicted in Figure 6.7, and selected bond lengths and angles are listed in the caption. In the cationic unit, two cyanides of two $[Fe(Tp)(CN)_3]^-$ metalloligands bridge two manganese ions, thus forming a [2+2] distorted centrosymmetric molecular square. The angles between the metal ions are close to orthogonality (Mn1···Fe1···Mn1' angle = 87.2° and Fe1···Mn1···Fe1' angle = 92.8°) while the edge lengths are almost identical (Fe···Mn distance = 5.26 Å and 5.20 Å). This is consistent with reported distances in other $\{Fe^{III}_2Mn^{II}_2\}$ discrete molecular squares.[87,88,115,117] The third non-bridging cyanides are orientated in *trans* in respect to the $\{Fe_2Mn_2\}$ plane.

The iron ions lie in a C_3N_3 distorted octahedral environment. The octahedral distortion amounts to 23.2°, which is about the octahedral distortion of the iron monomer. The Fe–$C_{cyanide}$ bond lengths average 1.922 Å. This value is indicative of a +III oxidation state for the iron ions. The Fe–N_{pz} bond lengths (mean Fe–N_{pz} distance = 1.971 Å) are a little smaller than in the building block, but are consistent with other reported Fe^{III}–N_{pz} distances (for example, **17**). All cyanide ligands bind the respective iron ions almost linearly with no significant difference between bridging and non-bridging cyanides (177.6(5)° ≤ Fe1-C-N angle ≤ 178.8(5)°).

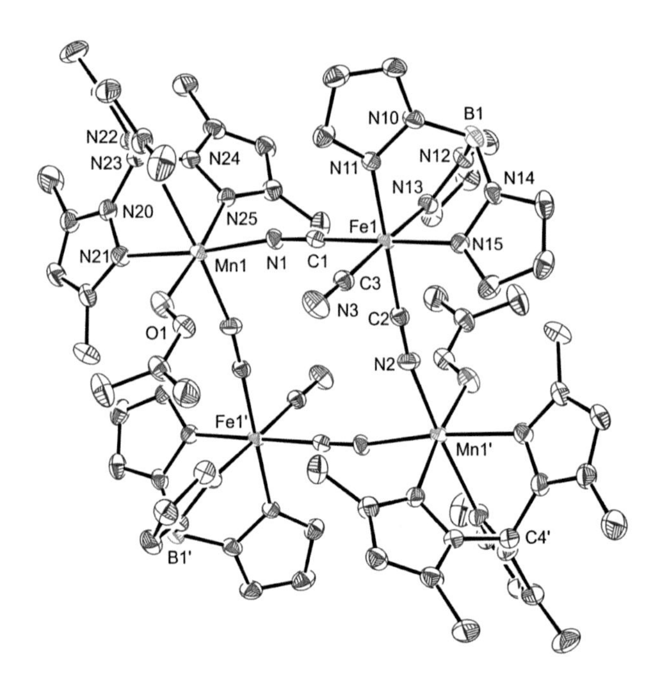

Figure 6.7: Perspective view of the cationic unit of **18**. Atoms are displayed as 30% probability ellipsoids. Hydrogen atoms, DMF, water molecules and the perchlorate counteranions are omitted for clarity. Equivalent atoms (noted with apostrophe) within the molecular square are generated with the following symmetry operations: 1-x, 1-y, 1-z. Selected bond lengths (Å) and angles (°) for **18**: Fe1–C1 1.930(6), Fe1–C2 1.902(5), Fe1–C3 1.935(6), Fe1–N11 1.975(4), Fe1–N13 1.967(5), Fe1–N15 1.971(4), Mn1–O1 2.171(4), Mn1–N1 2.199(4); Mn1–N2' 2.151(5), Mn1–N21 2.299(5), Mn1–N23 2.262(5), Mn1–N25 2.277(5), C1-Fe1-C2 88.3(2), C1-Fe1-C3 88.3(2), C2-Fe1-C3 87.0(2), C2-Fe1-N15 91.4(2); C3-Fe1-N15 93.5(2), C1-Fe1-N11 91.8(2), C3-Fe1-N11 91.1(2), N15-Fe1-N11 88.50(18), C2-Fe1-N13 90.4(2), C1-Fe1-N13 93.2(2), N13-Fe1-N15 86.28(19), N11-Fe1-N13 90.17(19), N1-Mn1-N23 96.53(17), N1-Mn1-N2' 93.49(18), N1-Mn1-N25 95.14(16), N23-Mn1-N25 80.75(18), N2'-Mn1-N25 90.94(18), N23-Mn1-N21 78.24(17), N2'-Mn1-N21 91.41(18), N21-Mn1-N25 81.86(17), N1-Mn1-O1 89.90(17), N23-Mn1-O1 95.85(19), N2'-Mn1-O1 91.62(19), N21-Mn1-O1 92.88(18), Fe1-C1-N1 178.8(5), Fe1-C2-N2 177.6(5), Fe1-C3-N3 177.9(5), C1-N1-Mn1 171.7(4), C2'-N2'-Mn1 170.5(5), Mn1-O1-C_{DMF} 130.1(4).

As in **17**, the coordination sphere of the manganese ions are completed by a neutral *fac*-coordinating *N*-donor Tpm* ligand. However, since the reaction did not take place in the same solvent as for **17**, the remaining coordination site is occupied by a coordinated DMF molecule, thus forming a N_5O octahedral environment around the manganese ions. The octahedral distortion is high, reaching *ca* 57.1°. The Mn1–$N_{cyanide}$ bond lengths are

2.199(4) and 2.151(5) Å respectively, while the Mn1–N_{pz} bond lengths are longer and range from 2.262(5) to 2.299(5) Å. This is consistent with a high-spin manganese(II) ion.[10,88,115,117,172,181] The cyanides bind the manganese ions with a slightly bent angle of 170.5(5)° and 171.7(4)° for C2≡N2 and C1≡N1, respectively.

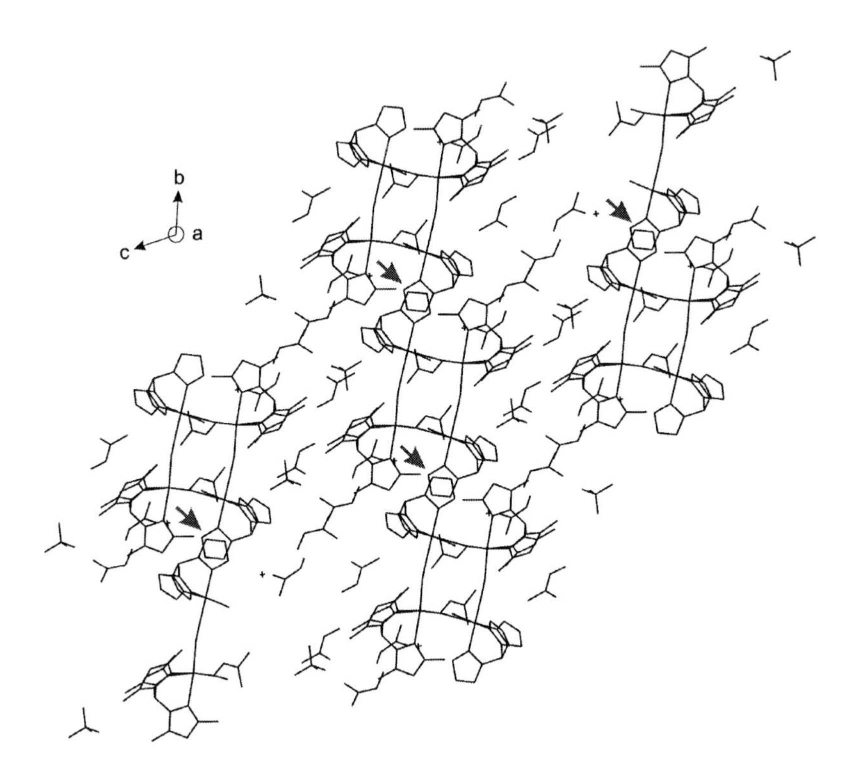

Figure 6.8: Crystal packing of **18** along the *a* axis. Piles of cations and anions are eclipsed. π-π interactions between the pyrazoles of adjacents piles along the *b* axis are indicated by the red arrows.

Unlike **17**, no hydrogen bond network connects the neighbouring molecular squares. However, careful observation of the crystal packing (Figure 6.8) indicates that π-π interactions between the pyrazolyl heterocycles of the Tp ligands take place, connecting the molecular squares along the b axis ($Centroid_{square1} \cdots Centroid_{square2}$ = 3.42 Å). Along this axis, the shortest intermolecular metal to metal distance amounts to 7.68 Å between to iron ions. Along the a and c axis, the cationic units are well isolated by DMF molecules and perchlorate counteranions.

6.1.3 Fourier Transform InfraRed spectroscopy

$\{[Fe^{III}(Tp)(CN)_3]_2[Co^{II}(Tpm^*)(MeOH)]_2\}(ClO_4)_2 \cdot 2MeOH$ (17)

The FT-IR absorption spectrum of freshly filtered **17** (see Figure 6.9) was recorded at room temperature using an ATR module. The presence of $\{Fe^{III}(Tp)(CN)_3\}$ units is displayed by its slightly shifted B–H stretching band at 2545 cm^{-1} (*cf* 2536 cm^{-1} in the starting material $[\mathbf{1}]^-$) and its pyrazolyl ring stretch at 1501 cm^{-1}, which is typical of non-methylated pyrazolyl species (in that case Tp). Furthermore, **17** displays two cyanide stretching vibrations well above 2100 cm^{-1}, which are characteristic of ferricyanides. The position of these two absorptions, at 2169 and 2149 cm^{-1}, indicate two different types of bridging cyanides ligands, in agreement with the X-ray diffraction data. A third stretch could be expected around 2123 cm^{-1} to match with the third terminal cyanide ligand. This is not the case; however, a closer look at the peak at 2149 cm^{-1} unravels a small shoulder at 2143 cm^{-1}. This is 20 cm^{-1} higher than typical non-bridging cyanides, but still too low to be a third type of cyanide bridge. In fact, this intermediary position can be explained by strong hydrogen interactions between the cyanide C3≡N3 and the lattice methanol molecules as already depicted in Figure 6.6. This was confirmed by recording a different spectrum of the same sample a few days after filtration: while the rest of the spectrum remains unchanged, the shoulder peak disappeared to be replaced by a weak absorption at 2129 cm^{-1}, which is in the normal range for non-bridging cyanide stretching vibrations.

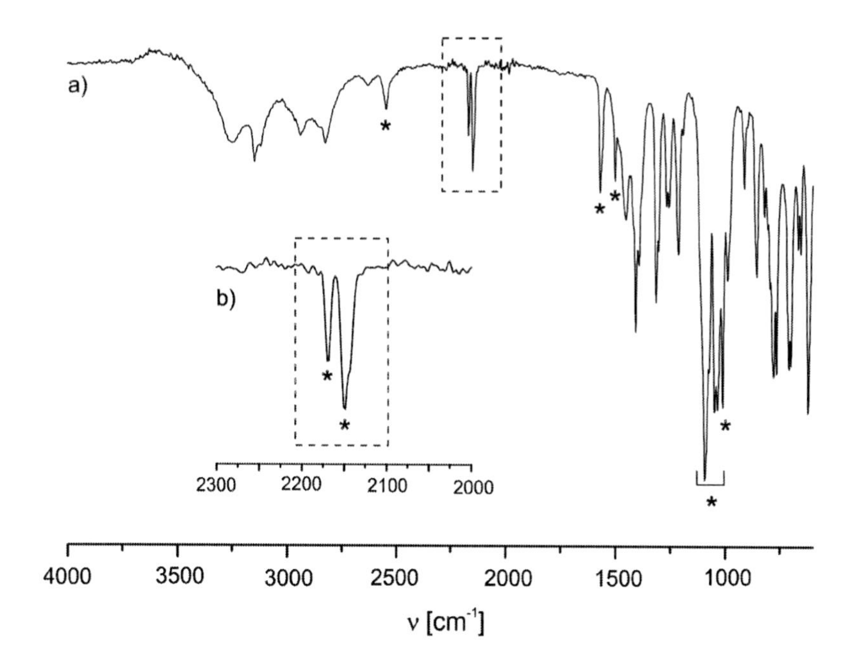

Figure 6.9: a) FT-IR (ATR) transmission spectrum of freshly filtered **17** between 4000 and 600 cm^{-1} with a 4 cm^{-1} resolution. Selected IR vibration bands in cm^{-1} and their intensities are marked with an asterisk: 988(m), 1013 (vs), 1036 (vs), 1050 (vs), 1074 (s), 1093 (vs), 1501 (w), 1569 (w), 2149 (w), 2143(sh), 2169 (vw), 2545 (vw). b) Zoom on the cyanide stretching band area of the same spectrum.

The presence of the $\{Co^{II}Tpm^*\}$ moiety can also be detected in the IR spectrum even though the oxidation state of the cobalt ion cannot be assessed with certainty using only infrared data. Indeed, pyrazole rings substituted at the 3- and the 5-positions display a vibration band (pyrazole ring breathing) at slightly higher frequency than their non-substituted analogues, in that case at 1569 cm^{-1}, but this frequency is relatively independent from the oxidation state of the coordinated metal ion. The spectrum also displays bands in the non-aromatic C–H stretching area, between 2850 and 3000 cm^{-1} which can be attributed to the methyl groups of the Tpm* ligands and the methyl moiety of the coordinated and lattice methanol molecules. The presence of non-coordinated perchlorate ions, which exhibit four consecutive characteristic bands at 1036, 1050, 1074 and 1093 cm^{-1}, points towards a cationic molecular square. Finally, methanol can also be detected in the sample through its C–O stretch at 1013 cm^{-1}.

$\{[Fe^{III}(Tp)(CN)_3]_2[Mn^{II}(Tpm^*)(DMF)]_2\}(ClO_4)_2 \cdot 3\ DMF \cdot 2\ H_2O$ (18)

The FT-IR spectrum of **18** was recorded at room temperature, with an ATR module. The obtained IR spectrum is depicted in Figure 6.10 and selected vibrations are listed in its caption.

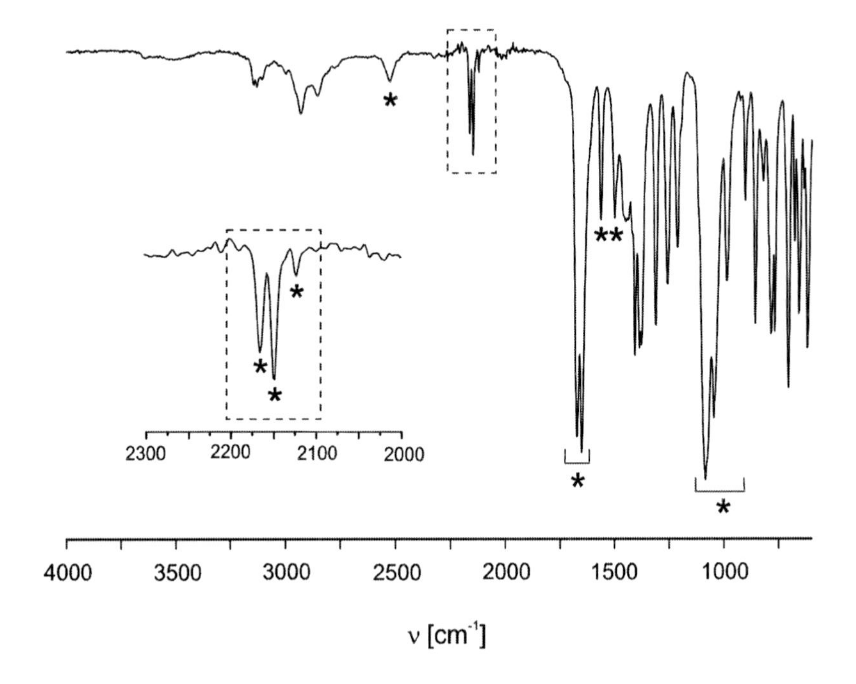

Figure 6.10: FT-IR (ATR) transmission spectrum of freshly filtered **18** between 4000 and 600 cm^{-1} with a 4 cm^{-1} resolution. Selected IR vibration bands in cm^{-1} and their intensities are marked with an asterisk: 988 (m), 1047 (vs), 1084 (br, vs), 1502 (w), 1565 (w), 1651 (vs), 1673 (vs), 2122 (vw), 2148 (w), 2164 (vw), 2525 (vw).

18 exhibits three cyanide stretching vibrations above 2100 cm^{-1}, two of them corresponding to bridging cyanides (2148 and 2164 cm^{-1}). The last one is clearly terminal, with a typical stretching band at 2122 cm^{-1}. This strongly supports the occurrence of the $\{Fe^{III}{}_2Mn_2{}^{II}\}$ oxidation state for **18**. The B–H stretching vibration of the $\{Fe^{III}(Tp)(CN)_3\}$ moiety is redshifted of about 20 cm^{-1} compared to that of **17**. The pyrazolyl ring stretch of the same moiety is, as usual, not affected by the coordination modes of the cyanides and is detected at 1502 cm^{-1}. The 3,5-dimethylpyrazolyl ring stretch band of the Tpm* ligand is also relatively uninfluenced by the nature of the coordinated metal (here manganese(II), compared to cobalt(II) in **17**) and arise at about the same frequency in both complexes (1565 cm^{-1} in **18**, 1569 cm^{-1} for **17**). Perchlorate counteranions provide three characteristic bands, at 988, 1047 and 1084 cm^{-1}. One would expect four bands, but the relative broad linewidth of some of them in the above spectrum does not allow sufficient resolution. Interestingly, the IR spectrum of **18** exhibits two intense absorption bands in the carbonyl region; one of them ($\tilde{\nu}$ = 1651 cm^{-1}) is typical of coordinated DMF molecule. The second absorption band, 22 cm^{-1} shifted towards higher frequency, corresponds to the uncoordinated lattice DMF molecules. The CH_3 moieties of both types of DMF present a vibration band at about the two same frequencies: $\tilde{\nu}$ = 2854 and 2932 cm^{-1}.

6.1.4 SQUID magnetometry

{[Fe^{III}(Tp)(CN)$_3$]$_2$[Co^{II}(Tpm*)(MeOH)]$_2$}(ClO_4)$_2$ · 2MeOH (17)

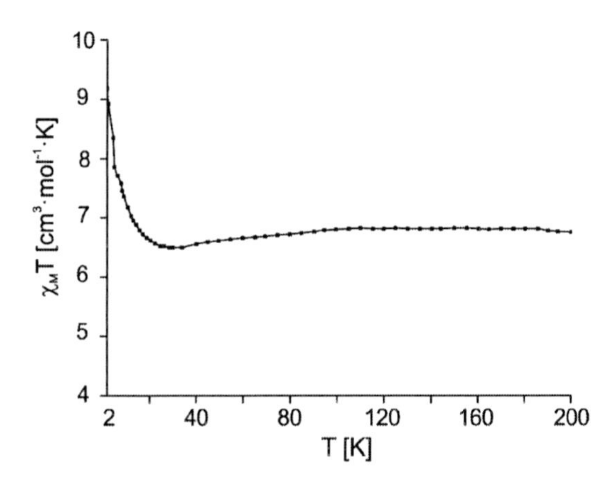

Figure 6.11: First provisional magnetic data of freshly filtered **17**. $\chi_M T$ product *vs* T between 2 K and 200 K, H = 2500 Oe. The sample was prepared as follows: m_{sample} = 29.19 mg, $m_{capsule}$ = 48.6 mg.

17 exhibits a paramagnetic behaviour over the whole temperature range. At 200 K, the $\chi_M T$ product of **17** reaches 6.75 cm^3·mol^{-1}·K, that is about the expected 6.6 cm^3·mol^{-1}·K for an independent set of two high-spin cobalt(II) ions ($\chi_M T \approx$ 2.7–3.6 cm^3·mol^{-1}·K) and two low-spin iron(III) ions (($\chi_M T \approx$ 0.6 cm^3·mol^{-1}·K). The $\chi_M T$ decreases slightly with the temperature, which is accounted to the spin-orbit coupling effects of the cobalt and iron ions. It reaches a minimum at T = 28 K ($\chi_M T$ = 6.50 cm^3·mol^{-1}·K), then increases to reach 9.18 cm^3·mol^{-1}·K at T = 2 K. The increase at low temperature is due to ferromagnetic interactions between the iron and cobalt ions, as already observed for compounds **10**, **11**, **13** and in the literature.[88]

$\{[Fe^{III}(Tp)(CN)_3]_2[Mn^{II}(Tpm^*)(DMF)]_2\}(ClO_4)_2 \cdot 3\ DMF \cdot 2\ H_2O$ (18)

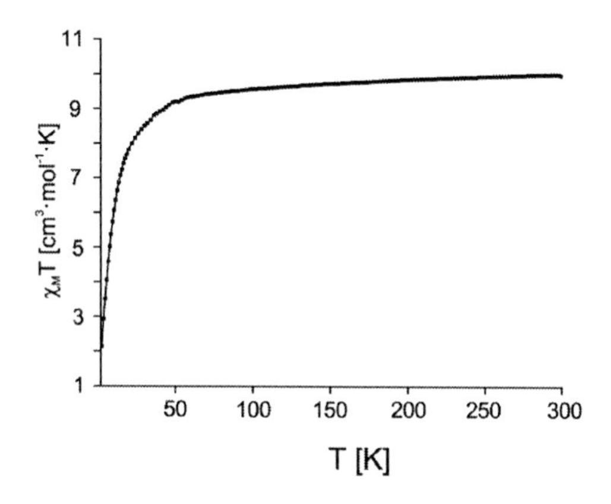

Figure 6.12: Magnetic properties of freshly filtered **18**. $\chi_M T$ product *vs* T between 2 K and 300 K, H = 250 Oe. The sample was prepared as follows: m_{sample} = 8.01 mg, m_{film} = 8.50 mg.

At high temperatures, the $\chi_M T$ product of **18** reaches 9.97 $cm^3 \cdot mol^{-1} \cdot K$ (Figure 6.12). This value is in the expected range for a set four independent paramagnetic ions: two high-spin manganese(II) ions (4.5 $cm^3 \cdot mol^{-1} \cdot K$ each) and two low-spin iron(III) ions (0.7 $cm^3 \cdot mol^{-1} \cdot K$ each). The curve exhibits a slightly descending slope as the temperature decreases between 300 K and 100 K, which is due to the spin-orbit coupling of the iron(III) ion. The strong increase below *ca* 80 K is likely due to antiferromagnetic interactions between the manganese(II) and the iron(III) ions.. At lower temperatures, the $\chi_M T$ product does not bounce back to a higher value, as it could have been expected for a S = 4 ground spin state. The π interactions between the Tp ligands of neighbouring cationic units found by X-ray analysis of **18** are probably responsible for an additional intermolecular antiferromagnetic interaction between the molecular square cationic units. Indeed, and as observed for similar $\{Fe^{III}Mn^{II}\}$ compounds of various nuclearity and topology,[88,117,171,172,181] the $\chi_M T$ product decreases rapidly under 50 K.

6.2 Cyanide-based hexanuclear complexes

6.2.1 Syntheses

An equivalent of Tpe ligand was pre-coordinated to one equivalent of $M(ClO_4)_2 \cdot x\ H_2O$ (M = Co (**19**), Mn (**20**)) in pure methanol (**20**) or in a methanol/water 5:1 mixture (**19**). The resulting solution, yellow in case of cobalt and colourless for manganese, was added dropwise to the stirred red solution of Li[**1**] in the same solvent. The resulting red (cobalt) and orange (manganese) solutions were filtered, covered with pierced paraffin film and stored for a month. The reaction with the cobalt salt produced directly red block-like crystals of **19** for suitable X-ray diffraction analysis. However, the reaction with the manganese ions first provided small black rods of a first species (**20'**), whose quality was not sufficient for X-ray diffraction analysis. The reaction mixture was covered hermetically to avoid further loss of solvent and was stored for another month, during which the small polycrystalline rods of **20'** were converted into bigger red block-like crystals of **20**, very similar in form and shape to **19**. It is noteworthy that using either the $[PPh_4]^+$ or the K^+ salt of $[\mathbf{1}]^-$ in the synthesis does not lead to crystallisation of the expected product but either to recrystallisation of $[\mathbf{1}]^-$ (in case of $[PPh_4]^+$) or precipitation of a red undefined compound (K^+).

6.2.2 Structural analyses

$\{[Fe^{III}(Tp)(CN)_3]_4[M^{II}(Tpe)]_2\}$ (M = Co (19), Mn (20))

19 and **20** are isostructural, and crystallise in the triclinic space group $P\bar{1}$, (Z = 1). Their crystal structures consist of an hexanuclear $\{Fe_4M_2\}$ (M = Co (**19**), Mn (**20**)) neutral complex and lattice water molecules. The structure of **20** is depicted in Figure 6.13, with selected bond lengths and angles included in the caption for both **19** and **20**.

The hexametallic structures of **19** and **20** can be described as {$Fe^{III}{}_2M^{II}{}_2$} molecular squares, quite similar to **17** and **18**, where the tripodal *N*-donor *fac*-coordinating Tpe replaces the Tpm* in the coordination sphere of the metal ion M. While the fourth and last coordination position is occupied by a coordinated methanol molecule in the structure of **17** and a DMF molecule in the structure of **18**, the coordination spheres of the M ions in **19** and **20** are completed by a nitrogen atom of a cyanide ligand from two supplementary {$Fe^{III}(Tp)(CN)_3$} moieties. Since they are singly negatively charged, they also can be described as coordinated counterions to the doubly positively charged {Fe_2M_2} molecule square, leading to an overall neutral molecule. Alike to the methanol (DMF) ligands in **17** (in **18**), the two satellite {$Fe^{III}(Tp)(CN)_3$} moieties are oriented in *trans* position relative to the mean plan of the molecular square.

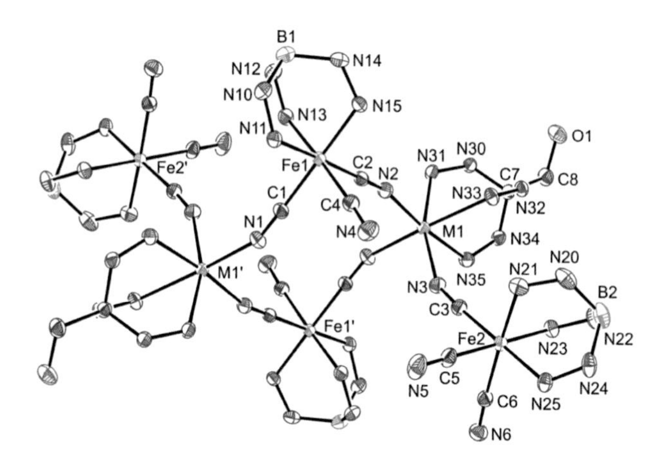

Figure 6.13: Perspective view of the hexanuclear unit of **20**. Atoms are displayed as 30% probability ellipsoids. Hydrogen atoms, pyrazolyl carbon atoms and solvent molecules are omitted for clarity. Equivalent atoms (noted with apostrophe) are generated with the following symmetry operations: 2-x, 1-y, -z (**19**) and 2-x, 1-y, -z (**20**).
Selected bond lengths (Å) and angles (°) for **19** (M = Co): Fe1–C1 1.920(8), Fe1–C2 1.893(9), Fe1–C4 1.909(9), Fe1–N11 1.960(7), Fe1–N13 1.944(6), Fe1–N15 1.975(6), Fe2–C3 1.912(8), Fe2–C5 1.911(10), Fe2–C6 1.915(9), Fe2–N21 1.961(8), Fe2–N23 1.973(6), Fe2–N25 1.963(7), Co1–N1' 2.078(6), Co1–N2 2.083(8), Co1–N3 2.093(6), Co1–N31 2.137(6), Co1–N33 2.148(6), Co1–N35 2.110(6), C7-C8 1.544(11), C8-O1 1.420(10), Fe1-C1-N1 176.1(7), Fe1-C2-N2 178.4(6), Fe1-C4-N4 176.1(8), Fe2-C3-N3 174.9(6), Fe2-C5-N5 177.5(10), Fe2-C6-N6 173.0(8), C2-N2-Co1 168.3(6), C1'-N1'-Co1 160.6(6), C3-N3-Co1 155.7(6), C1-Fe1-C2 87.8(3), C1-Fe1-C4 90.4(3), C2-Fe1-C4 88.9(3), C1-Fe1-N11 92.2(3), C1-Fe1-N13 88.3(3), C4-Fe1-N11 89.8(3), C4-Fe1-N15 92.3(3), C2-Fe1-N13 93.2(3), C2-Fe1-N15 91.6(3), N11-Fe1-N13 88.0(3), N11-Fe1-N15 88.4(3), N13-Fe1-N15 89.0(3), C3-Fe2-C5 89.8(3), C3-Fe2-C6 90.2(3), C5-Fe2-C6 86.4(4), C3-Fe2-N21 92.8(3),

C3-Fe2-N23 89.8(3), C5-Fe2-N21 92.6(4), C5-Fe2-N25 92.2(3), C6-Fe2-N23 92.5(3), C6-Fe2-N25 89.1(3), N21-Fe2-N23 88.5(3), N21-Fe2-N25 87.9(3), N23-Fe2-N25 88.2(3), N2-Co1-N1' 91.9(3), N2-Co1-N3 90.4(2), N1'-Co1-N3 94.4(3), N1'-Co1-N31 94.5(2), N1'-Co1-N35 89.3(3), N2-Co1-N31 97.2(2), N2-Co1-N33 95.9(2), N3-Co1-N33 90.1(2), N3-Co1-N35 90.6(2), N31-Co1-N33 80.1(2), N31-Co1-N35 81.6(2), N33-Co1-N35 82.8(2), C7-C8-O1 106.7(7)
Selected bond lengths (Å) and angles (°) for **20** (M = Mn): Fe1–C1 1.917(3), Fe1–C2 1.915(3), Fe1–C4 1.926(3), Fe1–N11 1.976(2), Fe1–N13 1.956(2), Fe1–N15 1.965(2), Fe2–C3 1.916(3), Fe2–C5 1.918(3), Fe2–C6 1.915(3), Fe2–N21 1.961(2), Fe2–N23 1.981(2), Fe2–N25 1.972(2), Mn1–N1' 2.178(2), Mn1–N2 2.160(2), Mn1–N3 2.173(2), Mn1–N31 2.267(2), Mn1–N33 2.289(2), Mn1–N35 2.268(2), C7-C8 1.541(4), C8-O1 1.412(4), Fe1-C1-N1 176.4(2), Fe1-C2-N2 179.4(3), Fe1-C4-N4 176.8(3), Fe2-C3-N3 175.6(2), Fe2-C5-N5 179.5(3), Fe2-C6-N6 175.7(3), C2-N2-Mn1 165.7(2), C1'-N1'-Mn1 157.7(2), C3-N3-Mn1 155.5(2), C1-Fe1-C2 86.84(11), C1-Fe1-C4 88.68(12), C2-Fe1-C4 88.75(12), C1-Fe1-N11 92.70(11), C1-Fe1-N13 89.72(11), C4-Fe1-N11 90.13(11), C4-Fe1-N15 92.84(11), C2-Fe1-N13 92,89(11), C2-Fe1-N15 92.12(11), N11-Fe1-N13 88.22(11), N11-Fe1-N15 88.36(10), N13-Fe1-N15 88.80(10), C3-Fe2-C5 88.72(12), C3-Fe2-C6 90.13(12), C5-Fe2-C6 87.94(14), C3-Fe2-N21 92.86(12), C3-Fe2-N23 90.55(11), C5-Fe2-N21 92.03(13), C5-Fe2-N25 92.02(11), C6-Fe2-N23 91.58(11), C6-Fe2-N25 88.63(11), N21-Fe2-N23 88.49(11), N21-Fe2-N25 88.38(11), N23-Fe2-N25 88.70(10), N2-Mn1-N1' 96.42(10), N2-Mn1-N3 92.78(10), N1'-Mn1-N3 97.63(10), N1'-Mn1-N31 93.99(9), N1'-Mn1-N35 88.11(10), N2-Mn1-N31 98.50(9), N2-Mn1-N33 96.03(9), N3-Mn1-N33 90.14(9), N3-Mn1-N35 90.35(10), N31-Mn1-N33, N31-Mn1-N35 77.36(9), N33-Mn1-N35, C7-C8-O1 108.6(3).

The $\{Fe_2M_2\}$ core of **19** and **20** is quite distorted and centrosymmetric. One of the quadratic core edges is slightly more elongated than the other (Fe···Co distance = 5.07 Å and 5.13 Å; Fe···Mn distance = 5.12 Å and 5.19 Å) but they are both slightly longer than the Fe2···M1 distance (Fe2···Co1 distance = 5.03 Å, Fe2···Mn1 distance = 5.09 Å). In **19**, the angles between the metal ions in the $\{Fe_2Co_2\}$ quadratic core slightly depart from orthogonality (Fe···Co···Fe angle = 93.2° and Co···Fe···Co angle = 86.8°) while in **20**, the $\{Fe_2Mn_2\}$ core structure is far more distorted (Fe···Mn···Fe angle = 96.02° and Mn···Fe···Mn angle = 83.98°). However, the two species display similar Fe2···M···Fe1 angles (90.82° and 98.16° for **19**, 90.88° and 98.88° for **20**). Fe1 and Fe2 are both part of $\{Fe(Tp)(CN)_3\}$ moieties and therefore have the same type of environment. Their distortion to perfect octahedron amount to 19.5°, 20.6° (**19**), 21.3° and 18.3° (**20**) respectively. These values are somehow lower than the octahedral distortion displayed by PPh_4[**1**] (25.8°). The mean Fe–$C_{cyanide}$ bond length values for **19** amount to 1.908 Å (Fe1) and 1.913 Å (Fe2). **20** exhibits slightly longer mean Fe–C bonds (1.919 Å for Fe1 and 1.916 Å for Fe2). This is coherent with a formal +III oxidation state for both Fe1 and Fe2. The Fe–N_{pz} bond lengths are slightly longer and range from 1.944(6) Å to 1.981(2) Å. They average 1.960 (Fe1) and 1.966 Å (Fe2) for **19** and 1.966 Å and 1.971 Å for **20** respectively. The cyanides bind Fe1 in the two complexes almost linearly with no

Fe1-C-N angle smaller than 176.1(8)°. In contrast, the Fe2 cyanides are far more bent than they are in the monomer [**1**]$^-$; the effect is more drastic in **19** with the smallest value being 173.0(8)° than in **20** (smallest angle value is 175.6(2)°). These cyanides are oriented inwards in an alternated configuration in respect to the cyanide bridges as depicted in Figure 6.14.

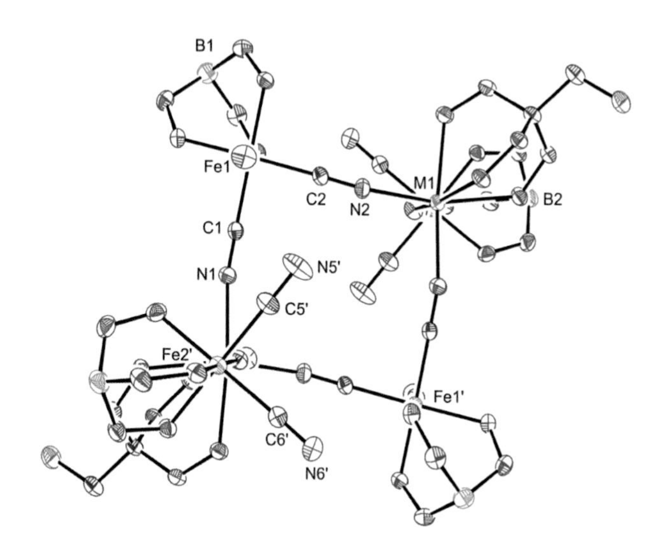

Figure 6.14: Molecular structure of **20** along another axis than in Figure 6.13.

The coordination spheres of the cobalt and manganese ions consist of a *fac*-coordinating tridentate *N*-donor Tpe and three *N*-capped cyanide bridges from the adjacent $\{Fe^{III}(Tp)(CN)_3\}$ moieties. The octahedron distortions around the cobalt and manganese ions are high, and reach 51.2° and 75.6°, respectively. In **19**, the $Co–N_{cyanide}$ bond lengths range from 2.078(6) to 2.093(6) Å, while the $Co–N_{pz}$ bond lengths average 2.132 Å. Those are typical values for high-spin cobalt(II) complexes in similar cyanide-bridged architectures.[87,88] The $Mn–N_{cyanide}$ bond lengths are longer than their cobalt analogues and average 2.170 Å, while the mean $Mn–N_{pz}$ bond length amounts to 2.275 Å. This is also in the range of reported values for manganese(II) high-spin complexes in comparable architectures found in the literature.[87,88,117,181] The cyanide bridges are notably bent on the cobalt/manganese side, with the C3-N3-M angles being the most bent (168.3(6) Å – 155.7(6) Å for **19**, 165.7(2) Å – 155.5(2) Å for **20**). One of the pyrazolyl rings of the Tpe is notably bent (C7-N32-N33-M1 torsion angle = 10.2° and 11.7° for **19** and **20** respectively) while the other arms undergo only minimal torsion. The distance between the metal ion and the apical C7 atom of its Tpe ligand is 3.24 Å for cobalt and 3.39 Å for manganese, which is quite long in respect to the building blocks (*ca* 3.08 Å). The -CH_2OH moieties point right between two pyrazolyl groups.

While the molecules are well isolated along the *b* and *c* axis in the crystal packing, intermolecular head-to-tail hydrogen bonding occurs along the *a* axis as depicted in Figure 6.15, between the O1 Tpe alcohol functions and the non-bridging C4-N4 cyanide moieties. Such a strong, direct interaction between two neighbouring molecules is expected to be able to mediate magnetic information along the *a* axis. The smallest intermolecular metal-metal distance concerns two Fe2 atoms and amounts to 7.851 Å for **20** and 8.35 Å for **19**.

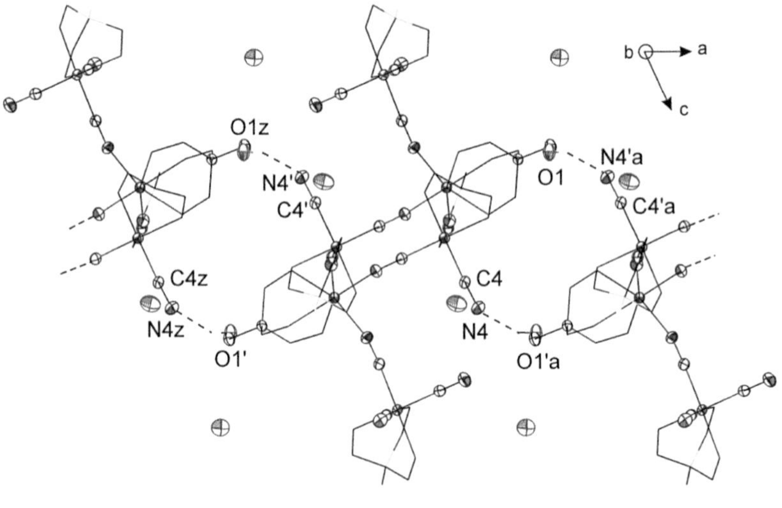

Figure 6.15: Perspective view of **20** with its network of hydrogen bonds (spaced dotted lines) along the *b* axis. Letters in atom labels refer to consecutive cationic units in the crystal packing.

6.2.3 Fourier Transform InfraRed spectroscopy

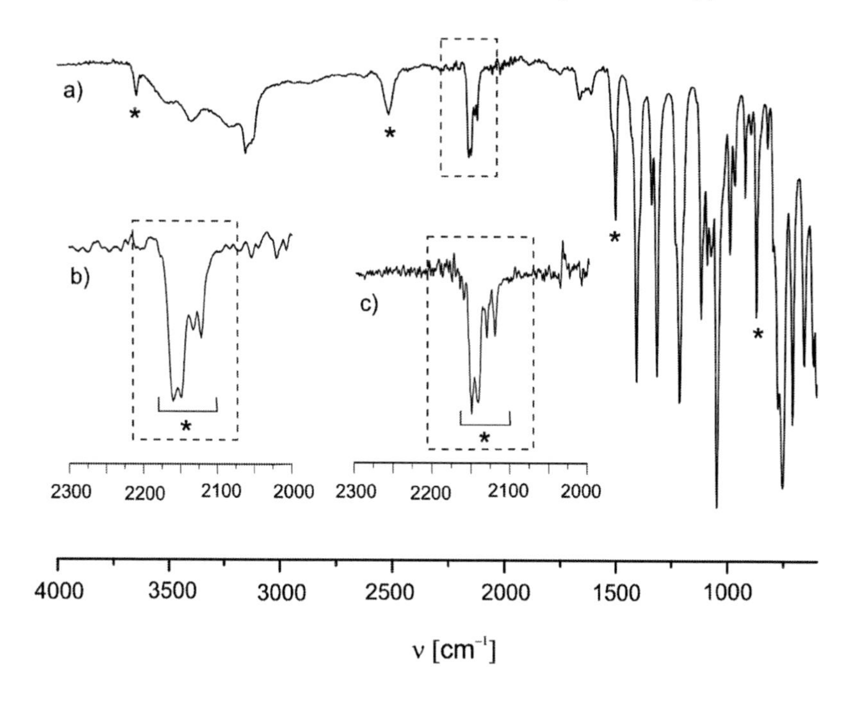

Figure 6.16: a) FT-IR (ATR) spectrum of **19** with a resolution of 4 cm^{-1}. A zoom on the cyanide stretching bands is depicted as b). The IR spectrum of **20** is identical to the one of **19**, and is therefore not displayed here. The only tiny difference between the two spectra is the cyanide stretches, which are zoomed in for **20** as c) (resolution of 1 cm^{-1}).
Selected IR vibrations bands in cm^{-1} and their intensities are marked with an asterisk: 871 (m), 1501 (w), 1518 (w, sh), 2516 (vw), 3645 (vw). For 22: 2122 (vw), 2133 (vw), 2149 (w), 2160 (w). For 23: 2122 (vw), 2132 (vw), 2152 (w), 2162 (w).

The FT-IR (ATR) spectra of **19** and **20** are almost identical, and only differ slightly for the cyanide stretching bands broadness which led to the acquisition of a spectrum with better resolution for **20** (1 cm^{-1} instead of 4 cm^{-1} as default) in order to have a better look at the four different cyanide stretches. The following description applies therefore to IR spectra of both cobalt and manganese compounds.

Even though their B–H stretching vibration band overlap at 2516 cm^{-1}, both types of $\{Fe^{III}(Tp)(CN)_3\}$ units are visible in the IR spectrum. The cyanide area displays four different cyanide vibrations, two of them assigned to bridging ligands (2149 and 2160 cm^{-1}), the two others accounting for non-bridging cyanides (2122 and 2133 cm^{-1}). It is quite difficult to make a precise attribution for the cyanide bridges: only two bridging cyanide stretching vibrations are observed, of about the same intensity, instead of the three expected ones. Furthermore, as already seen with **17**, both frequencies can apply for $\{Fe^{III}(Tp)(\mu\text{-}CN)_2(t\text{-}CN)\}$ units. One of the cyanide stretching vibrations is a typical for an iron(III) terminal coordinated cyanide. The second non-bridging cyanide stretching vibration located at slightly higher frequency ($\Delta\tilde{v}$ = 10 cm^{-1}) can be attributed to a cyanide ligand which is involved in a hydrogen interaction with the alcohol group of Tpe ligands. The ring stretches of the pyrazolyl rings for both $\{Fe^{III}(Tp)\}$ moieties also overlap at 1501 cm^{-1}. Hints for the occurrence of {M(Tpe)} (M = Co, Mn) can also be found in the spectrum: the pyzolyl rings of Tpe stretch at slightly higher frequency than their Tp counterparts and can be detected as a shoulder at 1518 cm^{-1}. This is coherent with literature values for $[Fe(Tpe)_2](X)_2$ (X = OTf).[10] The vibration band at 871 cm^{-1} is specific of the Tpe ligand, as it can be found in Tpe, PPh_4[**4**] and $[Fe^{II}(Tpe)_2](X)_2$ (X = $[BF_4]^-$, OTf) but not in Tp or Ttp infrared spectra. Finally, the vibration band at 3645 cm^{-1} cannot be attributed to the alcohol moiety of the Tpe ligand since the crystal data clearly states that this OH moiety is involved in a hydrogen bonding interaction with the non-bridging cyanide, which should shift its vibration to lower frequency. One of the water molecules in the crystal structure shows no hydrogen interaction with its environment and could be responsible for this peak.

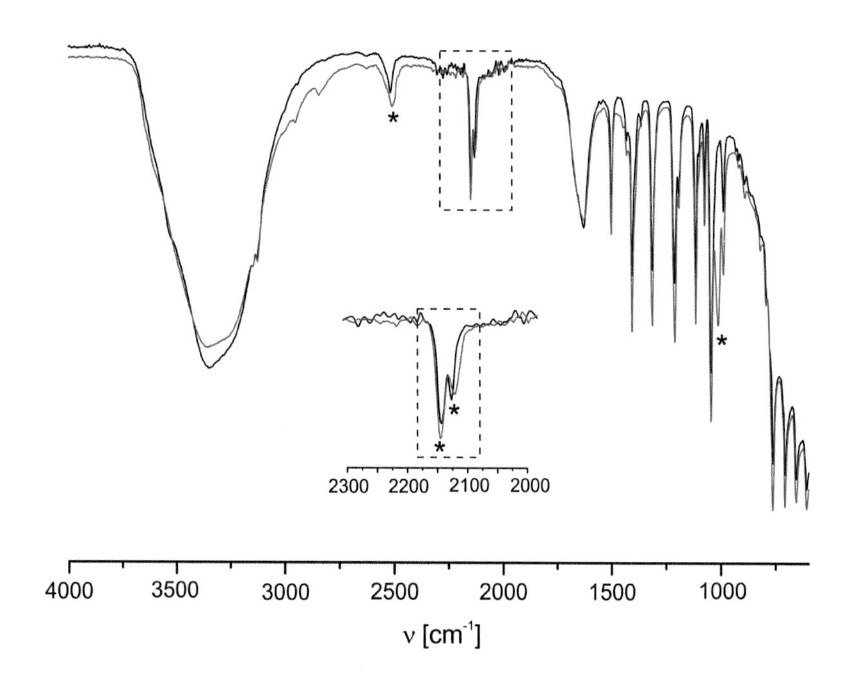

Figure 6.17: red curve: FT-IR (ATR) spectrum of **20'**. Black curve: FT-IR (ATR) spectrum of the literature-known {FeMn} double zigzag chain.[181] Differences between the two spectra are marked with an asterisk.
For **20'**: 1015 (m), 2129 (vw), 2151 (w), 2510 (vw).
For the {FeMn} double zig-zag chain: 2133 (vw), 2150 (w), 2518 (vw).

Even though the crystal structure of **20'** could not be obtained due to the bad quality of the crystals, the collected infrared data gave a hint about the nature of **20'**. The same reaction in an acetonitrile/water mixture (4:1) instead of methanol yields black rod-like crystals suitable for X-Ray diffraction, which were identified as a literature known double zigzag iron(III)-manganese(II) coordination polymer chain[181] in which two water molecules in *trans* position completes the manganese coordination sphere. The overlaid infrared spectra of **20'** and the iron-manganese chain are displayed in Figure 6.17. Both spectra are almost identical; the B–H vibration is slightly shifted towards higher frequency in the infrared spectrum of **20'**. The cyanide stretching vibrations do not completely overlap, but remain close to each other. Most notably, the vibration band at 1015 cm^{-1} in **20'** is completely absent in the black spectrum: it corresponds to coordinated methanol as already evoked for the IR analysis of **17**. It is reasonable to assume that **20'** is also a double zigzag iron(III)-manganese(II) chain, where methanol completes the manganese coordination sphere instead of water.

6.2.4 SQUID magnetometry

{[Fe^{III}(Tp)(CN)$_3$]$_4$[Co^{II}(Tpe)]$_2$} (19)

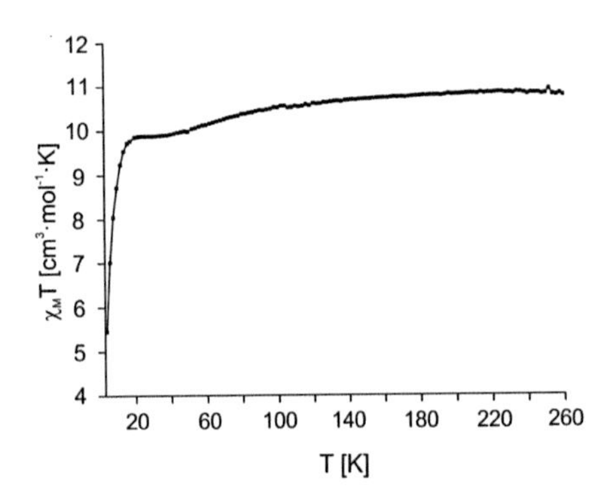

Figure 6.18: Magnetic properties of freshly filtered **19**. $\chi_M T$ product *vs* T between 260 K and 2 K, H = 2500 Oe. The sample was prepared as follows: m_{sample} = 7.70 mg, m_{film} = 5.5 mg.

The magnetic properties of **19** were investigated using SQUID magnetometry, by plotting the $\chi_M T$ product *vs* T (Figure 6.18.). The $\chi_M T$ product of **19** at 260 K reaches 10.83 cm^3·mol^{-1}·K. It is slightly higher than the expected value (8.4 to 10 cm^3·mol^{-1}·K) for an independent set of two cobalt(II) ions (2.8 – 3.6 cm^3·mol^{-1}·K each) and four iron(III) ions (0.7 cm^3·mol^{-1}·K each). However, the raw data were processed using the molecular mass of the solvated compound, while the introduction of the sample in the SQUID at 260 K instead of 200 K as usual might have partly removed the solvent. The curve exhibits a very slightly descending slope as the temperature decreases. This is due to the spin-orbit coupling of four low-spin iron(III) ions of **19**. The $\chi_M T$ reaches a pseudo plateau of 9.88 $cm^3 \cdot mol^{-1} \cdot K$ at 30 K. As for **18**, the $\chi_M T$ product then experiences a drastic decrease below 10 K. The lack of $\chi_M T$ increase at relatively low temperature (in contrast with some other related {FeCo} materials) could be ascribed to weaker intramolecular ferromagnetic interactions, while the decrease at very low temperatures (*ca* $T < 10$ K) could be due to intermolecular antiferromagnetic interactions. This is not surprising as such magnetic interactions are expected, due to direct hydrogen bonds between neighbouring molecules found in the crystal packing (see Figure 6.15). Further rationalising of the magnetic properties would require additional theoretical calculations (e.g. DFT calculations) that are beyond the scope of the present study.

{[$Fe^{III}(Tp)(CN)_3]_4[Mn^{II}(Tpe)]_2$} (20)

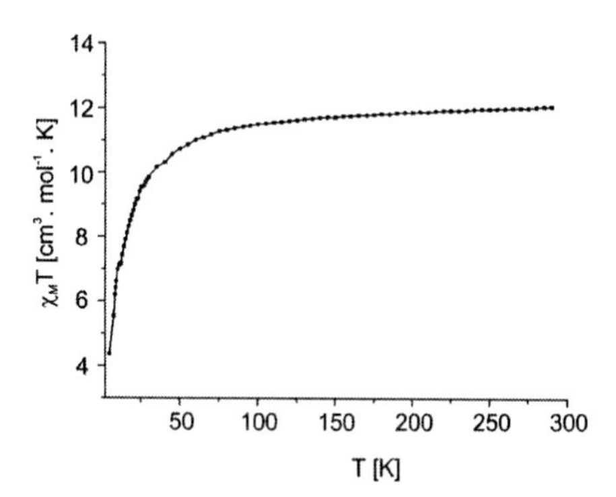

Figure 6.19: Magnetic properties of freshly filtered **20**. $\chi_M T$ product *vs* T between 4 K and 300 K, H = 10000 Oe. The sample was prepared as follows: m_{sample} = 18.06 mg, $m_{capsule}$ = 61.20 mg.

The $\chi_M T$ *vs* T curve of the hexanuclear {Fe_4Mn_2} compound **20** (see Figure 6.19.a) is strongly reminiscent of the magnetic curves obtained for the {Fe_2Mn}$_\infty$ polymeric chain

17 (by a factor 2) and the $\{Fe_2Mn_2\}$ molecular square **18**, but also very similar to the $\chi_M T$ *vs T* curve of the parent hexanuclear compound $\{[Fe^{III}(Tp)(CN)_3]_4[Mn^{II}(DMF)_2(H_2O)]_2\}$ reported in 2006 by Jiang *et al.*[171] The value of the $\chi_M T$ product obtained at 300 K ($12.07\ cm^3 \cdot mol^{-1} \cdot K$) is slightly higher than the expected one ($11.54\ cm^3 \cdot mol^{-1} \cdot K$) for the set of independent magnetic ions: two high-spin manganese(II) ions ($4.37\ cm^3 \cdot mol^{-1} \cdot K$ for $g_{Mn} = 2.0$) and four low-spin iron(III) ions ($0.7\ cm^3 \cdot mol^{-1} \cdot K$). Upon cooling, the $\chi_M T$ product decreases very slowly down to 50 K ($11.5\ cm^3 \cdot mol^{-1} \cdot K$), because of the spin-orbit coupling of the iron(III) ions. Below 50 K, the $\chi_M T$ product plummets to $4.37\ cm^3 \cdot mol^{-1} \cdot K$ (4.8 K), that is, about $1\ cm^3 \cdot mol^{-1} \cdot K$ lower than the expected value for a $S = 3$ system. This is probably due, as already explained for **18**, to the conjugated effect of the weak intramolecular antiferromagnetic interactions and the noticeable intermolecular antiferromagnetic interactions present in the crystal structure (see Figure 6.15).

6.3 Cyanide-based molecular octanuclear complexes

Many examples of cyanide-based molecular boxes were reported since the first cyanide example that was synthesised by Heinrich *et al.* in 1998.[182] The reported compounds can be homometallic $\{M_8\}$ with M being cobalt or iron ions,[120,175,182,183] but most of them are heterobimetallic species, which involve metal ions from the first row (M = Cr, Mn, Fe, Co , Ni, Zn), but also from the second row (M = Mo, Ru, Rh) and even from the third row (M = Re).[87,93,176,179,180,182,184–188] Heterotrimetallic $\{Co_4Ru_3M\}$ cubic structures, with M being a copper, silver or nickel ion were also reported by Boyer *et al.* in 2007.[186,187] In most of the cases, the capping ligands that coordinate the metal ions at the corner of the boxes (to prevent the polymerisation toward Prussian Blue Analogues (PBAs)) are from the scorpionate family. They are often either Tp or Tpm derivatives,[93,120,175,176,179,180] but other tridentate ligands include tacn,[182,183] Cp and Cp*,[186–188] as well as triphos ligands.[185]

The most striking property of some of these cyanide-bridged boxes is their ability to host a guest, which then could act as a templating agent. In this respect, alike the PBA compounds they are molecular models of, cyanide-bridged boxes often accommodate alkali metals within their cyanide–bridged core. While lithium seems to possess a Van der Waals radius too small to be accommodated inside the molecular cages, example of sodium,[176] potassium[188] and above all caesium ions[186–188] are reported in the literature. Other guests include solvent molecules,[183] or a tetrafluoroborate counteranion.[120]

6.3.1 Syntheses

$\{[Fe^{III}(Tp)(CN)_3]_4[Co^{II}(Tpe)]_4\}(ClO_4)_4$ (21)

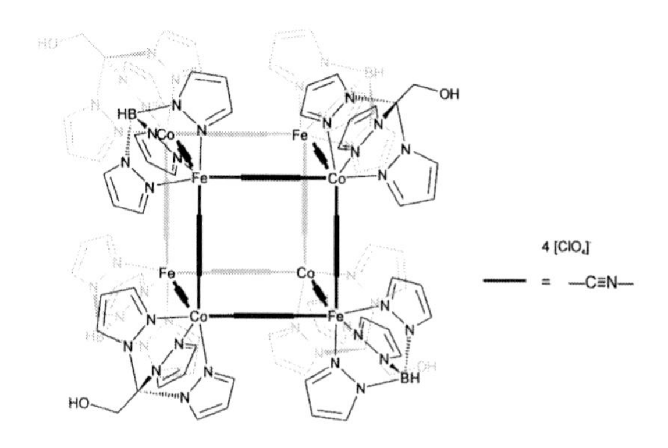

Figure 6.20: Schematic representation of **21**.

As previously shown in chapter 3, $[Fe^{III}(Ttp)(CN)_3]^-$ ($[\mathbf{9}]^-$) and $[\mathbf{1}]^-$ have quite similar electronic properties. These two tricyanido building blocks were previously shown to: (i) lead to the same kind of products when placed in the same reaction conditions, (ii) confer the same properties to the resulting compound[98] (*i.e.* photomagnetism, about the same half transition temperature $T_{1/2}$ for spin-state transitions). In 2008, Li *et al.* reported a photomagnetic $\{Fe_4Co_4\}$ cubic structure where Tpe ligands complete the coordination sphere of the cobalt ions while the iron ions are capped with a Ttp ligand.[93] Up to now, it is the only {FeCo} photomagnetic cubic molecule that has been reported. **21**, which is the related Tp derivative, was synthesised by following the same protocol and replacing $[NEt_4][Fe^{III}(Ttp)(CN)_3]$ by its Tp analogue.

The resulting compound **21** crystallises as red blocks by layering a DMF solution with diethyl ether. Apparition of the crystals is in a matter of days followed by precipitation of a green, diamagnetic compound. This slow decomposition of **21** in DMF solution is accelerated by exposition of the solution to light, but also occurs, albeit at a much slower rate, in darkness. This behaviour is a strong indicator that **21** exists only as part of an equilibrium in DMF, and is dissolvable, rather than soluble, in this solvent.

K@$\{[Fe^{II}(Tp)(CN)_3]_4[Co^{III}(Ttp)]_3[Co^{II}(Ttp)]\}$ (22)

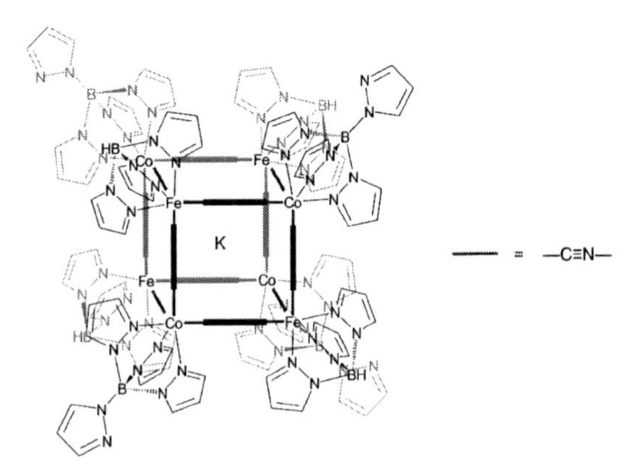

Figure 6.21: Schematic representation of **22**.

The synthesis of **22** is accomplished in two steps. The first step is the formation of a brick red cube precursor that is obtained by treating one equivalent of $[Et_4N]$[**1**] in DMF solution with one equivalent of solid $Co(ClO_4)_2 \cdot 6\ H_2O$ as in the reported synthesis of the photomagnetic molecular cube.[93] This first step works best in larger quantities (up to 580 mg) and the obtained species is storable at room temperature.

Based on IR data, the compound is assumed to be the cubic compound $\{[Fe^{III}(Tp)(CN)_3]_4[Co^{II}(DMF)_3]_4\}(ClO_4)_4$, and an arbitrary molecular mass of $M = 2900\ g{\cdot}mol^{-1}$ was attributed to the compound for further reactions.

22 was obtained by addition of 6.3 equivalents of solid, colorless K[Ttp] to a deep red concentrated solution of the cube precursor in DMF. The suspension instantly turned deep green, and very small particles of a dark blue solid precipitated. This colour change from red to dark blue is a good indicator that some redox processes take place during the reaction. The obtained powder is too fine to be filtered conventionally, and was therefore centrifuged. IR spectrum analysis of the yellow DMF solution content indicated that it primarily consists of [**1**]$^-$, that is the decomposition product of the brick red cube precursor. The dark blue solid was washed several times with a DMF/Et_2O 1:8 mixture to remove traces of [**1**]$^-$. The blue solid was then dissolved in diethyl ether and the off-white insoluble residue ($[Co^{III}(Ttp)_2]$) was filtered. Diethyl ether was slowly evaporated and a blue solid was obtained (61.5%, based on the brick red solid amount). **22** is highly soluble in dichloromethane, soluble in ether and ethyl acetate but not in DMF and pentane. It decomposes in cyanide chemistry common solvents such as acetonitrile, methanol or water.

ESI-MS analysis of **22** in CH_2Cl_2 provided a useful insight on the stability of the cube in solution: indeed, the obtained cationic molecular mass matches with the cubic cage ($M = 2740\ g{\cdot}mol^{-1}$), plus a potassium ion with the expected mass distribution due to the iron, boron and potassium isotopes. (This implies that the cube would have been oxidised by one electron in these experimental conditions). This tends to indicate that the cube is stable in dichloromethane and that the potassium would remain inside the cage in solution in CD_2Cl_2. The anionic molecular mass obtained in the same solution corresponds to the adduct between **22** and a chlorine from the solvent, leading to the singly negatively charged $m/z = 2814$ species. Except for a trace peak corresponding to the $[Co^{III}(Ttp)_2]^+$ cation coming from the last step of the workup, the ESI-MS spectra only display the two above mentioned molecular peaks. If equilibria were taking place in CH_2Cl_2, the other transient species should be detected as well. Their absence in ESI-MS, combined with NMR spectroscopy and cyclovoltametric studies strongly points toward a stable species in this solvent.

6.3.2 Structural analyses

{[Fe^{III}(Tp)(CN)$_3$]$_4$[Co^{II}(Tpe)]$_4$}(ClO_4)$_4$ (21)

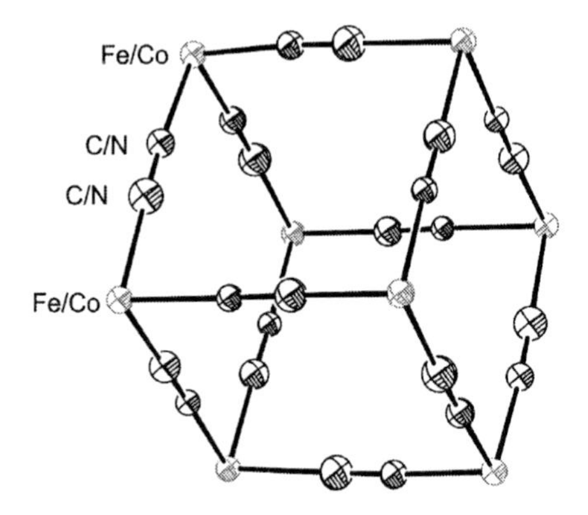

Figure 6.22: Molecular structure of the cationic unit of **21** at 260 K. The ligands capping the metal ions are omitted for clarity. Metal atoms (50/50 Fe/Co) are depicted in orange. Cyanide atoms, also 50/50 C/N disordered, are depicted in black. Atoms are displayed as 30% probability ellipsoids.

In contrast to its reported parent {[Fe^{III}(Ttp)(CN)$_3$]$_4$[Co^{II}(Tpe)]$_4$}(ClO_4)$_4$,[93] **21** crystallises in the hexagonal space group $R\bar{3}c$. This space group corresponds to the point group symmetry O_h for the cationic unit of **21**. However, by construction, the cationic unit of **21** necessarily belongs to the point group symmetry T_d. This results in a molecular cube, which is superposable upon itself and statistically disordered. Each metal atom is statistically half cobalt and half iron; just as the atoms of the cyanide bridges, which are statistically half carbon and half nitrogen. The capping ligands coordinating the iron ions are {HB(pz)$_3$} (Tp) whereas those coordinating the cobalt ions are {HOCH$_2$C(pz)$_3$} (Tpe). A statistical 50/50 disorder of both ligands with superposed {E(pz)$_3$} units (E = B, C) and half occupied CH_2OH chains is therefore observed. Only three out of the four perchlorate ions are visible in the X-ray diffraction analysis, the last one being lost in the diffuse residual electronic density in the lattices (see Figure 6.23) along with DMF lattice molecules. Because of the intrinsic properties of the crystal organisation, it is not possible to discuss in details bond length and angles in the structure.

Figure 6.23: View of the crystal packing of **21** alongside the lattices. The cubic units and perchlorate anions are eclipsed. No bond lengths are discussed due to high structural disorder leading to lower data quality.

$K@\{[Fe^{II}(Tp)(CN)_3]_4[Co^{III}(Ttp)]_3[Co^{II}(Ttp)]\} \cdot 12\ CH_2Cl_2$ (22 – Phase #1)

When a dichloromethane solution of **22** is layered with *n*-hexane, deep blue, partly intergrown plates of **22** are obtained.

At 200 K, **22** crystallises in triclinic space group $P\bar{1}$ with a whole formula unit in the asymmetric unit ($Z = 2$). The latter consists of an octanuclear $\{Fe_4Co_4\}$ cyanide-bridged molecular box containing an inserted potassium ion and twelve dichloromethane molecules. Because of the high amount of dichloromethane molecules per molecular cube, these crystals are extremely sensitive and lose their crystallinity in a matter of seconds. A perspective view of **22** is depicted in Figure 6.24, and selected bonds and angles are listed in the caption.

The octanuclear core structure of **22** consists of a slightly distorted, monoanionic cubic $\{Fe_4Co_4\}$ cage, in which iron and cobalt ions occupy alternate corners and are bridged by cyanide ligands along the cube edges. The iron-cobalt distances are about the same length and average 4.989 Å, the angles of the quadratic faces are close to orthogonality and the sum of their deviation to 90° amounts 39.8°. The potassium countercation is trapped inside the cage and probably acts as template. Unlike the crystal structure of **21**, the crystallographic phase #1 of **22** seems to be ordered, except for the inserted potassium ion, which is disordered on three sites. Indeed, potassium is slightly too small for the box: a caesium ion would have the optimal van der Waals radius for it and occupies preferentially cyanide-bridged cages, as testified by the use of Prussian Blue as a caesium-133 antidote in medicine.[189] Such a crystallographic disorder for inserted potassium and preference for Cs^+ ions have been observed in other cyanide-bridged $\{Rh_4Mo_4\}$ boxes.[188] The potassium ion interacts with the π system of three nearby cyanide bridges.

The iron ions are capped with anionic *fac*-coordinating Tp ligands, and thus lie in a typical C_3N_3 environment. In the four $\{Fe(Tp)(CN)_3\}$ moieties, the Fe–C bonds lengths are all below 1.900 Å (average 1.890 Å). This is indicative of the occurrence of low-spin iron(II) spin and oxidation states for the four iron ions. The Fe–N_{pz} bond lengths exhibit also nearly identical distances (2.009 Å). Their octahedral distortion amounts to 20.6° for

Fe1, 28.2° for Fe2, to 29.2° for Fe4 and to 22.0° for Fe4. These values are in the range of those found for tricyanido iron(II) and iron(III) complexes (see chapter 3). The cyanides bind the respective iron ions almost linearly, with bent angles ranging from 173.9(4)° to 178.8(4)°.

The coordination sphere of the four cobalt ions is completed by the *fac*-coordinating anionic scorpionate ligand Ttp, thus leading to a N_6 distorted octahedral environment for the four cobalt ions of **22**.

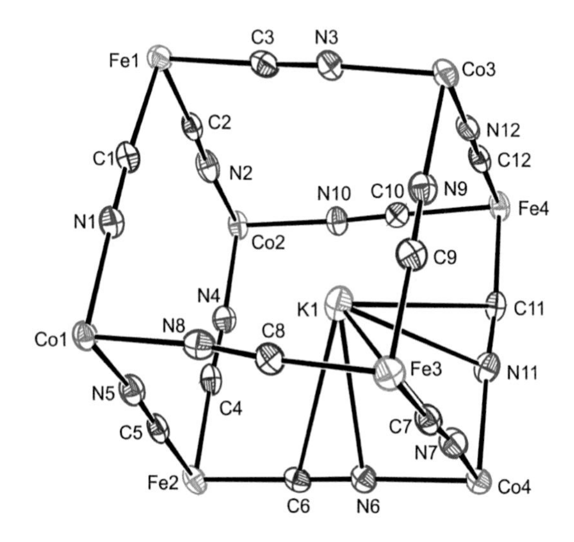

Figure 6.24: Molecular structure of the phase #1 of **22**. Atoms are displayed as 30% probability ellipsoids. Tp and Ttp capping ligands, solvent molecules and two of the potassium partially occupied positions are omitted for clarity.

Selected bond lengths (Å) for the phase 1 of **22** at 200 K: Fe1–C1 1.892(5), Fe1–C2 1.892(5), Fe1–C3 1.897(5), Fe1–N101 2.006(4), Fe1–N111 2.007(4), Fe1–N121 2.012(4), Fe2–C5 1.871(5), Fe2–C6 1.890(5), Fe2–C4 1.895(5), Fe2–N221 1.993(5), Fe2–N201 2.010(5), Fe2–N211 2.010(5), Fe3–C7 1.884(5), Fe3–C8 1.896(5), Fe3–C9 1.888(5), Fe3–N301 2.009(4), Fe3–N321 2.014(4), Fe3–N311 2.019(5), Fe4–C12 1.883(5), Fe4–C10 1.891(5), Fe4–C11 1.902(5), Fe4–N401 1.999(4), Fe4–N411 2.004(4), Fe4–N421 2.024(5), Co1–N5 1.921(4), Co1–N8 1.931(4), Co1–N1 1.936(4), Co1–N501 1.943(4), Co1–N511 1.945(4), Co1–N521 1.960(4), Co2–N601 1.921(5), Co2–N4 1.933(4), Co2–N10 1.937(4), Co2–N2 1.943(4), Co2–N621 1.950(5), Co2–N611 1.951(5), Co3–N9 1.917(4), Co3–N701 1.918(5), Co3–N3 1.922(4), Co3–N12 1.923(5), Co3–N711 1.932(4), Co3–N721 1.938(4), Co4–N11 1.998(4), Co4–N6 1.999(4), Co4–N7 2.001(5), Co4–N811 2.006(5), Co4–N821 2.017(5), Co4–N801 2.027(5).

Selected angles (°) for the phase #1 of **22**: N5-Co1-N8 87.94(17), N5-Co1-N1 91.40(17), N8-Co1-N1 91.03(16), N5-Co1-N501 90.25(18), N8-Co1-N501 90.18(17), N8-Co1-N511 93.79(18), N1-Co1-N511 90.45(18), N501-Co1-N511 87.87(19),

N5-Co1-N521 91.54(18), N1-Co1-N521 90.45(16), N501-Co1-N521 88.35(18), N511-Co1-N521 86.68(18), C1-Fe1-C2 88.2(2), C1-Fe1-C3 89.72(19), C2-Fe1-C3 92.2(2), C1-Fe1-N101 92.70(19), C3-Fe1-N101 90.23(19), C2-Fe1-N111 91.4(2), C3-Fe1-N111 90.67(19), N101-Fe1-N111 87.64(19), C1-Fe1-N121 92.37(18), C2-Fe1-N121 90.71(18), N101-Fe1-N121 86.82(17), N111-Fe1-N121 87.26(17), C5-Fe2-C6 92.77(19), C5-Fe2-C4 87.2(2), C6-Fe2-C4 88.4(2), C6-Fe2-N221 91.1(2), C4-Fe2-N221 93.5(2), C5-Fe2-N201 89.0(2), C4-Fe2-N201 93.70(19), N221-Fe2-N201 87.1(2), C5-Fe2-N211 93.1(2), C6-Fe2-N211 89.79(19), N221-Fe2-N211 86.4(2), N201-Fe2-N211 88.08(19), N601-Co2-N4 88.8(2), N4-Co2-N10 90.36(18), N601-Co2-N2 91.6(2), N4-Co2-N2 91.30(18), N10-Co2-N2 88.89(18), N601-Co2-N621 88.5(2), N4-Co2-N621 89.8(2), N10-Co2-N621 91.04(19), N601-Co2-N611 88.9(2), N10-Co2-N611 91.91(18), N2-Co2-N611 91.42(19), N621-Co2-N611 87.5(2), N9-Co3-N701 88.71(19), N9-Co3-N3 90.63(17), N701-Co3-N3 92.24(19), N9-Co3-N12 91.28(17), N3-Co3-N12 88.73(17), N701-Co3-N711 88.4(2), N3-Co3-N711 91.44(18), N12-Co3-N711 91.53(19), N9-Co3-N721 89.31(18), N701-Co3-N721 88.48(19), N12-Co3-N721 90.55(18), N711-Co3-N721 88.7(2); C7-Fe3-C9 87.9(2), C7-Fe3-C8 94.3(2), C9-Fe3-C8 88.1(2), C7-Fe3-N301 91.68(19), C8-Fe3-N301 91.90(18), C7-Fe3-N321 90.55(19), C9-Fe3-N321 93.17(19), N301-Fe3-N321 86.87(18), C9-Fe3-N311 92.99(19), C8-Fe3-N311 87.79(19), N301-Fe3-N311 87.44(18), N321-Fe3-N311 87.30(18), N11-Co4-N6 90.88(17), N11-Co4-N7 91.48(17), N6-Co4-N7 87.83(17), N11-Co4-N811 90.51(18), N7-Co4-N811 94.49(18), N11-Co4-N821 90.42(18), N6-Co4-N821 90.95(17), N811-Co4-N821 86.69(18), N6-Co4-N801 93.20(17), N7-Co4-N801 91.80(17), N811-Co4-N801 85.28(18), N821-Co4-N801 86.39(18), C12-Fe4-C10 91.7(2), C12-Fe4-C11 88.1(2), C10-Fe4-C11 88.49(19), C12-Fe4-N401 90.53(19), C10-Fe4-N401 90.18(18), C12-Fe4-N411 91.5(2), C11-Fe4-N411 93.96(19), N401-Fe4-N411 87.44(18), C10-Fe4-N421 90.47(19), C11-Fe4-N421 92.7(2), N401-Fe4-N421 88.68(19), N411-Fe4-N421 86.35(19), N1-C1-Fe1 176.0(4), N2-C2-Fe1 177.2(4), N3-C3-Fe1 177.7(4), N4-C4-Fe2 173.9(4), N5-C5-Fe2 175.6(4), N6-C6-Fe2 178.2(4), N7-C7-Fe3 176.8(4), N8-C8-Fe3 174.2(4), N9-C9-Fe3 175.1(4), N10-C10-Fe4 178.1(4), N11-C11-Fe4 174.4(4), N12-C12-Fe4 178.7(4), C1-N1-Co1 178.8(4), C2-N2-Co2 176.1(4), C3-N3-Co3 175.6(4), C4-N4-Co2 173.2(4), C5-N5-Co1 172.2(4), C6-N6-Co4 178.1(4), C7-N7-Co4 172.9(4), C8-N8-Co1 173.0(3), C9-N9-Co3 171.0(4), C10-N10-Co2 176.6(4), C11-N11-Co4 175.5(4), C12-N12-Co3 178.8(4).

Three cobalt ions (Co1, Co2 and Co3) exhibit Co–N bond lengths ranging from 1.960(4) Å to 1.917(4) Å (average Co–N bond distances for each cobalt ion = 1.925 Å – 1.934 Å). These bond distances are clearly longer than those reported for diamagnetic $\{Fe_2Co_2\}$ molecular squares at 200 K (*ca* 1.900-1.910 Å),[96–98] in the chapter 5 of this work (*ca* 1.89 Å at 200 K) and in the diamagnetic phase of the already mentioned reported photomagnetic $\{Fe_4Co_4\}$ molecular cube measured at 90 K (< 1.900 Å).[93] However, they are by far shorter than the typical 2.1 Å Co–N bond lengths in high-spin cobalt(II) complexes.[41,93,98,118] The octahedral distortion experienced by these cobalt ions remains low (15.3–16.2°) even though they are higher than those found for the cobalt ions of **15** and **16**. Overall, these data indicate that a low-spin cobalt(III) spin and oxidation state for these cobalt ions is more plausible. By contrast, the remaining cobalt ion (Co4) exhibits longer Co–N bond distances (average 2.008 Å) than the three other cobalt ions. This value is however rather short in regard to typical Co–N bond lengths for high-spin cobalt(II) species (*ca* 2.090-2.105 Å).[41,93,97,98,118] Moreover, the octahedral environment of Co4 is far more distorted than those of the three other cobalt ions although the distortion remains lower than those observed in other high-spin cobalt(II) species: 27.5°, to be compared to 37.5° (compound **17** – at 200 K), 51.2° (compound **19** – at 200 K) and 44° – 49° for the reported {Fe_4Co_4} photomagnetic cube measured at 260 K.[93] The cyanides *N*-bind the respective cobalt ions with bent angles ranging from 171.0(4)° (quite bent) to 178.8(4)° (linear) but without clear difference between the cobalt(II) and the cobalt(III) ions.

In view of these data, the phase #1 of **22** may not be as "disorder-free" as it seems to be on first sight. It rather exhibits some cobalt(II)-cobalt(III) disorder with however a preferential cobalt(II) site (Co4). This preferred cobalt(II) site is to be correlated with the potassium ion location in the cage: as shown in Figure 6.24, the main potassium position is displaced toward Co4 (occupancy 50%).

The molecules of **22** are well isolated from each other by the twelve dichloromethane molecules. The shortest intermolecular metal-metal distance, 9.71 Å, is between the two Co4 of neighbouring molecules.

K@{[Fe^{II}(Tp)(CN)$_3$]$_4$[Co^{III}(Ttp)]$_3$[Co^{II}(Ttp)]} · 3 DMF (22 – Phase #2)

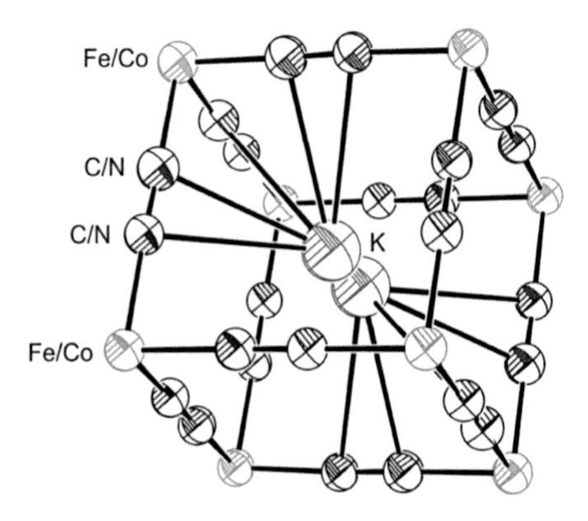

Figure 6.25: Perspective view of the cubic units of the phase #2 of **22**. The ligands capping the metal ions are omitted for clarity. Metal atoms (50/50 Fe/Co) are depicted in orange. Cyanide atoms, also 50/50 C/N disordered, are depicted in black. The cage is inhabited by one potassium ion, which is disordered on two positions (purple). Atoms are displayed as 30% probability ellipsoids.

As for **21**, **22** crystallises in CH_2Cl_2/DMF in the hexagonal space group $R\bar{3}c$ as deep blue blocks. **22** presents the same symmetry problem as **21**: the molecular cubic skeleton of the phase #2 is highly disordered so the iron and cobalt metal atom are not discernable, as the carbon and nitrogen atoms of the cyanides bridge. Similarly, each blocking ligand is statistically half Tp and half Ttp. As for the phase #1 of **22**, the guest potassium ion seems to occupy preferentially only two positions in the cage, which might be an indication that the phase #2 of **22** is not statistically disordered like **21**, but rather along a preferred C_3 axis. Because of the intrinsic properties of the crystal organisation, it is not possible to discuss in details the bond lengths and angles in the structure.

6.3.3 EDX spectroscopy

The presence of potassium was checked by measuring EDX (Energy Dispersive X-ray spectroscopy) spectra of crystals of the phase #1 and phase #2 of **22**. Our preliminary qualitative results on both phases are similar and confirm the presence of K, Co and Fe as "heavy" elements ($Z > 10$) (Figure 6.26). In each case, photons corresponding to the Kα (L shell to K shell) and Kβ (M shell to K shell) transitions energies of potassium are detected, confirming the nature of the atom trapped inside the cage. Quantitative data will be performed soon to check the exact amount of K/Fe/Co in the material.

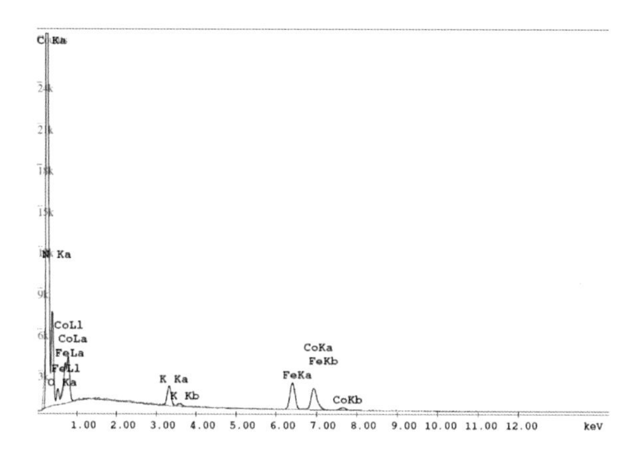

Figure 6.26: Qualitative EDX spectrum of the phase 1 of **22**, with a tension of 20 kV.

6.3.4 NMR spectroscopy

In order to shed light on the solution properties/behaviour of **22**, its ^{1}H NMR spectrum was recorded at different temperatures between T = 298 K and T = 183 K in dichloromethane. The spectra are reported in Appendix (Figure 12.1, Figure 12.2 and Figure 12.3). The replacement of one of the low-spin cobalt(III) ions by a high-spin cobalt(II) ion has a huge impact on the overall symmetry of the cube, transforming the expected T_d symmetry for a cubic structure into a C_{3v} one. This implies the loss of equivalence of the pyrazolyl rings of both Tp and Ttp ligands coordinating the iron(II) and cobalt(III) ions that do not lie on the 3-fold rotation axis, as illustrated by Figure 6.27. Indeed, 24 four pyrazolyl signals are expected for such a compound:

- One pyrazolyl set (3 signals) for the C_3-symmetric Tp ligand of the iron(II) lying on the 3-fold axis corresponding each to 3 protons. Set: 3, 3, 3.
- One pyrazolyl set (3 signals) for the C_3-symmetric Ttp ligand binding the cobalt(II) ion on the 3-fold rotational axis, and integrating for 3 protons each. The fourth pyrazolyl ring is expected to freely rotate, giving three additional signals integrating for one proton each. Set : 3, 3, 3, 1, 1, 1.

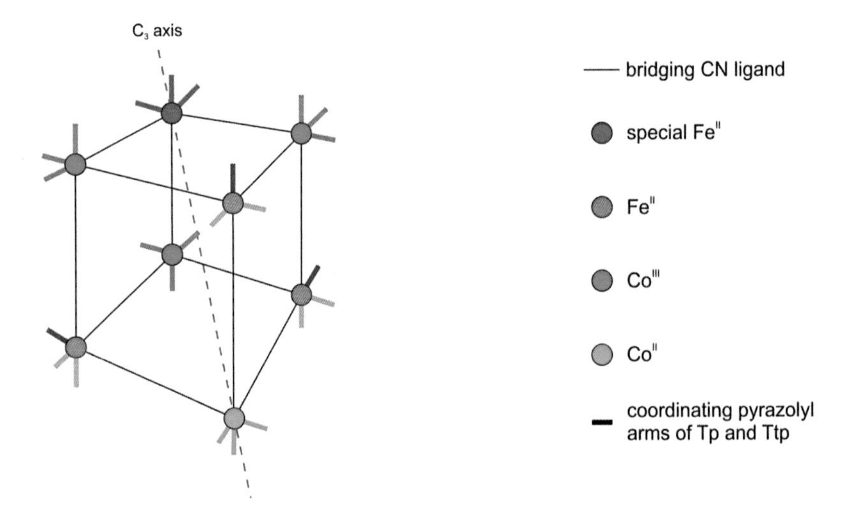

Figure 6.27: Schematic representation of the cyanide-bridged molecular cube **22**. The blocking ligands coordinating the iron (Tp) and the cobalt (Ttp) ions are represented by their three binding pyrazolyl arms. Each Ttp ligand possesses a fourth, non-binding pyrazolyl heterocycle, which is omitted for clarity, as well as the potassium ion normally present in the cube. Pyrazole heterocycles of the same colour are chemically and magnetically equivalent in NMR.

In the case of the metal ions that are not lying on the 3-fold axis, the pyrazolyl rings of each unit are no more equivalent:

- Two of the pyrazolyl heterocycles of the Tp ligands binding the remaining iron(II) ions are facing the cobalt(II) ion, while the third is facing the special-position iron(II) ion; this leads to two equivalent pyrazolyl rings by σ_v-symmetry and a third non-equivalent pyrazolyl heterocycle. Due to the 3-fold axis, the three iron(II) ions (in orange in the Figure 6.27) are equivalent; thus six signals with the following intensities are expected: sets 6, 6, 6, 3, 3, 3.
- The Ttp ligands of the cobalt(III) ions exhibit the same signal pattern, except that the relative positions of each set is exchanged. Furthermore, one additional set of signals is expected for its fourth, non-binding pyrazolyl moieties; thus nine signals with the following intensities are expected: sets 3, 3, 3, 6, 6, 6, 3, 3, 3.

At room temperature, only 19 signals are clearly visible; two of them (at 91.4 ppm and 37.5 ppm) are only "suggested" by the distortion of the baseline and three are completely invisible. At 233 K, however, all signals are present and are integrated as expected (*vide supra*). The partial attributions are summarised in the spectrum of Figure 6.28.

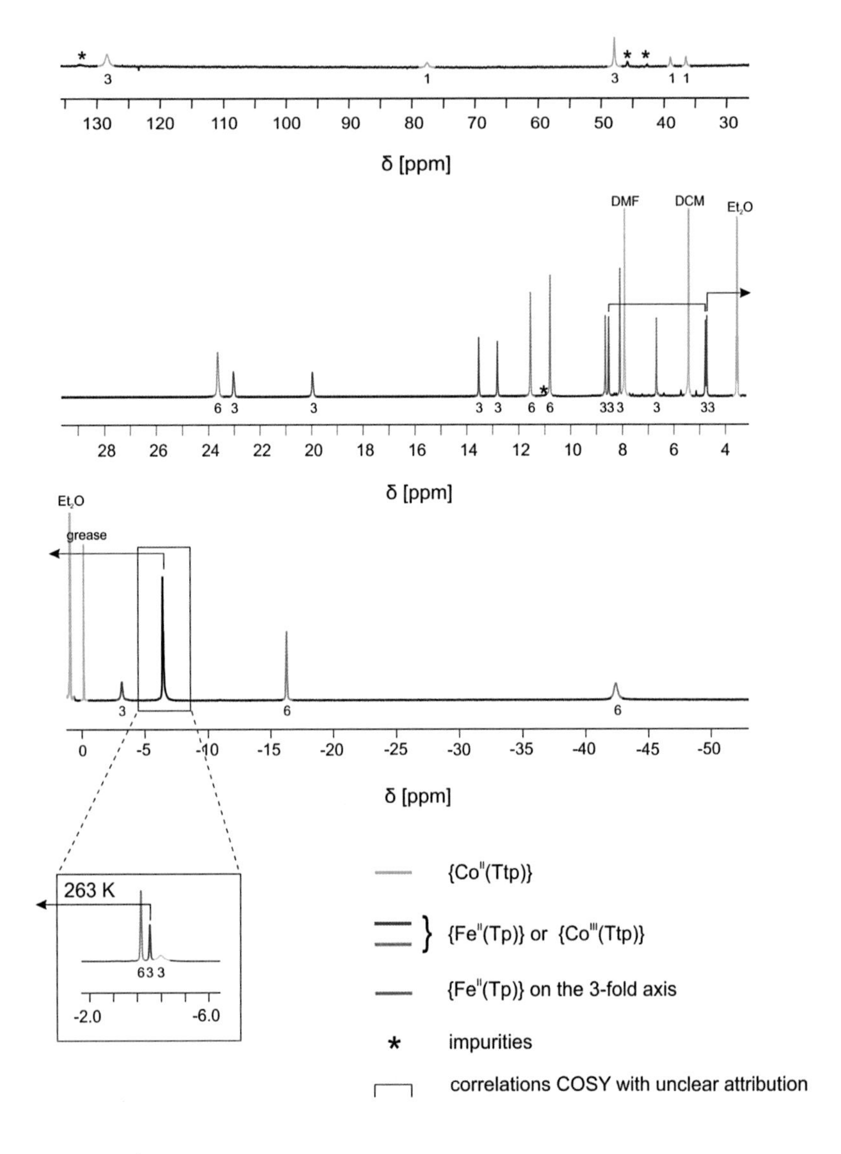

Figure 6.28: ^{1}H NMR spectrum at T = 233 K of **22** in CD_2Cl_2. Since at this temperature, three signals overlap at $\delta \approx$ -4 ppm, a zoom of the same spectral region at higher temperature (T = 273 K) is depicted in the inset. Partial information about connectivity within pyrazolyl rings when precise attribution is not possible (obtained by gCOSY) are depicted as brackets.

Assignment of the different signal sets

“Co^{II}(Ttp)} moiety, paramagnetic royal blue set”

As expected, the signals the ^{1}H NMR chemical shifts assigned to the {Co^{II}(Ttp)} moiety (in royal blue) are strongly shifted outside the “normal” diamagnetic range, and except for the signal at δ = -2.64 ppm (T = 298 K), they are all absent/barely noticeable at room temperature. Indeed, they are significantly broader than the other signals (e.g. δ = -2.64 ppm, $\Delta\nu_{1/2}$ = ~90 Hz at 298 K). Their chemical shift is also strongly dependent on the temperature. They shift up to 1 ppm per K, (see in Appendix the Figure 12.1, Figure 12.2 and Figure 12.3).

“{Fe^{II}(Tp)} moiety lying on the 3-fold axis, red set”

These signals belong to the {Fe^{II}(Tp)} moiety, which is the farthest away from the paramagnetic centre, and lies on the 3-fold axis. They appear in the aromatic diamagnetic region of the ^{1}H NMR spectrum: δ = 8.34, 7.65 and 6.57 ppm at 298 K. They are almost temperature independent as they shift from 0 to 0.5 ppm over the whole temperature range (115 K). They exhibit indeed less temperature dependency than the DMF methyl group and water impurity signals. The three {Fe^{II}(Tp)} signals correlate with each other in the ^{1}H, ^{1}H gCOSY spectrum, and even show an unresolved fine structure. Despite showing strong diamagnetic behaviour (which is not surprising considering that the paramagnetic centre is more than 7 Å away), they unmistakably belong to **22** as stated by the diffusional NMR studies (see below).

“pink set of signals integrating for 6H”

As they do not exhibit crosspeaks in the ^{1}H, ^{1}H gCOSY spectrum, it is not possible to sort out the six signals highlighted in pink (integration: 6 protons each) into two distinct spin systems; however, three of them are strongly shifted and are temperature dependent: δ = 18.44, -25.05 and -8.45 ppm (298 K), while the three others, at δ = 9.68, 9.71

and -1.49 ppm (298 K) tend to show less temperature dependency. The first three signals probably belong to the six pyrazolyl groups facing the cobalt(II) ions and located on the three {Fe^{II}(Tp)} moieties that are directly connected to the paramagnetic ions (Figure 6.27). The three last signals could be ascribed to the six equivalent pyrazolyl groups belonging to the {Co^{III}(Ttp)} moieties, that are farther away from the paramagnetic centre (Figure 6.28).

"navy blue set of signals"

The remaining nine signals (in navy blue in Figure 6.28), corresponding to three protons each, can be assigned either to the {Fe^{II}(Tp)} moiety that is away from the 3-fold axis (three sets of 3H signals expected) or to the {Co^{III}(Ttp)} moieties (three sets of 3H signals for the cobalt-coordinated pyrazolyl groups and three sets of 3H signals for the uncoordinated pyrazolyl of the Ttp ligand). Only partial assignment can be achieved by the ^{1}H, ^{1}H gCOSY analysis: the pairs of peaks at δ = 7.94 and 5.69 ppm, and those at δ = -2.01 and 5.65 ppm belong to the same two spin systems, but it is impossible to further ascribe the five other signals at δ = 17.91, 15.91, 11.39, 10.44 and 0.99 ppm (298 K) without more information.

In theory, two additional signals corresponding to the apical protons of the Tp ligands should be visible in the ^{1}H NMR spectrum: one signal (3H) belonging to the {Fe^{II}(Tp)} moiety linked to the paramagnetic cobalt(II) ion and one (1H) belonging to the {Fe^{II}(Tp)} on the C_3 axis. However, the ^1J coupling to the quadrupolar boron splits and broadens significantly the signals already broadened by paramagnetism, preventing their detection.

At room temperature, **22** exhibits three signals in ^{11}B NMR spectrum. The relatively sharp signal at 196.5 ppm (298 K) exhibits a strong temperature dependency. It does not appear in the usual ^{11}B NMR frequency range and can therefore be ascribed to the paramagnetic {Co^{II}(Ttp)} moiety. Another sharp signal, thrice as big in integral as the first one, appears at 1.64 ppm. It only shows a small temperature dependency and is therefore attributed to the low-spin {Co^{III}(Ttp)} moieties. The remaining signals of the four {Fe^{II}(Tp)} units appear at -13.5 ppm as a very broad signal that cannot be sharpened by proton decoupling. Its chemical shift, on the top of the boron glass signal, does not allow reliable integration.

Diffusion ^{1}H NMR spectroscopy

Diffusional ^{1}H NMR spectroscopy was performed on a 10 mM solution of **22** in CD_2Cl_2 with a diffusion parameter of $\Delta = 50$ and 75 ms at room temperature. Since the T_1 values are short (paramagnetic species) and in the same order of magnitude as Δ, longer Δ values gave rise to relaxation problems, with noticeable issues on all the peaks. For the most diamagnetic signals, which are not (or less) affected by this relaxation problem at small enough Δ, the diffusion coefficient was found to be $D = 6.874\times10^{-10}$ $m^2 \cdot s^{-1}$ ($\pm$2%). This corresponds to a hydrodynamic radius of $R_H = 7.6$ Å. Considering the M-CN-M' distance of 5 Å (cube edge), and the size of the capping Tp ligands (if the free pyrazolyl rings of the Ttp ligands are not taken into account), with Fe···B distances of about 3.1 Å, the estimated value for hydrodynamic radius would be $R_H = 7.43$ Å. It is thus clearly in line with the value from the diffusion experiment.

6.3.5 Fourier Transform InfraRed spectroscopy

$\{[Fe^{III}(Tp)(CN)_3]_4[Co^{II}(Tpe)]_4\}(ClO_4)_4$(21)

21 exhibits only one sharp cyanide stretching absorption band (*ca* 2169 cm^{-1}), whose frequency corresponds to a bridging $\{Fe^{II}$-μCN-Co$\}$ moiety (see Figure 6.29). The absence of stretching band below 2100 cm^{-1} is consistent with the absence of electronic transfer and a $\{Fe^{III}_4Co^{II}_4\}$ oxidation state for **21**. The B–H moiety absorbs at 2522 cm^{-1}, that is at higher frequency than the B–H absorption in $PPh_4[Fe^{III}(Tp)(CN)_3]$. The pyrazole rings of Tp and Tpe exhibit two, barely resolved from each other absorption bands at 1500 and 1516 cm^{-1}; the same applies for their C–H pyrazolyl rings, which only give one set of very weak absorption at 3109, 3131 and 3151 cm^{-1}. However, the -CH_2OH moiety of the Tpe ligands is responsible for the broad OH peak around 3475 cm^{-1}. As indicated by the intense broad absorption at 1654 cm^{-1}, the sample contains uncoordinated DMF; it is also responsible for the two C–H stretches at 2866 and 2930 cm^{-1}. This group of signals also shows two additional shoulders at 2885 and 2960 cm^{-1} which can be ascribed to the ν-CH_2 vibration mode of the –CH_2OH Tpe moiety.

21 is four-fold positively charged and was isolated as perchlorate ions. Three of their characteristic stretching vibrations are observed at 1049, 1073 and 1085 cm^{-1}.

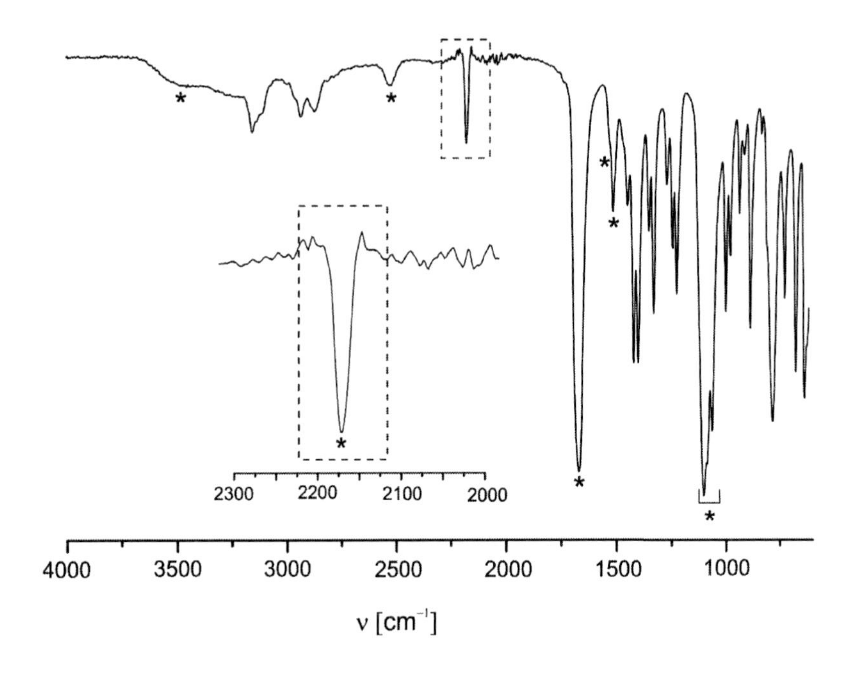

Figure 6.29: FT-IR (ATR) transmission spectrum of fresh filtered **21** between 4000 and 600 cm^{-1} with a 4 cm^{-1} resolution. Selected IR vibration bands in cm^{-1} and their intensities are marked with an asterisk: 1049 (vs), 1073 (vs), 1085 (vs), 1500 (w), 1516 (sh, w), 1654 (br, vs), 2169 (w), 2522 (vw), 3475 (br, vw).

K@{[Fe^{II}(Tp)(CN)$_3$]$_4$[Co^{III}(Ttp)]$_3$[Co^{II}(Ttp)]} (22)

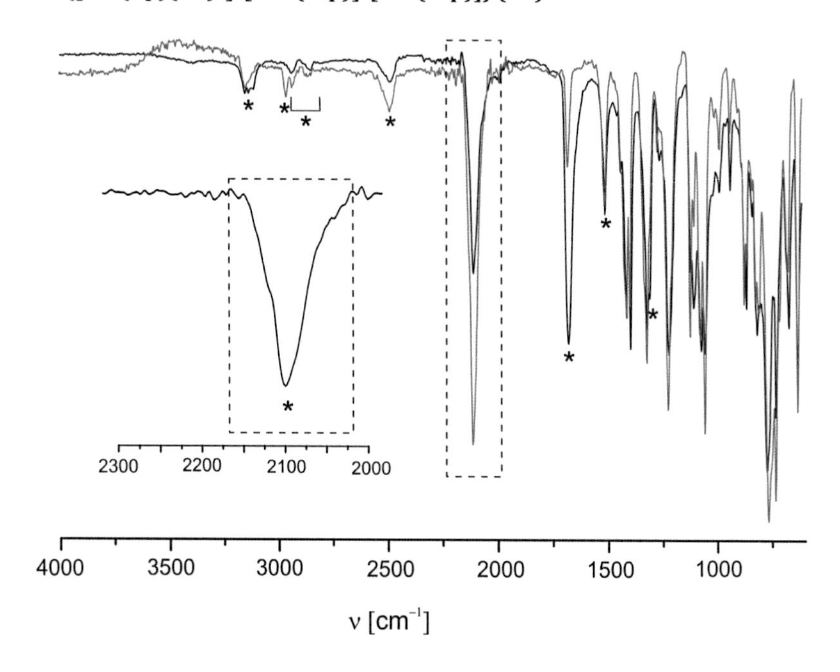

Figure 6.30: FT-IR (ATR) transmission spectra of fresh filtered **22** between 4000 and 600 cm^{-1} with a 4 cm^{-1} resolution:
Red curve: compound **22** crystallised from CH_2Cl_2/*n*-pentane (phase #1).
Black curve: compound **22** crystallised from DMF/CH_2Cl_2 (phase #2).
Selected IR vibration bands in cm^{-1} and their intensities are marked with an asterisk: 1178 (m – phase #1), 1503 (w), 1669 (br, s), 2103 (br, m), 2480 (vw), 2845 (vw), 2932 (vw), 2960 (vw – phase #1), 3108 (vw), 3131 (vw), 3146 (vw).

The FT-IR spectra of the two phases of **22** are almost identical; the only difference between these two lies in the presence in the phase #1 spectrum (sample recrystallised by layering a CH_2Cl_2 solution of **22** with *n*-pentane) of two additional vibration bands at the following wavenumbers: 1178 and 2960 cm^{-1}. These are characteristic vibrations of CH_2Cl_2 and disappear when the sample is left a few seconds out of its mother liquor before the acquisition of the spectrum. The intense peak at 1669 cm^{-1} and the two very weak peaks at 2845 cm^{-1} and 2932 cm^{-1} correspond to the lattice DMF molecules. According to the crystal structures, the phase #2 of **22** contains DMF molecules but not the phase #1, (or at least those are not detected by the X-ray diffraction study). Despite containing both Tp and Ttp ligands, the FT-IR spectra of **22** display only one ring stretch above 1500 cm^{-1}, as well as only one non-*C* substituted pyrazolyl C–H pattern with the three, very weak absorptions at 3108, 3131 and 3146 cm^{-1}. However, a look at the IR table for the precursors (see Table 3.1) indicates that the signals of Tp and Ttp appear at about the same frequency, and in the case of **22**, overlap. The Tp distinctive B–H stretching band appears at 2480 cm^{-1}. This is only *ca* 8 cm^{-1} blueshifted compared to $K_2[Fe^{II}(Tp)(CN)_3]$, but 22 cm^{-1} redshifted compared to PPh_4[**1**]. Most interesting is the broad, strong stretching band due to the stretching cyanide moieties. With a resolution of 4 cm^{-1}, the maximal absorption happens at 2103 cm^{-1}, but is slightly shifted toward lower frequencies when a better resolution is used because of the incidence on the relative intensities of the different contributions. This is consistent with several unequivalent bridging cyanides, all *C*-bound to iron(II) ions.

6.3.6 Electrochemistry

K@{[Fe^{II}(Tp)(CN)$_3$]$_4$[Co^{III}(Ttp)]$_3$[Co^{II}(Ttp)]} (22)

In order to shed some light on the solubility and electronic properties of **22**, cylclovoltammetric studies were performed in dichloromethane solution at room temperature. Crystals of phase #2 (DMF/CH_2Cl_2) were used as a sample. **22** provides a rather complicated cyclovoltammogram over the whole E = +1100 / -1200 mV potential range (Figure 6.31. b). When no potential lower than -1200 mV is applied to the system (see Figure 6.31.a), only the first cycle of Figure 6.31.b is obtained. **22** undergoes first a seemingly irreversible, one-electron oxidation at E_{pa} = -23 mV, whose intensity decreases upon cycling at a sufficiently high scan rate. It is reasonable to ascribe this process to the one-electron oxidation of the only cobalt(II) ion of the cube to cobalt(III), producing the neutral cyanide-bridged cage $\{[Fe^{II}(Tp)(CN)_3]_4[Co^{III}(Ttp)]_4\}$. The cube remains intact in solution, as demonstrated by the four consecutive one-electron quasi-reversible redox processes at $E°_{1/2}$ = +349 mV (ΔE_p = 75 mV), +490 mV (ΔE_p = 86 mV), +637 mV (ΔE_p = 65 mV) and +815 mV (ΔE_p = 76 mV).

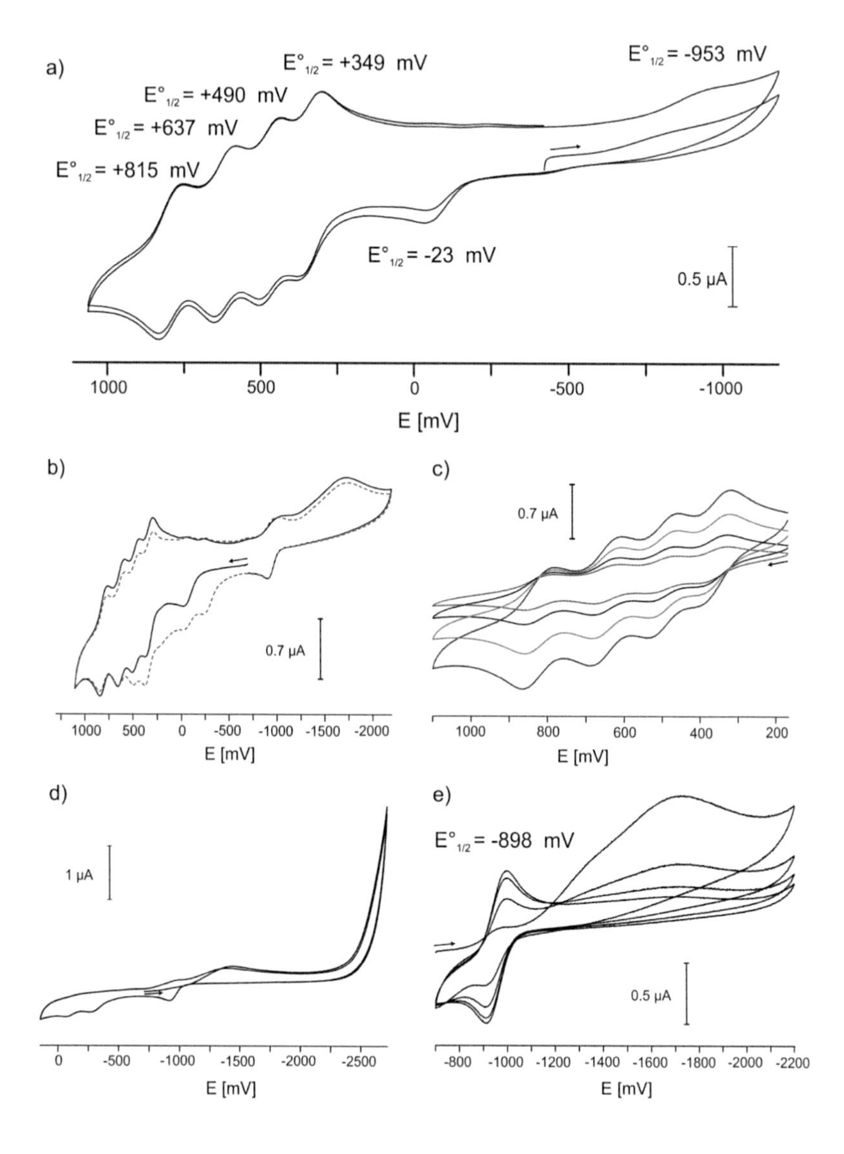

Figure 6.31: Cyclovoltammograms of **22** in dry CH_2Cl_2. at room temperature against $[Fc]/[Fc]^+$, Pt/[n-Bu_4N][PF_6]/Ag.
a) Two potential scans between 1100 mV and -1200 mV. Scan rate v = 250 mV·s^{-1}.
b) Two potential scans over the whole potential range. The first scan is depicted in black, while the second one is represented by the red dotted line. Further scans were identical to the second potential scan. Scan rate v = 250 mV·s^{-1}.
c) Cyclovoltammogram of the four quasi-reversible oxidation processes at $E°_{1/2}$ = 349 mV, $E°_{1/2}$ = 490 mV, $E°_{1/2}$ = 637 mV and $E°_{1/2}$ = 815 mV. Scan rates: 50 (red), 100 (black), 250 (green)

and 500 (blue) mV·s^{-1}. For clarity, and as both cycles are identical, only one per scan rate is depicted here.
d) Two potential cycles over the reductive potential range. Scan rate v = 100 mV·s^{-1}.
e) Four potential scans of the irreversible reduction process at E_{pc} = -1722 mV and the quasi-reversible oxido-reduction process at $E°_{1/2}$ = -958 mV.

The four quasi-reversible processes can be assigned to the following successive oxidations of the four iron(II) of **22** to iron(III):

First, $\{Fe^{II}_4Co^{III}_4\} \rightarrow \{Fe^{III}Fe^{II}_3Co^{III}_4\}^+$ at $E°_{1/2}$ = +349 mV, then, $\{Fe^{III}Fe^{II}_3Co^{III}_4\}^+ \rightarrow \{Fe^{III}_2Fe^{II}_2Co^{III}_4\}^{2+}$ at $E°_{1/2}$ = +490 mV. This is followed by the oxidation process $\{Fe^{III}_2Fe^{II}_2Co^{III}_4\}^{2+} \rightarrow \{Fe^{III}_3Fe^{II}Co^{III}_4\}^{3+}$ at $E°_{1/2}$ = +637 mV and finally by $\{Fe^{III}_3Fe^{II}Co^{III}_4\}^{3+} \rightarrow \{Fe^{III}_4Co^{III}_4\}^{4+}$ at $E°_{1/2}$ = +815 mV. As shown in Figure 6.31.c, they remain quasi-reversible at different scan rates (v = 50, 100, 250 and 500 mV·s^{-1}), which indicates a fast oxidation and reduction process and thus, confirms the quasi-reversibility. This electrochemical behaviour is similar to what is reported for a $\{Fe_8\}$ and two $\{Fe_4Ni_4\}$ cubic cyanidometallate cages by Oshio *et al.*[175,176] and for a $\{Re_4Fe_4\}$ cubic cyanidometallate cage by Schelter *et al.*[185]

A seemingly irreversible reduction process takes place at E_{pc} = -953 mV in Figure 6.31.a. It only takes place after the oxidation process at E_{pa} = -23 mV, independently from further oxidation. It can be assigned to the reduction of the $[Fe^{II}_4(Tp)_4Co^{III}_4(Ttp)_4]$ species to $[Fe^{II}_4(Tp)_4Co^{III}_3Co^{II}(Ttp)_4]^-$. The peak to peak potential difference between the two half waves amounts to 930 mV, and is due to the structural reorganisation accompanying the oxidation of a high-spin cobalt(II) into a low-spin cobalt(III) species, as it was reported for $[Co^{II}(Tpmd)_2]$ by Kuzu *et al.*[41] and for $[Co^{II}(Tpm)_2](BF_4)_2$ by Sheets and Schultz.[190]

Further investigation of the reductive part of the spectrum between +150 mV and -2700 mV (see Figure 6.31.d) at a scan rate of 100 mV·s^{-1} reveals that the irreversible reduction process at E_{pc} = -1419 mV generates several new electro-active species that are oxidised at E_{pa} = 259 mV and E_{pa} = -898 mV. The comparison of Figure 6.31.d with other cycles at different scan rates (not shown) reveals a scan rate dependence of the half-wave potential for all electro-active species between +150 mV and -2700 mV.

The reversibility of the oxidation process at E_{pa} = -898 mV is increased at higher scan rates until reaching quasi-reversibility from $v > 500\ mV \cdot s^{-1}$ on. Furthermore, Figure 6.31.e demonstrates that, at higher scan rates than the diffusion of **22** in dichloromethane, the scan after scan decrease of the intensity the reduction wave at E_{pc} = -1722 mV due to consumption of the nearby **22** molecules is directly linked to the progressive increase of anodic and cathodic currents of the decay product. This is consistent with the facts that (i) this species is generated by the dissociation of **22** and (ii) diffuses away quite rapidly from the electrode, which ascertains a rather small hydrodynamic radius. A possible candidate is [**1**]$^{-}$, which is known to be one of the dissociation products of **22** in DMF, water and acetonitrile and exhibits a similar electrochemical potential of $E°_{1/2}$ = -824 mV in acetonitrile (see chapter 3). Potential scanning between -1100 and -700 mV did not show any process at all, which is consistent with the fact that **22** does not undergo dissociation in dichloromethane at room temperature.

6.3.7 UV-visible Spectroscopy

K@{[Fe^{II}(Tp)$(CN)_3$]$_4$[Co^{III}(Ttp)]$_3$[Co^{II}(Ttp)]} (22)

The UV-visible absorption spectrum of a dichloromethane solution of **22** was recorded at room temperature (see Figure 6.32). Crsytals of phase #2 (DMF/CH_2Cl_2) were used as a sample.

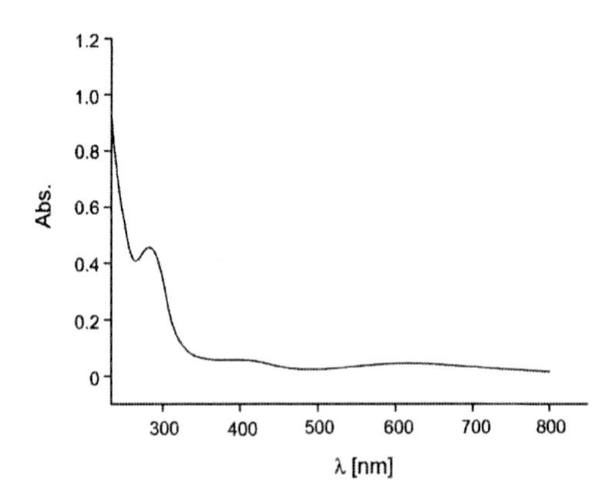

Figure 6.32: UV-visible spectrum of **22** in dichloromethane at room temperature, at c = 14.07 μM.

In dichloromethane, **22** features three absorption bands at $\lambda = 282$ nm ($\varepsilon_{282} = 32480$ L·mol^{-1}·cm^{-1}), 404 nm ($\varepsilon_{404} = 4050$ L·mol^{-1}·cm^{-1}) and 618 nm ($\varepsilon_{618} = 3270$ L·mol^{-1}·cm^{-1}). The absorption at 282 nm is attributed to intra-ligand transition due to the pyrazole rings of the coordinated Tp and Ttp ligands. The $PPh_4[Fe^{II}(L)(CN)_3]$ (L = Tp, Ttp, Tt) building blocks also exhibit an absorption band at $\lambda \approx 404$ nm,[98] which is ascribed to a metal-to-ligand charge transfer (MLCT). Finally a very broad absorption is detected between 500 and 750 nm, with a maximal absorption at 618 nm. This absorption is responsible for the blue colour of **22** and is assigned to the intervalence charge transfer (MMCT), in analogy with the attribution of the 690 nm absorption band in the Prussian Blue UV-visible spectrum analysis.[191]

6.3.8 SQUID magnetometry

$\{[Fe^{III}(Tp)(CN)_3]_4[Co^{II}(Tpe)]_4\}(ClO_4)_4$ (21)

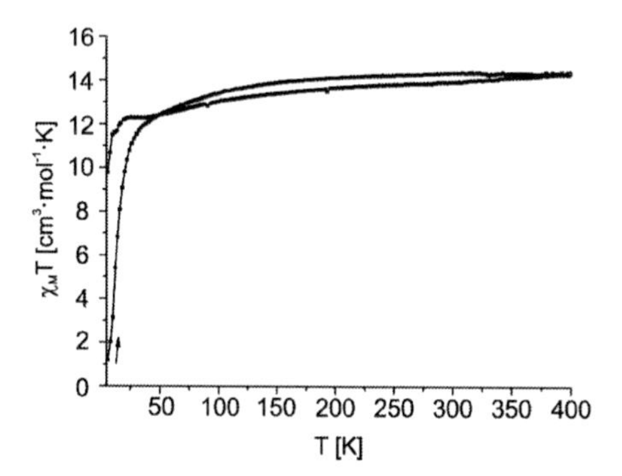

Figure 6.33: Magnetic properties of freshly filtered **21**: $\chi_M T$ product *vs* T between 2 K and 400 K, then between 400 K and 2 K, H = 5000 Oe. The sample was prepared as follows: m_{sample} = 7.7 mg, $m_{capsule}$ = 49.7 mg.

Magnetic measurements were performed on 7.7 mg of freshly filtered **21** and the obtained $\chi_M T$ product in function of the temperature is depicted in Figure 6.33.

In contrast with the analogue photomagnetic cube $\{[Fe^{III}(Ttp)(CN)_3]_4[Co^{II}(Tpe)]_4\}(ClO_4)_4$ reported by Li *et a*l. in 2008,[93] and despite similar electronic properties of the $[Fe^{III}(Tp)(CN)_3]^-$ and $[Fe^{III}(Ttp)(CN)_3]^-$ tricyanido iron(III) complexes, **21** remains in a paramagnetic state $\{Fe^{III}{}_4Co^{II}{}_4\}$ over the whole temperature range. The $\chi_M T$ product at 300 K amounts to 13.92 $cm^3 \cdot mol^{-1} \cdot K$ for the fresh sample and 14.34 $cm^3 \cdot mol^{-1} \cdot K$ after *in situ* desolvation at 400 K. This is exactly what is expected for the following set of eight magnetically independent ions: four low-spin iron(III) ions (0.7 $cm^3 \cdot mol^{-1} \cdot K$, four units) and four high-spin cobalt(II) ions (2.8 $cm^3 \cdot mol^{-1} \cdot K$, four units). Both curves exhibit a slightly decreasing slope toward low temperatures due to the iron and cobalt spin-orbit coupling. At low temperatures (~50 K), the $\chi_M T$ product of the solvated sample decreases rapidly, probably due to antiferromagnetic intra- and/or intermolecular interactions. It is worth noticing that the low temperature behaviour of **21** is significantly affected by the desolvation because (i) desolvatation may lead to changes in the geometry of the {Fe-CN-Co} bridges and thus in the intensity of the moderate intramolecular exchange interaction (which is very dependent on the cyanide bridge geometry) (ii) intermolecular interactions can be modified upon desolvation (e.g. if solvent mediated).

K@{[Fe^{II}(Tp)$(CN)_3$]$_4$[Co^{III}(Ttp)]$_3$[Co^{II}(Ttp)]} (22)

The magnetic properties of both crystal phases of **22** were studied in the SQUID magnetometer. Since the disordered crystal phase #2 contains only DMF molecules as lattice solvent, the corresponding sample is much less sensitive to the desolvatation that can occur during the sample preparation. We can thus consider that the magnetic experiments were performed on a 'fresh' solvated compound. The ordered phase #1, however, contains a significant number of highly volatile dichloromethane molecules per molecular cube. As they leave the crystal lattice as soon as the crystals are removed from solution, it was not possible to measure a true fully solvated phase. In the following section, phase #1of **22** refers to partially desolvated crystals of the non-disordered crystal phase, with no dichloromethane left, but still containing DMF lattice molecules (*cf* IR).

Thermo-induced ECTST

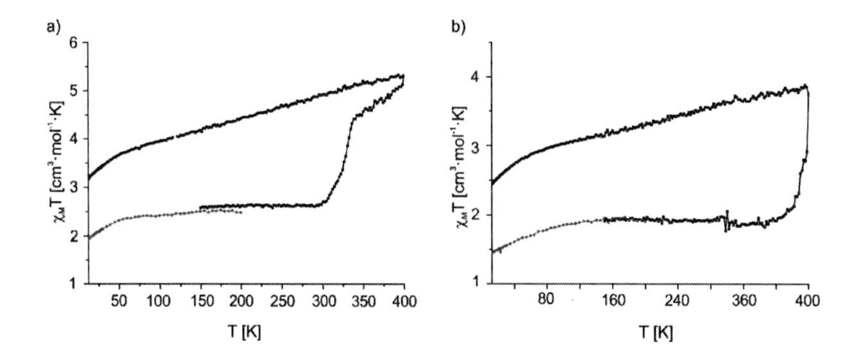

Figure 6.34: $\chi_M T$ *vs* temperature plot of a) phase #1 and b) phase #2 of **22** between 12 K and 400 K.
a) m_{sample} = 5.55 mg, m_{film} = 8.10 mg, H = 10000 Oe,
b) m_{sample} = 5.80 mg, m_{film} = 13.20 mg, H = 10000 Oe.

The two phases of **22** exhibit quite similar magnetic behaviours. The $\chi_M T$ product *vs* temperature curves for both phases are depicted in Figure 6.34. Freshly filtered samples of both phases display the expected curves for a single high-spin cobalt(II) ion between 2 and 300 K. At 300 K, the $\chi_M T$ product of the phase #1 is 2.70 $cm^3 \cdot mol^{-1} \cdot K$, which is in the expected range of 2.8–3.6 $cm^3 \cdot mol^{-1} \cdot K$ for an isolated high-spin cobalt(II) ion. At 12 K, the $\chi_M T$ product is smaller due to spin-orbit coupling and amounts to 1.98 $cm^3 \cdot mol^{-1} \cdot K$. The $\chi_M T$ product of the phase #2, however, amounts only to 1.9 $cm^3 \cdot mol^{-1} \cdot K$ at 300 K, and 1.48 $cm^3 \cdot mol^{-1} \cdot K$ at 12 K. At higher temperature, both phases exhibit a significant increase of $\chi_M T$ that could be due to an ETCST. This transition in phase #1 is somehow smoother, starts at 300 K and presents an unexplained inflexion point at T = 338 K ($\chi_M T$ = 4.41 $cm^3 \cdot mol^{-1} \cdot K$). The $\chi_M T$ value at 400 K reaches 5.32 $cm^3 \cdot mol^{-1} \cdot K$. This is lower than the expected minimum 6.2 $cm^3 \cdot mol^{-1} \cdot K$ if one of the diamagnetic $\{Fe^{II}_{LS}Co^{III}_{LS}\}$ pairs of **22** is converted in a paramagnetic $\{Fe^{III}_{LS}Co^{II}_{HS}\}$ one, pointing to a partial ETCST at 400 K. Similarly, it clearly appears that only the beginning of the ETCST transition is detected in the phase #2, the $\chi_M T$ product reaching only

3.85 $cm^3 \cdot mol^{-1} \cdot K$ at 400K. It also starts at a higher temperature (T = 360 K) than for phase #1 and it is very abrupt. The in-situ desolvated compounds of both phases exhibit similar behaviour. No transition is observed, but the $\chi_M T$ product is linearly decreasing with the temperature, and reaches 3.24 $cm^3 \cdot mol^{-1} \cdot K$ for phase #1 and 2.46 $cm^3 \cdot mol^{-1} \cdot K$ for phase #2, which is indicative of residual {$Fe^{III}{}_{LS}Co^{II}{}_{HS}$} pairs in the system.

The $\chi_M T$ *vs* T curve of phase #1 was simulated at low temperature. In these conditions, **22** can be approximated to a single cobalt(II) high-spin complex. The total Hamiltonian of the system can, like for **8**, also be expressed as Equation (2). However, the adequate spin-orbit coupling Hamiltonian is expressed in Equation (15). The Zeeman Hamiltonian is:

$$\mathcal{H}_{Ze} = \left(-\frac{3\alpha}{2} L_\nu + 2S_\nu\right) \beta\, H_\nu \tag{19}$$

In order to obtain a better fit of the magnetic properties of **22**, a rhombohedricity parameter was introduced in the distortion Hamiltonian:

$$\mathcal{H}_{dist} = \Delta\left(L_z^2 - \frac{1}{3}L^2\right) + E(L_x^2 - \frac{1}{3}L^2) \tag{20}$$

The best estimate was obtained with an orbital reduction parameter α of 0.82, a spin-orbit coupling of -167 cm^{-1} and a TIP parameter of c = $45 \cdot 10^{-6}$ $cm^3 \cdot mol^{-1} \cdot K$. The axial distortion Δ was found to be -2089 cm^{-1} and the rhombohedricity E amounts to 290 cm^{-1}. Even though the quality of the data is quite low, a very good agreement factor was obtained for this fit: $6.9 \cdot 10^{-5}$. α, λ and c are in the same order of magnitude as the found values for the simulation of the magnetic data of **10** and **11**, as well as in the literature.[88] Δ is twice as high as for the mentioned molecular squares, and is also negative. An hypothesis for that resides in the strong axial structural symmetry of **22**, compared to molecular squares.

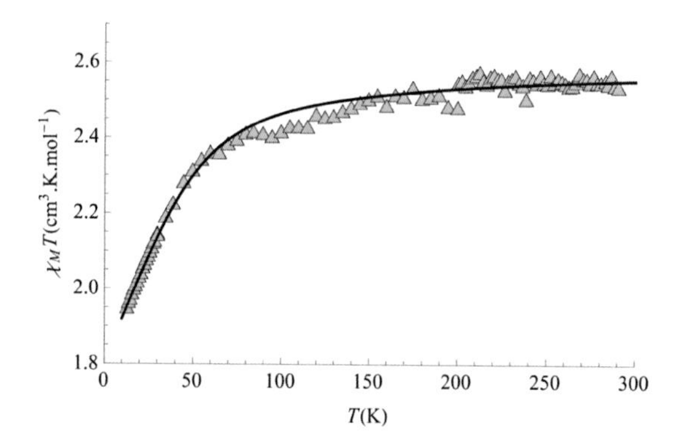

Figure 6.35: Experimental (blue triangles) and simulated (black curve) $\chi_M T$ *vs* temperature curves of the phase #1 of **22**. m_{sample} = 5.55 mg, m_{film} = 8.10 mg, H = 10000 Oe.

ON mode – Photo–induced ETCST

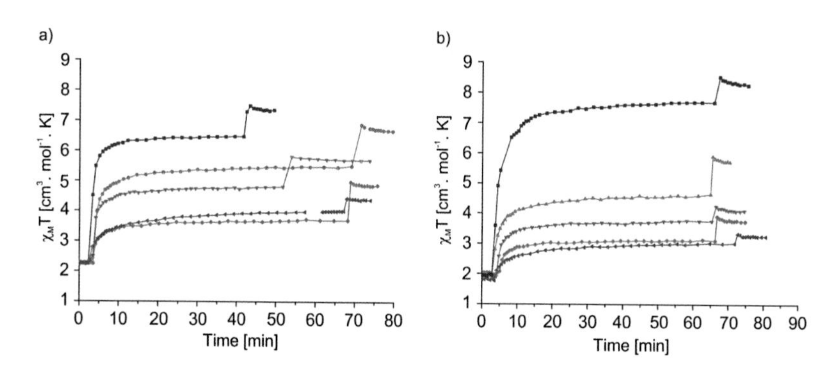

Figure 6.36: $\chi_M T$ *vs* time plot of a) phase #1 and b) phase #2 of **22** under 405 (blue), 532 (green), 635 (red), 808 (wine red) and 900 (grey) nm laser light irradiation at 20 K.
a) m_{sample} = 0.54 mg, H = 10000 Oe,
b) m_{sample} = 0.49 mg, H = 10000 Oe.

Both phases of compound **22** show significant photomagnetic effects at 20 K under light irradiation in the visible and near infrared range, at 405, 532, 635, 808 and 900 nm (see Figure 6.36). However they are both insensible to the 1313 nm wavelength.

The 808 nm wavelength is in both cases the most efficient one, with the highest photo-conversion rate, followed by the 900 nm wavelength. In phase #2, the 808 nm is much more efficient compared to the other wavelengths but the gap in efficiency between the different wavelengths is not as big for phase #1. Interestingly, the 808 nm wavelength is also the most efficient wavelength for photo-conversion in the parent photomagnetic $\{Fe_2Co_2\}$ molecular square reported by the Parisian research group.[96–98] It falls indeed within the MMCT band of the $\{Fe(Tp)–CN–Co(bik)_2\}$ pair. In terms of kinetics, the photo-conversion of phase #1 and phase #2 are similar (20 min *vs* 30 minutes to reach saturation). and slightly faster than that observed for the above mentioned square measured in the same conditions.[96–98] The desolvated phase #1 of **22**, albeit more paramagnetic at the beginning, also shows photomagnetic effects for the same wavelength. Since it reaches the same values than the solvated phase, the curves are not displayed here.

OFF mode – reverse LIETCST and ON/OFF cycling

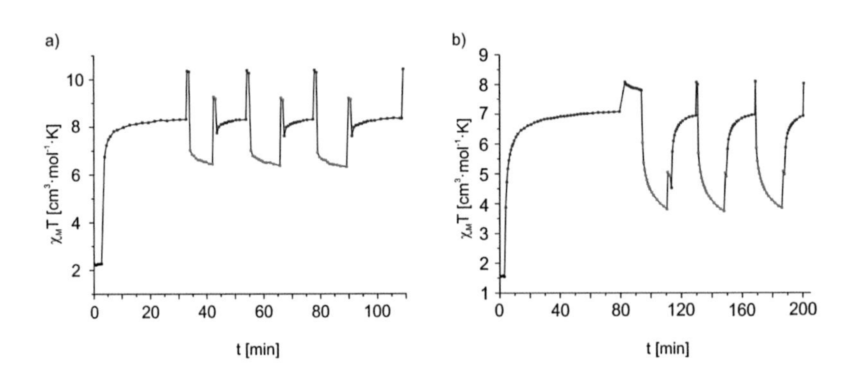

Figure 6.37: Time dependence of the $\chi_M T$ product of a) phase #1 and b) phase #2 of **22** under successive laser irradiations at 808 (wine red) and 532 (green) nm.
a) m_{sample} = 0.54 mg, H = 10000 Oe,
b) m_{sample} = 0.49 mg, H = 10000 Oe.

The reproducibility of the partially reversible photo-induced phenomena have been investigated for both phases by recording their $\chi_M T$ product over time, under successive irradiations at 808 (wine red) and 532 (green) nm for 110 minutes (phase #1) and 200 minutes (phase #2) respectively (see Figure 6.37). In both cases, each on-off cycle is similar to the first one, with identical $\chi_M T$ product values when the laser is switched off (for a same wavelength), showing thus no aging effect under these experimental conditions. The reverse process seems to be more effective for the disordered phase #2 than for the ordered phase #1: (13.8%, and 48.1% decrease of $\chi_M T$ under these experimental conditions). The OFF mode is way less efficient for both phases compared to the photomagnetic $\{Fe_2Co_2\}$ squares reported by Mondal *et al.* from the Parisian research group, which exhibits up to 90% recovery of the diamagnetic ground state with the same 532 nm laser light.[96–98]

In terms of kinetics, in ON mode (808 nm laser light) the $\chi_M T$ product reaches its final value faster than in OFF mode (532 nm laser light) for both phases. This difference in photo-conversion rate is most visible for phase #2 (see Figure 6.37.b) as for about the same time span, the ON photo-conversion reaches $\chi_M T$ saturation while it is not the case for the OFF retro-photo-conversion.

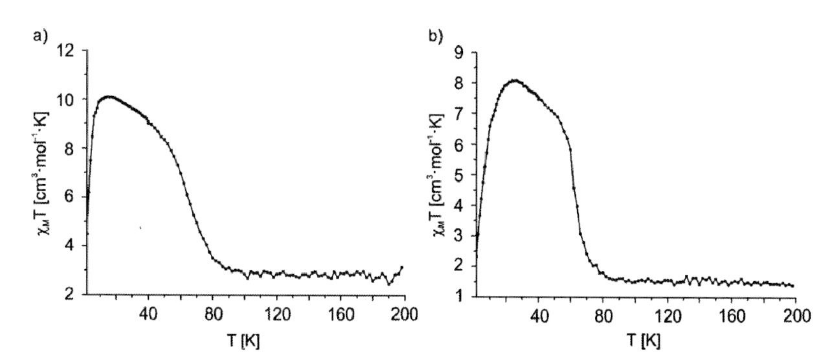

Figure 6.38: Temperature dependence of the $\chi_M T$ product for a) phase #1 and b) phase #2 of **22**, after irradiation at 808 nm (at 20 K).
a) m_{sample} = 0.54 mg, H = 10000 Oe, 0.4 K·min^{-1}.
b) m_{sample} = 0.49 mg, H = 10000 Oe, 0.4 K·min^{-1}.

7 Conclusions and outlook

The target of this work was to extend the family of the $[Fe(Tp)(CN)_3]^-$ building blocks with the preparation (or improved preparation) of new *fac*-tricyanido iron complexes, and to explore their reactivity towards paramagnetic complexes in order to obtain new molecular magnetic materials. A special focus was given to the preparation of low dimensional systems: chains that could behave as SCM (single chain magnet) or polynuclear complexes that could exhibit photomagnetic effect.

In chapter 3, we report on the synthesis and extensive characterisation of several new octahedral iron(II) and iron(III) building blocks based on cyanide and scorpionate ligands of the form $[Fe(L)(CN)_3]^{n-}$, where L is a tris(pyrazolyl)methane derivative (L = Tpm, Tpe, Tpm*). Indeed, the electronic and structural properties of these tricyanido building blocks govern the magnetic properties of the polymetallic species thereof. The synthesis of some of these complexes were already reported in the literature (L = Tp*, Ttp and Tpms), however, the missing spectroscopic and structural data were obtained in this work, and their properties were compared with those of the new tricyanido building blocks. All of these complexes show a C_{3v} symmetry where the iron ion occupies a C_3N_3 environment formed by three *C*-bound cyanides and the three imine-type moieties of the respective scorpionate ligand.

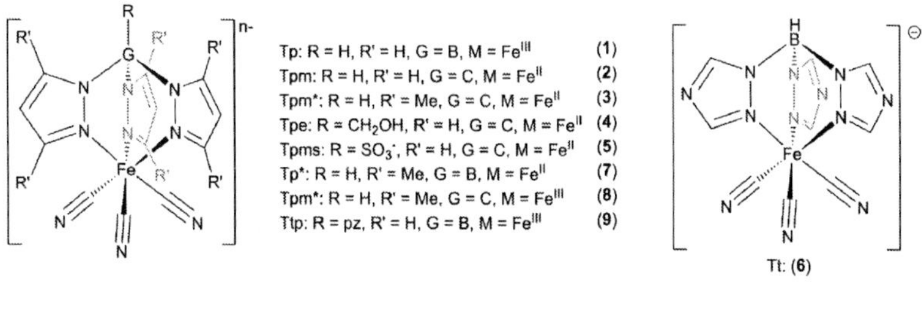

Tp: R = H, R' = H, G = B, M = Fe^{III} (1)
Tpm: R = H, R' = H, G = C, M = Fe^{II} (2)
Tpm*: R = H, R' = Me, G = C, M = Fe^{II} (3)
Tpe: R = CH_2OH, R' = H, G = C, M = Fe^{II} (4)
Tpms: R = SO_3^-, R' = H, G = C, M = Fe^{II} (5)
Tp*: R = H, R' = Me, G = B, M = Fe^{II} (7)
Tpm*: R = H, R' = Me, G = C, M = Fe^{III} (8)
Ttp: R = pz, R' = H, G = B, M = Fe^{III} (9)
Tt: (6)

Figure 7.1: Synthesised iron(II) and iron(III) complexes based on cyanide and scorpionate ligands.

Cyclic voltammetry studies showed that **1**, **3/8**, **5**, **6** and **7** undergo quasi-reversible iron-centred redox processes in acetonitrile. Comparison of their infrared spectra allowed the identification of several key spectral features, whose frequencies and intensities contain specific electronic and structural information so that the structure of unknown new polymetallic species can be deduced from infrared analysis.

The g values of **1** and **7** were extracted by EPR. For **8** a g_{eff} value was extracted from the magnetic data obtained by SQUID magnetometry.

In order to shed some light on the spin density distribution along the cyanide bridges, compound **1**, **7** and **8** were measured by MAS-NMR. It was shown that the mediation of the magnetic information primarily occurs as a spin polarisation phenomenon leading to strongly negative spin density in the 2s orbitals of the carbon atoms. The total spin density found for **1** and **7** corresponds to the DFT calculations. The spin density detected at the nitrogen atoms is positive, and is the result of spin delocalisation from the metal ion to 2p nitrogen orbitals.

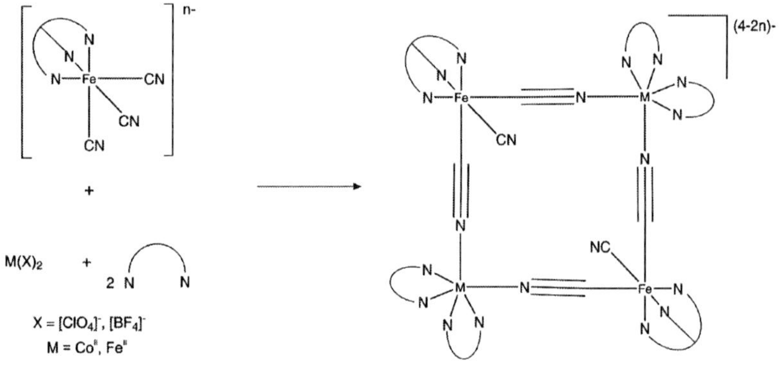

Figure 7.2: Schematic representation of the reaction of $[Fe^{III}(L)(CN)_3]^{n-}$ with partially blocked $\{M(L')_2(S)_y\}^{2+}$ units by self-assembly.

In chapter 4, reaction of **7** with partially blocked $\{M(bik)_2(S)_2\}^{2+}$ subunits ($M = Co^{II}$ or Fe^{II}, S = solvent) led to the formation of the molecular squares **10**, **11** and **12**

(M = Co^{II} (**10** and **11**), Fe^{II} (**12**)) (see Figure 7.2). **10** and **11** consist of paramagnetic $\{Fe^{III}_{LS}$-C≡N-$Co^{II}_{HS}\}$ bridges and do not undergo thermally induced spin transitions. For these complexes it was shown that the iron(III) and cobalt(II) ions experience ferromagnetic interactions. Their $\chi_M T$ *vs* T curves were modelled in order to get an estimate of the electronic parameters governing the magnetic properties (coupling constants, anisotropy parameters, orbital reduction parameters, Δ and g values). The mixed-valence $\{(Fe^{III}_{LS})_2(Fe^{II}_{HS})_2\}$ molecular square **12** shows a thermo-induced spin transition from high-spin to low-spin on the divalent iron centres, with $T_{1/2}$ = 227 K. It was also shown that a photo-induced spin-state switch (LIESST effect) can also be triggered at 20 K by laser light irradiation for 405-1313 nm wavelength, with an optimum efficacy in the 700-900 nm range. The metastable high-spin state of **12** is stable up to 35 K.

Reaction of **8** with cobalt(II) and manganese(II) ions led to new magnetic one-dimensional cyanide bridged double-zigzag chains **13** (M = Co^{II}) and **14** (M = Mn^{II}) (chapter 5). The magnetic study shows that the interactions between the iron(III) and the cobalt(II) ions in **13** are ferromagnetic, while antiferromagnetic interactions take place between the iron(III) and manganese(II) ions in **14**. Unfortunately, no SCM (Single Chain Magnet) behaviour was observed. Reaction of **8** with partially blocked $\{Co^{II}(L')_2(S)_2\}^{2+}$ subunits (L' : bik or bim ligands and S = solvent molecules) led to *in-situ* redox reaction between the metal ions and produced the molecular squares **15** and **16** containing the diamagnetic $\{Fe^{II}_{LS}$-C≡N-$Co^{III}_{LS}\}$ pairs. These two molecular squares remained diamagnetic over the whole temperature range, and did not show any change in their magnetic properties under laser light irradiation at low temperature. This is assigned to the mismatch of the redox potentials of the two building blocks that clearly favours the diamagnetic electronic state.

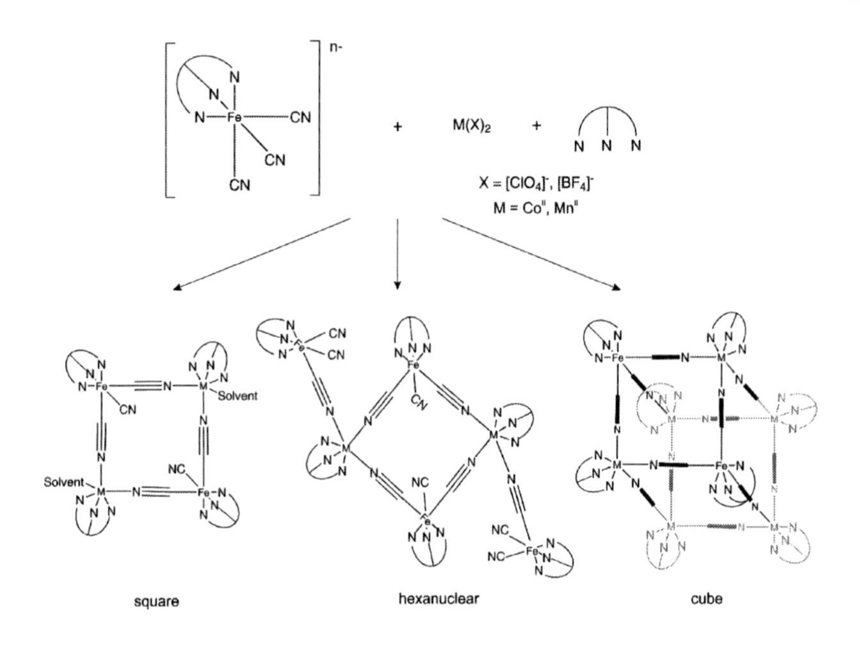

Figure 7.3: Schematic representation of the reaction of $[Fe^{III}(L)(CN)_3]^{n-}$ with partially blocked $\{M(L)(S)_3\}^{2+}$ units by self-assembly.

In chapter 6, reaction of **1** with partially blocked $\{M(L)(S)_3\}^{2+}$ subunits (M = Co^{II}, Mn^{II}, L = scorpionate ligand, S = solvent molecule) under various conditions was explored (see Figure 7.3). Interestingly, changing the synthetic conditions allows to tune the nuclearity and the architecture of the obtained polymetallic assemblies. The reaction of **1** with $\{M(Tpm^*)(S)\}^{2+}$ units allowed the preparation of the molecular squares **17** (M = Co^{II}) and **18** (M = Mn^{II}). The reaction of **1** with $\{M(Tpe)(S)\}^{2+}$ units in acetonitrile/water mixtures produced the isostructural, paramagnetic hexanuclear $\{Fe_4M_2\}$ neutral compounds **19** (M = Co^{II}) and **20** (M = Mn^{II}). Interestingly, the same reagents in DMF produced the four-fold cationic, paramagnetic $\{Fe_4Co_4\}$ molecular cube **21**. Ferromagnetic interactions between the metal ions take place in the {FeCo} species **17**, **19** and **21**. These interactions are antiferromagnetic for the {FeMn} species **18** and **20**.

Reaction of **1** with $\{Co(Ttp)(S)_3\}^{2+}$ subunits in DMF led to the formation of the anionic $\{Fe_4Co_3^{III}Co^{II}\}$ molecular box **22**. EDX and structural analysis revealed that the potassium countercation could serve as template inside the box so that the overall charge of the compound is zero. As a consequence, a fully stable, undissociated compound **22** is obtained in solution, which was highlighted by five consecutive redox processes in cyclic voltammetry studies and ESI-MS analysis in dichloromethane. This quite rare property in cyanide chemistry allowed full characterisation of **22** in solution, and, notably, the determination of the hydrodynamic radius of **22** in dichloromethane by diffusion ^{1}H NMR spectroscopy. Freshly filtered samples of **22** show typical $\chi_M T$ *vs* T curves up to room temperature. At higher temperature, **22** undergoes a transition that could be due to a thermo-induced electron transfer, with a drastic increase of the $\chi_M T$ product. Laser light irradiation of **22** at 20 K triggered a strong increase of the magnetisation for wavelength from 405 nm to 900 nm. No effect was observed for the 1313 nm wavelength. Optimal response from **22** was obtained for irradiation at 808 nm in the iron(II)-cobalt(III) charge transfer band. A partial reverse effect could be obtained by irradiating the metastable state of **22** by 532 nm. Such systems could be used as molecular models of the well-known photomagnetic Prussian Blue Analogues. Further investigations will include the examination of the role played by the alkali ion in the photomagnetic properties of these systems.

Overall, the cyanide scorpionate chemistry allowed the synthesis of various new cyanide-bridged polynuclear systems including some showing thermo-induced and photo-induced switching of their magnetic properties. Further research on these systems will include:

- In-depth physical studies (X-ray diffraction under irradiation at low temperature, time-resolved spectroscopy) in order to get a better comprehension of the metastable state and therefore be able to better rationalise the switchable properties of these systems.

- The processing of solution-stable photomagnetic switching systems into hybrid materials: inclusion in polymer films, immobilisation on surfaces or on nanoparticles to produce multifunctional materials.

8 Zusammenfassung und Ausblick

Ziel dieser Arbeit war es, die Familie der $[Fe(Tp)(CN)_3]$-Einheiten durch die Synthese (oder die Optimierung bestehender Synthesen) neuartiger *fac*-tricyanido Eisenkomplexe zu erweitern und ihre Reaktivität gegenüber paramagnetischen Komplexen zu untersuchen, um somit neue magnetische Molekülmaterialen zu erzeugen. Besonderer Fokus liegt hierbei auf niedrig dimensionalen Systemen wie Ketten, die als SCMs (single chain magnets) fungieren, oder polynukleare Komplexe, die photomagnetische Effekte zeigen können.

In Kapitel 3 wurde die Synthese und ausführliche Charakterisierung einiger neuen oktaedrischen Eisen(II)- und Eisen(III)-Einheiten der Form $[Fe(L)(CN)_3]^{n-}$ auf Basis von Cyaniden und Scorpionat-Liganden (wobei L ein Derivat des Tris(pyrazolyl)methan ist; L = Tpm, Tpe, Tpm*) beschrieben.

In der Tat bestimmen die elektronischen und strukturellen Eigenschaften dieser Tricyanido-Untereinheiten die magnetischen Eigenschaften ihrer polymetallischen Folgeverbindungen. Die Synthesen einiger solcher Komplexe wurde bereits in der Literatur beschrieben (L = Tp*, Ttp and Tpms) wobei die fehlenden spektroskopischen und strukturellen Daten im Rahmen dieser Arbeit ergänzt werden konnten. Außerdem wurden die Eigenschaften der Komplexe mit denen der neu vorgestellten Tricyanido-Untereinheiten verglichen. Alle diese Komplexe zeigen annähernd C_{3v}-Symmetrie, wobei die Eisenionen in einer C_3N_3 Umgebung, von drei über die *C*-Atome gebundenen Cyaniden und den drei Imin-Einheiten der entsprechenden Scorpionat-Liganden, koordiniert werden.

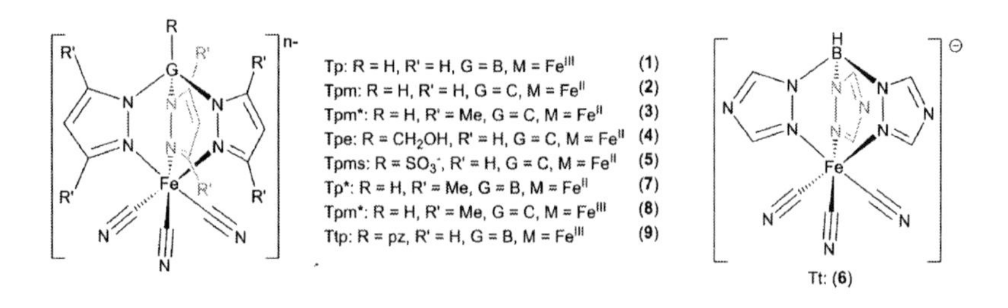

Abbildung 8.1: Synthetisierte Eisen(II)-und Eisen(III)komplexe auf Basis von Cyanido- und Scorpionat-Liganden.

Cyclovoltammetrische Untersuchungen zeigten, dass **1**, **3/8**, **5**, **6** und **7** in Lösung (Acetonitril) quasi-reversible eisenzentrierte Redoxprozesse durchlaufen. Durch Vergleichen der Infrarot-Spektren konnten charakteristische Banden identifiziert werden, deren Frequenzen und Intensitäten spezifische elektronische und strukturelle Informationen enthalten, wodurch sich die grundlegende Struktur unbekannter neuer polymetallischer Verbindungen durch Analyse dieser Daten herleiten lässt.

Die g-Werte von **1** und **7** wurden aus EPR- Experimenten erhalten. Für **8** wurde g_{eff} aus mittels SQUID-Messungen erhaltenen Daten bestimmt.

Um die Verteilung der Spindichten entlang der Cyanidbrücken eingehend zu untersuchen, wurden an den Verbindungen **1**, **7** und **8** MAS-NMR-Experimente durchgeführt. Es konnte gezeigt werden, dass die Übertragung der magnetischen Information hauptsächlich über Spinpolarisationsphänome geschieht, die zu ausgeprägten negativen Spindichten in den 2s Orbitalen der Kohlenstoffatome führt. Die gesamte gefundene Spindichte für **1** und **7** entspricht den Ergebnissen der DFT Berechnungen. Die an den Stickstoffatomen lokalisierte Spindichte ist positiv, was eine Folge der Spindelokalisierung vom Metallion in die 2p Orbitale der Stickstoffatome darstellt.

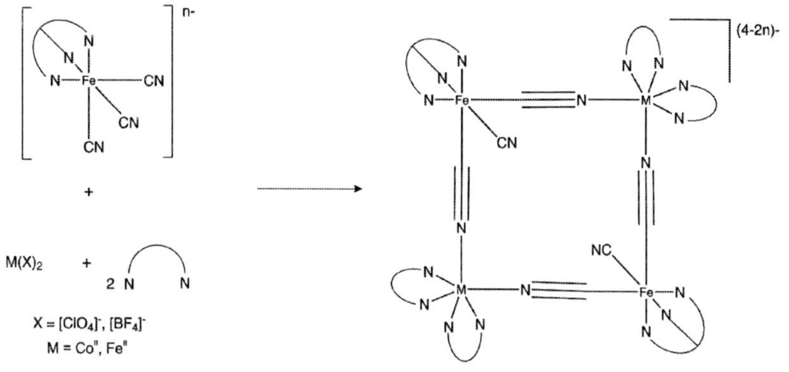

Abbildung 8.2: Schematische Darstellung der Reaktion von $[Fe^{III}(L)(CN)_3]^{n-}$ mit teilweise blockierten $\{M(L')_2(S)_y\}^{2+}$ Einheiten durch Selbstorganisation.

In Kapitel 4 wurde die Reaktion von **7** mit teilweise blockierten $\{M(bik)_2(S)_2\}^{2+}$ Untereinheiten (M = Co^{II} or Fe^{II}, S = Lösemittel) beschrieben. Es konnte die Bildung der quadratischen vierkernigen Verbindungen **10, 11** und **12** (M = Co^{II} (**10** und **11**), Fe^{II} (**12**)) (siehe Figure 7.2) gezeigt werden. **10** und **11** beinhalten paramagnetische $\{Fe^{III}_{LS}\text{-}C{\equiv}N\text{-}Co^{II}_{HS}\}$-Brücken und zeigen keine thermisch induzierten Spinübergänge. Für diese Komplexe konnte gezeigt werden, dass die Eisen(III)- und Cobalt(II)ionen ferromagnetisch interagieren. Um die den magnetischen Eigenschaften (Kopplungskonstanten, anisotropische Parameter, Reduzierung des Orbital-Drehimpulses, Δ and g Werte) zugrundeliegenden elektronischen Parameter abschätzen zu können, wurden die entsprechenden $\chi_M T$ gegen T Kurven modelliert. Der gemischtvalente vierkernige quadratische Komplex $\{(Fe^{III}_{LS})_2(Fe^{II}_{HS})_2\}$ **12** zeigt einen temperaturinduzierten Spinübergang von high-spin nach low-spin an den divalenten Eisenatomen, mit $T_{1/2}$ = 227 K. Es konnte ebenso gezeigt werden, dass ein photo-induzierter Wechsel des Spinzustandes (LIESST Effekt) bei 20 K durch Bestrahlung mit Licht der Wellenlängen 405-1313 nm angeregt werden kann. Die höchste Effizienz liegt dabei im Bereich von 700-900 nm. Der metastabile high-spin Zustand von **12** ist dabei bis 35 K stabil.

Die Reaktion von **8** mit Cobalt(II)- und Mangan(II)ionen führt zu den neuen magnetisch aktiven eindimensionalen cyanidverbrückten „zickzack" Doppelketten **13** (M = Co^{II}) und **14** (M = Mn^{II}) (Kapitel 5). Die magnetischen Untersuchungen belegen, dass die Interaktionen zwischen den Eisen (III)- und den Cobalt(II)ionen in **13** ferromagnetischer Natur sind, während zwischen den Eisen(III)- und Mangan(II)ionen in **14** antiferromagnetischer Austausch stattfindet. Bei diesen Verbindungen konnte jedoch kein SCM (Single Chain Magnet) Verhalten beobachtet werden. Die Reaktion von **8** mit dem teilweise blockierten $\{Co^{II}(L')_2(S)_2\}^{2+}$ Untereinheiten (L': bik oder bim Liganden und S = Lösemittelmoleküle) führte zur *in-situ* Redoxreaktion der Metallionen und damit zu den Viereck-Strukturen **15** und **16** mit den diamagnetischen Untereinheiten $\{Fe^{II}_{LS}\text{-}C{\equiv}N\text{-}Co^{III}_{LS}\}$. Diese zwei Viereck-Strukturen zeigen über den experimentellen Temperaturbereich diamagnetisches Verhalten wobei auch die Bestrahlung mit Licht bei niedrigen Temperaturen keine Veränderung der magnetischen Eigenschaften bewirkte. Dieses Verhalten wird einer ungeeigneten Abstimmung der Redoxpotentiale der Untereinheiten zugeschrieben, die einen diamagnetischen Zustand bevorzugen.

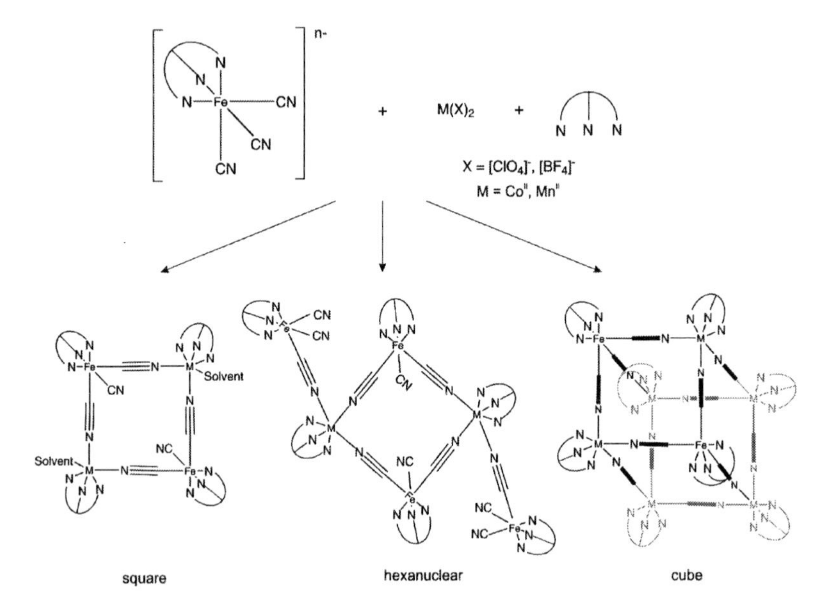

Abbildung 8.3: Schematische Darstellung der Reaktion von $[Fe^{III}(L)(CN)_3]^{n-}$ mit teilweise blockierten $\{M(L)(S)_3\}^{2+}$ Einheiten durch Selbstorganisation.

In Kapitel 6 wurde die Reaktion von **1** mit den teilweise blockierten $\{M(L)(S)_3\}^{2+}$ Untereinheiten ($M = Co^{II}$, Mn^{II}, L = Scorpionat Ligand, S = Lösemittelmolekül) unter verschiedenen Reaktionsbedingungen beschrieben. (Siehe Abbildung 8.3). Interessanterweise erlaubt die Veränderung der Reaktionsbedingungen eine Einflussnahme auf die Zahl der verknüpften Untereinheiten und die Architektur der erhaltenen polymetallischen Systeme. Die Reaktion von **1** mit $\{M(Tpm^*)(S)\}^{2+}$ Einheiten führte zu den vierkernigen quadratischen Komplexen **17** ($M = Co^{II}$) und **18** ($M = Mn^{II}$). Die Reaktion von **1** mit $\{M(Tpe)(S)\}^{2+}$ Einheiten in Acetonitrile/Wasser Mischungen ergab die isostrukturellen, paramagnetischen hexanuklearen $\{Fe_4M_2\}$ Neutralverbindungen **19** ($M = Co^{II}$) und **20** ($M = Mn^{II}$). Es ist bemerkenswert, das die gleichen Edukte in DMF zum vierfach katonischen, paramagnetischen $\{Fe_4Co_4\}$ Heterocuban **21** reagieren. Ferromagnetische Wechselwirkungen zwischen den Metallionen konnten für die {FeCo} Spezies **17**, **19** und **21** beobachtet werden, während die entsprechenden Wechselwirkungen für die {FeMn} Spezies **18** und **20** antiferromagnetischer Natur sind.

Die Reaktion von **1** mit $\{Co(Ttp)(S)_3\}^{2+}$-Untereinheiten in DMF führte zur Bildung der würfelförmigen anionischen $\{Fe_4Co_3^{III}Co^{II}\}$ Verbindung **22**. EDX-Messungen und die Röntgenstrukturanalyse zeigten auf, dass sich das Gegenion (Kalium) als Templat im Zentrum von **22** befindet und somit die Gesamtladung der Verbindung 0 beträgt. Folglich lässt sich **22** stabil und undissoziiert in Lösung bringen, was sich über die fünf aufeinanderfolgenden Redox-Prozesse im gemessenen Cyclovoltammogramm und in den ESI-MS-Untersuchungen in Dichlormethan belegen lässt. Diese Löslichkeit ist in der Cyanidchemie selten und erlaubte die vollständige Charakterisierung von **22** in Lösung, insbesondere die Bestimmung des hydrodynamischen Radius in Dichlormethan mittels Diffusions-^{1}H-NMR-Experimenten. Frisch abfiltrierte Proben von **22** zeigen typische $\chi_M T$ gegen T Kurven bis zur Raumtemperatur. Bei höheren Temperaturen vollzieht **22** einen Übergang, der auf einen thermisch induzierten Elektronentransfer zurückzuführen sein könnte und zu einer drastischen Steigerung des $\chi_M T$-Produkts führt. Die Belichtung der Verbindung bei 20 K resultierte in einem starken Zuwachs der Magnetisierung bei Wellenlängen von 405 nm bis 900 nm. Bei 1313 nm konnte kein Effekt festgestellt werden. Der stärkste Effekt konnte bei einer Belichtung mit 808 nm in der

Eisen(II)-Cobalt(III)-Charge-Transfer-Bande beobachtet werden. Ein teilweise umgekehrter Effekt konnte durch belichten des metastabilen Zustands von **22** bei 532 nm erhalten werden.

Solche Systeme könnten als molekulare Modelle der bekannten photomagnetischen Berliner Blau-Analoga dienen. Weiterführende Studien sollen die Rolle des Alkalikations im Zustandekommen der photomagnetischen Eigenschaften solcher Systeme beleuchten.

Zusammenfassend war es möglich im Rahmen der Cyanid-Scorpionatchemie eine Reihe neuer Cyanid-verbrückter polynuklearer Systeme zu synthetisieren. Einige dieser Verbindungen zeigen thermische bzw. photoinduzierte Schaltbarkeit ihrer magnetischen Eigenschaften. Weiterführende Untersuchungen an diesen Systemen legt Augenmerk auf:

- Eingehende physikalische Untersuchungen (Röntgenstrukturanalyse unter Belichtung und bei tiefen Temperaturen, zeitaufgelöste Spektroskopie) um ein besseres Verständnis der metastabilen Zustände aufzubauen und somit die Möglichkeit zu erhalten die Schalteigenschaften solcher Verbindungen besser steuern zu können.

- Die Verarbeitung von in Lösung stabilen photomagnetischen Schaltern zu Hybridmaterialien, zum Beispiel durch Inklusion in Polymerfilme, Fixierung auf Oberflächen oder auf Nanopartikeln, um neuartige Multifunktionsmaterialien zu erzeugen.

9 Conclusion et perspectives

L'objectif de ces travaux consitait à étendre la famille des composés $[Fe(Tp)(CN)_3]^-$ grâce à la préparation (ou grâce à une synthèse améliorée) de nouveaux complexes *fac*-tricyanidoferrates, ainsi qu'à explorer leur réactivité face à des complexes paramagnétiques, ce afin d'obtenir de nouveaux matériaux moléculaires magnétiques. L'accent a été particulièrement mis sur la préparation de systèmes de faibles dimensions, à savoir des chaînes pouvant se comporter comme des chaînes aimants (SCM = single chain magnet) ou des complexes polynucléaires photomagnétiques.

Au chapitre 3, nous avons décrit la synthèse et la caractérisation extensive de plusieurs nouvelles briques octahédriques de fer(II) et fer(III), à base de ligands cyanures et scorpionates de la forme $[Fe(L)(CN)_3]^{n-}$, dans laquelle L est un dérivé de tris(pyrazolyl)méthane (L = Tpm, Tpe, Tpm*). En effet, les propriétés électroniques et structurelles de ces composés tricyanurés dirigent les propriétés magnétiques des espèces polymétalliques dont ils constituent le squelette. La synthèse de certains de ces complexes a déjà été évoquée dans la littérature (L = Tp*, Ttp and Tpms) ; néanmoins, les données manquantes spectroscopiques et structurelles ont été obtenues au cours de ces travaux, et leurs propriétés ont été comparées avec celles des nouveaux composés tricyanurés. Tous ces complexes présentent une symétrie $C_{3v,}$ dans laquelle le fer occupe un environnement C_3N_3 formé de trois cyanures *C*-coordinés et de trois fragments imines, correspondant au ligand scorpionate respectif.

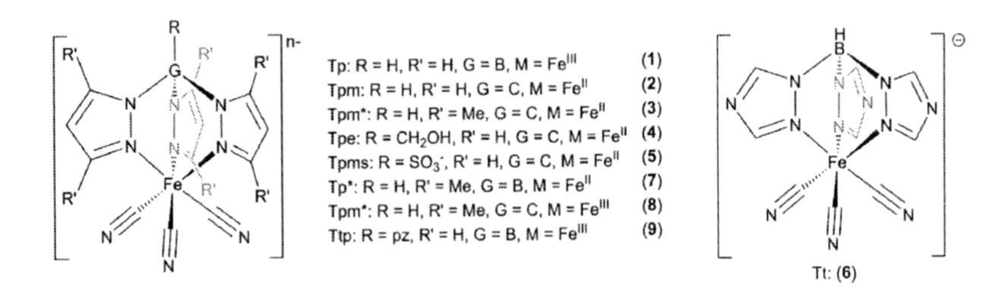

Figure 9.1: Complexes de fer(II) et fer(III) synthétisés, à base de ligands cyanures et scorpionates.

Les études de cyclovoltammétrie ont démontré que **1**, **3/8**, **5**, **6** et **7** subissent des processus redox quasi-irréversibles au niveau du fer en solution dans l'acétonitrile. Une comparaison de leurs spectres infrarouge a permis d'identifier plusieurs éléments-clefs du spectre, dont la fréquence et l'intensité contiennent des informations électroniques et structurelles spécifiques, si bien que la structure des nouvelles espèces polymétalliques inconnues peut être déduite par analyse infrarouge.

Les valeurs *g* de **1** et **7** ont été extraites pas RPE. Pour **8**, une valeur g_{eff} a été extraite des données magnétiques obtenues par magnétométrie SQUID.

Afin de mettre en lumière la distribution de densité du spin le long des ponts cyanures, les composés **1**, **7** et **8** ont été mesurés grâce à la RMN-MAS. Il a été démontré que la transmission des informations magnétiques s'effectue d'abord par polarisation du spin, ce qui conduit à des densités de spin fortement négatives dans les orbitales 2s des atomes de carbone. La densité totale de spin trouvée pour **1** et **7** correspond aux calculs de DFT. La densité du spin détectée dans les atomes d'azote est positive, et résulte d'une délocalisation de spin de l'ion métallique vers les orbitales 2p des azotes.

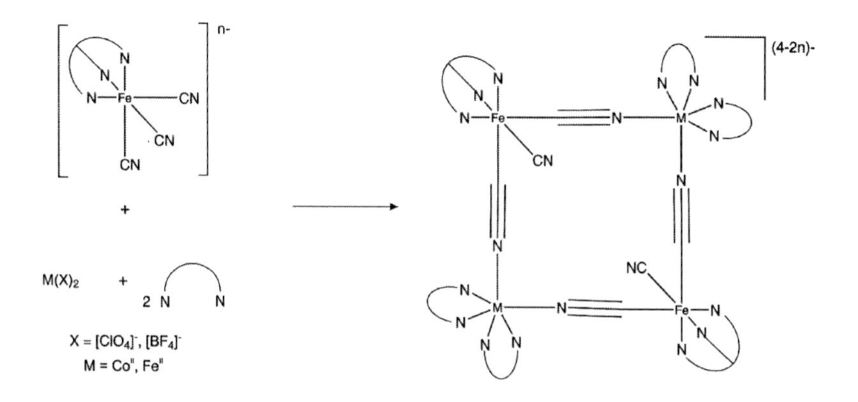

Figure 9.2: Représentation schématique de la réaction de $[Fe^{III}(L)(CN)_3]^{n-}$ avec des briques $\{M(L')_2(S)_y\}^{2+}$ partiellement bloquées par un auto-assemblage.

Au chapitre 4, la réaction de **7** avec les briques moléculaires $\{M(bik)_2(S)_2\}^{2+}$ (M = Co^{II} ou Fe^{II}, S = solvant) partiellement bloquées a conduit à la formation des carrés moléculaires **10**, **11** et **12** (M = Co^{II} (**10** et **11**), Fe^{II} (**12**)) (voir Figure 7.2). **10** et **11** sont constitués de ponts paramagnétiques $\{Fe^{III}_{LS}\text{-C}\equiv\text{N-}Co^{II}_{HS}\}$ et ne subissent pas de transitions de spin induites thermiquement. Il a été démontré que, pour ces complexes, les ions de fer(III) et de cobalt(II) sont soumis des interactions ferromagnétiques. Leurs courbes $\chi_M T$ *vs* T ont été modélisées pour obtenir une estimation des paramètres électroniques qui régissent les propriétés magnétiques (constantes de couplage, paramètres d'anisotropie, paramètres de réduction orbitalaire, valeurs Δ et g). Le carré moléculaire à valence mixte **12** $\{(Fe^{III}_{LS})_2(Fe^{II}_{HS})_2\}$ présente une transition de spin induite thermiquement, d'un état haut-spin vers un état bas-spin, avec une température à mi-transition de $T_{1/2}$ = 227 K. Il a été également démontré qu'un changement d'état de spin photo-induit (effet LIESST) peut également être déclenché à 20 K grâce à une irradiation laser avec des longueurs d'onde entre 405 et 1313 nm. Une efficacité optimale est obtenue entre 700 et 900 nm. L'état métastable haut-spin de **12** est stable jusqu'à 35 K.

Les réactions de **8** avec les ions cobalt(II) et manganèse(II) produisent les nouvelles chaînes 1D magnétiques à ponts cyanures en double zigzag **13** (M = Co^{II}) et **14** (M = Mn^{II}) (chapitre 5). L'étude magnétique montre que les interactions entre les ions fer(III) et cobalt(II) dans **13** sont ferromagnétiques, tandis que des interactions antiferromagnétiques se produisent entre les ions fer(III) et manganèse(II) dans **14**. Malheureusement, aucun comportement de chaîne aimant (SCM : Single Chain Magnet) n'a été observé. La réaction de **8** avec des sous-unités $\{Co^{II}(L')_2(S)_2\}^{2+}$ partiellement bloquées (L' : ligands bik ou bim et S = molécules de solvant) ont conduit à une réaction d'oxydo-réduction *in-situ* entre les ions métalliques, et ont produit les carrés moléculaires **15** et **16**, qui contiennent les paires diamagnétiques $\{Fe^{II}_{LS}\text{-C}\equiv\text{N-}Co^{III}_{LS}\}$. Ces deux carrés moléculaires sont restés diamagnétiques sur toute la gamme de température, et aucun changement de leurs propriétés magnétiques n'a été observé sous irradiation laser à basse température. Cela est attribué au fait que les potentiels redox des deux composés ne coïncident pas, ce qui favorise clairement l'état électronique diamagnétique.

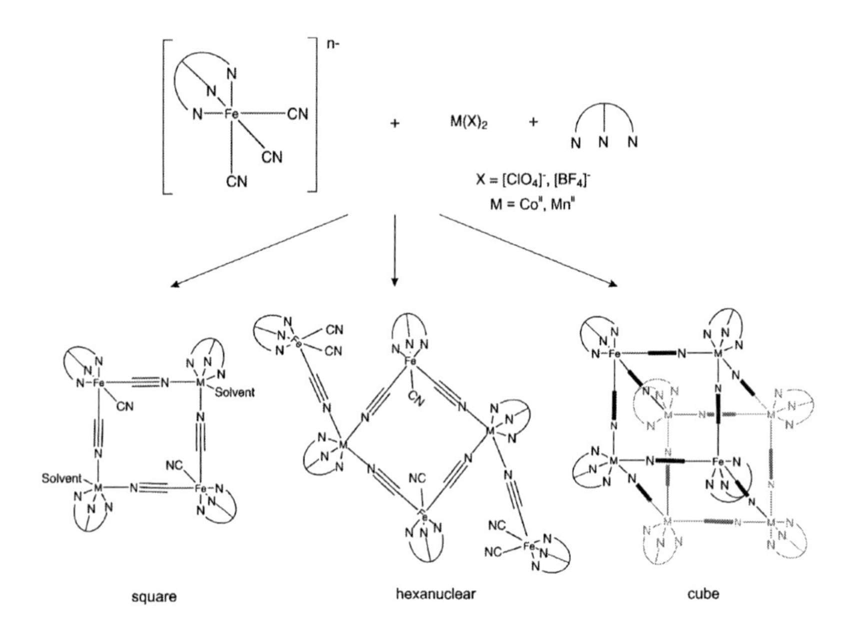

Figure 9.3: Représentation schématique de la réaction de $[Fe^{III}(L)(CN)_3]^{n-}$ avec les unités $\{M(L)(S)_3\}^{2+}$ partiellement bloquées par auto-assemblage.

Au chapitre 6, la réaction de **1** avec les briques moléculaires (M = Co^{II}, Mn^{II}, L = ligand scorpionate, S = molécule de solvant) partiellement bloquées $\{M(L)(S)_3\}^{2+}$ a été explorée dans différentes conditions (voir Figure 9.3). Il est intéressant de constater que le changement des conditions synthétiques permet de moduler la nucléarité et l'architecture des assemblages polymétalliques obtenus. La réaction de **1** avec les briques partiellement bloquées $\{M(Tpm^*)(S)\}^{2+}$ dans des mélanges d'acétonitrile et d'eau a produit les composés hexanucléaires isostructuraux $\{Fe_4M_2\}$ **19** (M = Co^{II}) et **20** (M = Mn^{II}), qui sont neutres et paramagnetiques. De façon intéressante, les mêmes réactifs dans le DMF ont produit le cube moléculaire paramagnétique $\{Fe_4Co_4\}$ **21**, quatre fois positivement chargé. Des interactions ferromagnétiques entre les ions métalliques se produisent dans les composés {FeCo} **17**, **19** et **21**. Ces interactions sont antiferromagnétiques dans les composés {FeMn} **18** et **20**.

La réaction de **1** avec des sous-unités $\{Co(Ttp)(S)_3\}^{2+}$ dans le DMF a conduit à la formation de la boîte moléculaire anionique $\{Fe_4Co_3^{III}Co^{II}\}$ **22**. L'analyse EDX et l'analyse structurelle ont révélé que le contre cation potassium pourrait servir de template à l'intérieur de la boîte, si bien que la charge totale du composé est nulle. Par conséquent, le composé **22** est stable en solution et ne se dissocie pas ; cette stabilité a été révélée par la présence de cinq processus d'oxydo-réduction les uns à la suite des autres en cyclovoltammétrie et par l'analyse du spectre de masse ESI dans le dichlorométhane. Cette propriété, assez rare dans la chimie des cyanures, a permis de caractériser entièrement **22** en solution, et en particulier de déterminer le rayon hydrodynamique de **22** dans le dichlorométhane grâce à la spectroscopie RMN diffusionnelle du proton. Des échantillons fraichement filtrés de **22** présentent des courbes typiques de $\chi_M T$ *vs* T jusqu'à température ambiante. A plus haute température, **22** passe par une transition qui pourrait être due à un transfert d'électron thermo-induit, tandis que la valeur du produit $\chi_M T$ augmente considérablement. L'irradiation au laser de **22** à 20 K déclenche une forte augmentation de l'aimantation, pour des longueurs d'onde allant de 405 à 900 nm. Il n'y a pas d'effets observables pour la longueur d'onde de 1313 nm. La réponse optimale sous irradiation de **22** a été obtenue à 808 nm dans la bande de transfert de charge fer(II)-cobalt(III). Un effet inverse partiel a pu être obtenu par l'irradiation de l'état métastable de **22** à 532 nm. Ces systèmes pourraient être utilisés en tant que modèles moléculaires pour les Analogues de Bleus de Prusse photomagnétiques. Des analyses ultérieures porteront notamment sur l'examen du rôle joué par l'ion alcalin dans les propriétés photomagnétiques de ces systèmes.

Plus généralement, la chimie des scorpionates et des cyanures a permis de synthesiser de plusieurs nouveaux systèmes polynucléaires cyanurés, parmi lesquels certains présentent une commutabilité thermo-induite et photo-induite de leurs propriétés magnétiques. Les prochaines recherches sur ces systèmes comporteront notamment :

- des études physiques approfondies (diffraction aux rayons X sous irradiation à basse température, spectroscopie résolue dans le temps), afin de mieux comprendre l'état métastable, et par conséquent de mieux rationaliser les propriétés de commutabilité de ces systèmes.
- la mise en forme de système commutables stables en solution et photomagnétiques en des matériaux hybrides : inclusion dans des films de polymères, immobilisation sur des surfaces ou sur des nanoparticules afin de produire des matériaux multifonctionnels.

10 Experimental section

If not stated otherwise, all syntheses were carried out under the hood without any inert atmosphere. Air- and moisture-sensitive compounds were synthesised using Schlenk-line techniques under extra purified argon atmosphere (concentrated sulphuric acid and phosphor pentoxide as drying agent). If needed, air- and moisture sensitive compounds were stored in a Glovebox under argon atmosphere (MB150B-G-II and Labmaster 130, Fa. M. Braun types).

10.1 Reagents and solvents

If not stated otherwise for syntheses carried out without any inert atmosphere, all chemicals were used as received.

If necessary, the solvents were dried using standard protocols and kept under inert atmosphere.[192] Diethyl ether, *n*-pentane, *n*-hexane and THF were refluxed over potassium (for toluene, over sodium) several days and benzophenone was used as indicator. Acetonitrile was treated the same way, with CaH_2 as a drying agent, and was stored over activated molecular sieve (3 Å).

The conduction salt (tetrabutylammonium hexafluorophosphate) used for the cyclic voltammetry was dried with neutral aluminium oxide (Al_2O_3, Brockmann I), several times recrystallised from absoluted ethanol and dried several hours in high vacuum.

The following ligands are literature-known and were synthesised using the literature protocols given as reference: Tp,[1–3] Tp*,[1–3] Ttp,[1–3] Tt,[193,194] Tpm,[14] Tpm*,[14] Tpe,[26] Tpms,[20,195] bik[98,196] and bim.[98,196] The following metal complexes are literature-known and were synthesised using the literature protocols given as reference: $PPh_4[Fe^{III}(Tp)(CN)_3]$,[98,114] $[Fe^{II}(Tpe)_2](OTf)_2$,[10,37] $(PPh_4)_2[Fe^{II}(Tpms)(CN)_3]$.[116]

10.2 Analytic and spectroscopic methods

Elemental Analysis

Elemental analyses (C, H, N, S) were performed using a "Elementar Vario EL" instrument by sample burning analysis. The values are given in mass percentages.

Melting point

When possible, decomposition was monitored with a ThermoFischer Scientific device and the values are uncorrected. For compounds with too intensive colour to be able to detect a change, the highest temperature reached in the SQUID magnetometer without change in the magnetic data was indicated.

Mass spectroscopy

The Electrospray ionisation (ESI) mass spectrometry was measured with a FTICR (Fourier Transform Ion Cyclotron Resonance) IonSpec mass spectrometer with magnets of 7 Tesla (Cryomagnetics, Inc). The sample inlet of the ESI-source was set to a potential of 3.20 kV while the quarz capillary covered with metal was set to the same potential. In order to get a better signal-to-noise, the ions which were produced were collected for 4 seconds in an hexapol before transfer to the ICR cell.

InfraRed spectroscopy

The infrared spectra were collected between at least 600 and 4000 cm^{-1} at room temperature using a Tensor 27 Bruker instrument (Paris), a VERTEX 70 Bruker spectrometer (Karlsruhe) or an Alpha Bruker spectrometer placed in the glovebox (air-sensitive compounds – Karlsruhe). All above mentioned spectrometer are working in ATR "Attenuated Total reflexion" mode. The intensity of the bands are reported using the following subdivisions: very strong (vs), strong (s), middle strong (m), weak (w), very weak (vw). When necessary, the mentions shoulder (sh) and broad (br) were also employed.

Nuclear Magnetic Resonance spectroscopy

Solution NMR samples were prepared in 5 mm o. d. glass tubes, using deuterated solvents as received. When necessary, this operation was carried out under inert atmosphere using purified deuterated solvents (see page 223). In this case, the samples are blowtorch sealed. Solution NMR spectra were recorded on two Bruker Avance 300 and a Bruker Avance 400 (300 MHz, 300 MHz and 400 MHz). The chemical shifts δ are expressed in ppm (parts per million) and are referenced, following IUPAC recommendations,[157] in respect to TMS (^{1}H, ^{13}C), $CFCl_3$ (^{19}F), H_3PO_4 (^{31}P) and NH_3 (^{15}N). When possible, the solvent signals were used as internal secondary references.[197,198] The multiplicity of the NMR signals are given using the following abbreviations: s = singlet, d = doublet, t = triplet, q = quartet, m = multiplet, dd = doublet of doublets, dt = doublet of triplets and br = broad signal. The value of the coupling constant *J* is given in Herz (Hz) as an absolute value.

High and low temperature experiments were performed on calibrated spectrometer with a 4% methanol in MeOD-d_4 sample for low temperature corrections and a 80% glycol in DMSO-d_6 sample for high temperature corrections. Particular attention was given to paramagnetic samples so that they are given time to reach thermal equilibrium.

All diffusion processing and molecular size estimations were performed by using the DiffAtOnce software package available at www.diffatonce.com. Gradients of the SMSQ10.100 form were used, and were calibrated using HDO in D_2O ($D = 1.902 \cdot 10^{-9}$ $m^2 \cdot s^{-1}$).

Solid-state MAS-NMR samples were prepared from microcrystalline samples packed in zirconium oxide rotors. The size of the rotor was adapted to the spectrometer probe head, it was tilted at the magic angle ($\theta \approx 54.7°$) and spinned at spectrometer-dependant high frequency. Data were collected on a Bruker Avance 500 spectrometer and on a Bruker Avance 400 spectrometer equipped each with a 4-BL MAS-NMR probe head (4 mm diameter rotor – 400 and 500 MHz – max. 14 kHz), on a Bruker Avance 300 spectrometer equipped with a 4-MQ MAS-NMR probe head (4 mm diameter rotor – 300 MHz – max. 14 kHz), or on a Bruker Avance 700 equipped with a 1.3-BL probe head (1.3 mm diameter rotor – 700 MHz – max. 67 kHz), depending on the nature of the nucleus and the properties of the samples.

For paramagnetic compounds, some nickelocene was added as internal temperature probe and sample temperature was tuned using a BCU Xtreme cooling unit. About 100 mg of sample was needed in case of a 4 mm rotor, but only 10 mg for a 1.3 mm rotor.

The Herzfeld-Berger Analysis was carried out using the module "Solids Line Shape Analysis" from Bruker's software package Topspin. It allows the extraction of experimental isotropic shifts $\delta_{T,iso}^{exp} = (\delta_{T,xx}^{exp} + \delta_{T,yy}^{exp} + \delta_{T,zz}^{exp})/3$, where $\delta_{T,xx}^{exp}$, $\delta_{T,yy}^{exp}$ and $\delta_{T,zz}^{exp}$ are the principal components of the chemical shift tensor (ordered as $|\delta_{T,zz}^{exp} - \delta_{T,iso}^{exp}| \geq |\delta_{T,xx}^{exp} - \delta_{T,iso}^{exp}| \geq |\delta_{T,yy}^{exp} - \delta_{T,iso}^{exp}|$). Further tensor describing parameters include the anisotropy $\Delta\delta_T^{exp} = \delta_{T,zz}^{exp} - (\delta_{T,xx}^{exp} + \delta_{T,yy}^{exp})/2$, and the asymmetry $\eta_T^{exp} = (\delta_{T,yy}^{exp} - \delta_{T,xx}^{exp})/(\delta_{T,zz}^{exp} - \delta_{T,iso}^{exp})$. The line broadening factor (LB) accounts for the broadness of the signals.

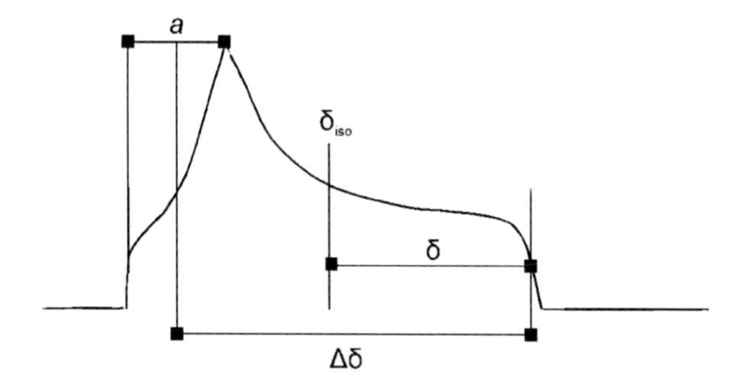

Figure 10.1: Haeberlean convention.

Cyclovoltametry

The cyclovoltammetric studies were performed with a setup from the firm METROHM using a potentiostat PGSTAT101 controled by the software NOVA within a glovebox with argon atmosphere. The working electrode consisted in a platinum rod (surface 0.785 cm^2). The auxiliary electrode was a platinum wire with a 1 mm diameter. The (pseudo)reference electrode was a silver wire. All given potentials are internally referenced *vs* ferrocene/ferrocenium (0.352 V *vs* Ag/AgCl). The conduction salt was tetrabutylammonium hexafluorophosphate ($[Bu_4N][PF_6]$). The programm ORIGIN PRO 10 was used to analyse the data.

UV/Visible Spectroscopy

The UV-visible spectra were acquired with a UV-Visible Spectrophotometer Varian Cary 100 Scan in solution contained in quartz cuvettes (l = 1 cm). The spectra of the compound were obtained by subtraction of the pure solvent spectrum.

EPR Spectroscopy

EPR measurements were performed using a Bruker ESP300 E spectrometer at a working frequency of 9.42 GHz (X-band) and a 33 GHz (Q-band). Calibrated silica tubes (suprasil quality grade) were filled with dry ground sample. The EPR spectra were recorded at 4 K, using a cooled helium flow device.

SQUID Magnetometry

All magnetic and photomagnetic data were collected with Quantum Design SQUID Magnetometers (MPMS-5S and MPMS XL-7). Variable temperature experiments were performed over a 2 – 400 K temperature range and the molar succeptibility χ_M was recorded. M is the magnetisation of the sample, while H is the applied magnetic field. Magnetic fields from 250 to 10000 Oe, depending on the mass of the sample, were applied and are mentioned in the caption of the figures. In order to prevent the loss of lattice solvent molecules, fresh samples were introduced at 200 K under helium flow and frozen before purging the airlock. The measurements were performed from 200 K to 2 K, then from 2 K to higher temperature with a sweep rate of 2 $K \cdot min^{-1}$.

Photomagnetic measurements were performed using a sample holder equipped with an optical fiber. In a typical experiment, a very small amount of sample (0.1 – 0.5 mg of ground crystals) was deposited on an adhesive pad. Laser sources were in the visible range at 405, 532, 635, 808 nm and in the Near InfraRed (1313 nm). The end of the optical fiber was located at 50 mm above the sample. In these experimental conditions, the estimated light powers were 5 (405 nm), 10 (532 nm), 12 (635 nm) and 6 $mW \cdot cm^{-2}$ (808 nm). The temperature was set to 20 K to minimise sample heating by light irradiation. A correction corresponding to the diamagnetic contribution of the constituent atoms and the residual diamagnetic signal from the sample holder was applied to the experimental data.

X-ray diffraction analysis

Unless written otherwise, all structural data were obtained at a temperature of 200 K.

A single crystal of each compound was selected, mounted onto a Hamilton cryoloop using Paratone N oil and glue to avoid solvent loss (Paris) or mounted on a glass capillary using perfluorined polyether oil (Karlsruhe) and placed in the cold flow produced with an Oxford Cryocooling device.

In Paris, the intensity data were collected with a Bruker Kappa APEX II with graphite-monochromated Mo Kα radiation source (λ = 0.71073 Å). Data collection was performed with APEX2 suite. Unit cell parameters refinement, integration and data reduction were carried out with the SAINT program. SADABS was used for multi-scan absorption corrections. The structure were solved by direct methods with SHELXS 97 and refined by full-matrix-least-squares methods using SHELXL 97. Almost all non-hydrogen atoms were refined anisotropically; only atoms of solvent molecules or disordered parts were refined isotropically. Hydrogen atoms were placed at calculated positions and refined with a "riding model".

In Karlsruhe, the intensity data were collected with a STOE IPDS II or a STOE STADI 4 diffractometer with a monochromatic radiation source Mo Kα (λ = 0.71073 Å) and a cooling device (200 K). The structures were solved using the SHELXTL (version 6.12) software packet using either the direct method or the Patterson method and step-by-step interpretation of the Fourier map with the full-matrix-least-square refinement method (against F or F^2).

The following quality factors were used:

$$R1 = \frac{\Sigma||F_0|-|F_c||}{\Sigma|F_0|};\ wR2 = \sqrt{\frac{\Sigma w(F_0^2-F_c^2)}{\Sigma wF_0^2}}$$

With $w = (\sigma^2 F_0^2 + a^2P^2 + bP)^{-1}$ and $P = \frac{1}{3}(F_0^2 + 2F_c^2)$

10.3 Syntheses of building blocks

$PPh_4[Fe^{II}(Tpm)(CN)_3]$ (PPh_4[2])

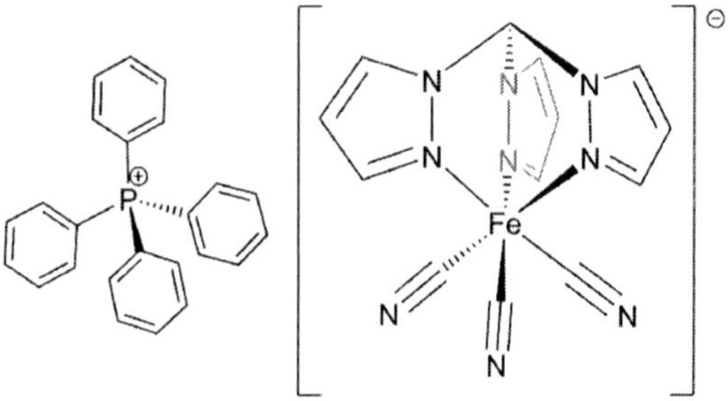

$[Fe^{II}(Tpm)_2](SO_4)$ (0.684 mg, 1.23 mmol) was dissolved in H_2O (20 mL). A solution of three equivalents of NaCN (0.179 mg, 3.69 mmol) in 5 mL water was added dropwise to the stirred resulting purple solution at room temperature. During the addition, the solution turned orange, and a solid precipitated. This solid redissolved before the end of the cyanide addition to produce an orange solution which was further stirred overnight at room temperature. Addition of one equivalent of tetraphenylphosphonium chloride (0.462 mg, 1.23 mmol) to the aqueous solution followed by slow evaporation of said solution produced crystals suitable for X-ray diffraction analysis. Yield: 65 mg (8%)

^{1}H NMR (400.1 MHz, 298 K, methanol-d_4, $[PPh_4]^+$): δ (ppm) = 6.46 (t, J_{HH} = 2 Hz, 3H, 4-pz-C*H*), 7.82 (m, 18H, P*Ph*$_4^+$), 7.98 (t, 4H, P*Ph*$_4^+$), 8.26 (d, J_{HH} = 2 Hz, 3H, pz-C*H*), 8.28 (s, 3H, pz-C*H*), 9.26 (s, 1H, $C^{apical}H$). In D_2O as Na^+ (300.1 MHz): δ = 6.41 (dd, J_{HH} = 2.9 and 2.2 Hz, 3 H, 4-pz-C*H*), 8.11 (dd, J_{HH} = 2.2 and 0.6 Hz, 3 H, 5-pz-C*H*), 8.18 (dd, J_{HH} = 2.9 and 0.7 Hz, 3 H, 3-pz-C*H*), 9.14 (s, disappearing, 1H).

^{13}C NMR (100.6 MHz, 298 K, methanol-d_4, $[PPh_4]^+$): δ (ppm) = 74.9 (s, 1C, C^{apical}), 107.5 (s, 3C, 4-pz-*C*H), 118.0 (d, $^1J_{CP}$ = 89.9 Hz, 4C, P*Ph*$_4^+$), 130.2 (d, $^3J_{CP}$ = 12.9 Hz, 8C, P*Ph*$_4^+$), 132.6 (s, 3C, 3-pz-*C*H), 134.5 (d, $^2J_{CP}$ = 10.5 Hz, 8C, P*Ph*$_4^+$), 135.3 (d, $^4J_{CP}$ = 3.1 Hz, 4C, P*Ph*$_4^+$), 148.4 (s, 3C, 3-pz-*C*H), 170.9 (s, 3C, *C*N). In D_2O as Na^+ (300.1 MHz, 298 K): δ (ppm) = 148.3 (s, 3C, 5-pz-*C*H), 108.2 (s, 3C, 4-pz-*C*H), 133.6 (s, 3C, 3-pz-*C*H), 175.5 (s, 3C, *C*N).

^{31}P NMR (161.9 MHz, 298 K, methanol-d_4): 23.2 ppm ([*P*Ph$_4$]$^+$).

Elemental analysis (%): calculated for $C_{37}H_{30}N_9FeP \cdot 3.25\, H_2O$: C 59.57, H 4.93, N 16.90; found: C 59.72, H 4.53, N 16.56.

IR (ATR, ν, cm^{-1}): 406 (w), 431 (w), 456 (w), 525 (vs), 557 (w), 608 (w), 646 (vw), 689 (m), 723 (s), 741 (m), 756 (w), 765 (w), 784 (w), 842 (vw), 863 (vw), 880 (vw), 984 (vw), 997 (vw), 1047 (w), 1088 (m), 1110 (m), 1162 (vw), 1183 (vw), 1224 (vw), 1238 (vw), 1284 (w), 1316 (vw), 1341 (vw), 1402 (w), 1440 (w), 1480 (vw), 1514 (vw), 1587 (vw), 1654 (vw), 2045 (m), 2054 (w), 2064 (w), 2997 (vw), 3052 (vw), 3135 (vw), 3150 (vw), 3394 (vw), 3461 (vw).

Melting point: ~180°C (decomposition).

Na[Fe^{II}(Tpm*)$(CN)_3$] (Na[3])

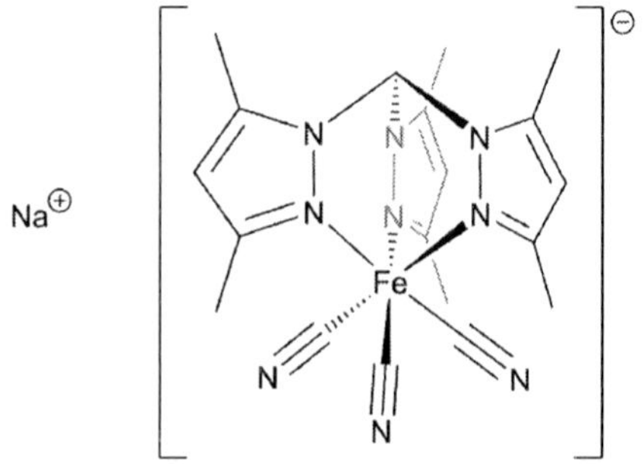

Solid Tpm* (0.298 g, 1.0 mmol) was added to a solution of $FeCl_2$ (0.126 g, 1.0 mmol) and ascorbic acid (a pinch of spatula) in 20 mL methanol. The resulting brown suspension was further stirred one hour to give a brown solution. It was added dropwise to a protected from light solution of NaCN (0.162 g, 3.3 mmol) in 10 mL methanol. The solution turned immediately yellow-red, and some solid precipitated. The reaction mixture was then stirred 16 hours. The solvent was then removed under low pressure and the residue dissolved in 30 mL H_2O. The unsoluble green solid was filtered off (iron-cyanide oligomers). Water was removed from filtrate and the residue was redissolved in 50 mL ethanol. The grey insoluble compound (hexacyanidometallate, NaCl) was filtered off. Ethanol was removed from filtrate to afford a yellow compound. Suitable crystals of compound [**3**]$^-$ for X-ray diffraction analysis were obtained as $[PPh_4]^+$ salt by cation metathesis in water and slow evaporation of the resulting solution.

Yield (Na^+ species): 0.355 g, 78%.

^{1}H NMR (300.1 MHz, 298 K, methanol-d_4): δ (ppm) = 2.59 (s, 9H, 5-pz-CH_3), 2.79 (s, 9H, 3-pz-CH_3), 6.06 (s, 3H, 4-CH_{pz}), 7.86 (s, 1H, C$^{apical}H$).

^{13}C NMR (75.5 MHz, 298 K, methanol-d_4): δ (ppm) = 9.2 (s, 3C, 5-pz-*C*H3); 14.6 (s, 3C, 3-pz-*C*H3); 67.1 (s, 1C, C^{apical}), 108.1 (s, 3C, 4-*C*H_{pz}), 140.5 (s, 3C, 5-C_{pz}-Me), 158.7 (s, 3C, 3-C_{pz}-Me), 170.5 (s, 3C, -*C*N).

^{15}N NMR (40.5 MHz, 298 K, methanol-d_4): no signal detected.

IR (ATR, ν, cm^{-1}): 632 (s), 673 (m), 703 (vs), 794 (m), 802 (m), 812 (m), 865 (s), 920 (vw), 978 (w), 1043 (m), 1091 (w), 1137 (w), 1156 (vw), 1262 (s), 1308 (s), 1396 (s), 1411 (s), 1449 (m), 1462 (s), 1568 (m), 2048 (vs), 2070 (s), 2925 (vw), 2968 (vw), 3144 (w) 3266 (br, w), 3373 (br, w).

Melting point: ~170°C (destruction)

$PPh_4[Fe^{II}(Tpe)(CN)_3]$ ($PPh_4[4]$)

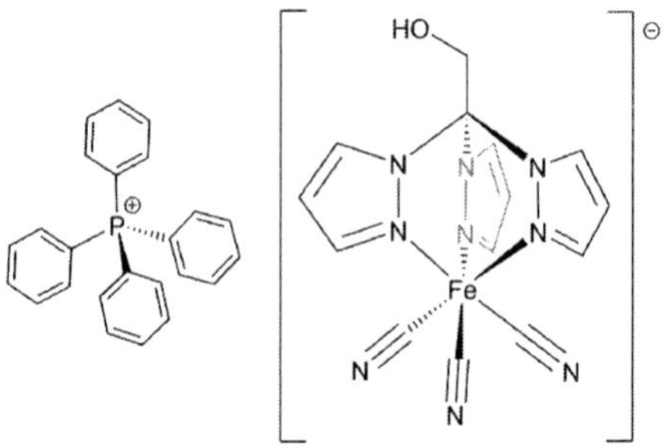

$[Fe^{II}(Tpe)_2]$ (3.00 g, 3.56 mmol) and NaCN (0.611 g, 12.46 mmol) were suspended in 30 mL isopropanol under light exclusion. The resulting red suspension was refluxed 16 hours. The resulting sand yellow solid was filtered off and washed with isopropanol and acetonitrile. The resulting yellow powder was then dissolved in 30 mL H_2O and reacted with one equivalent of PPh_4Cl salt to afford several crops of yellow microcrystals of $PPh_4[Fe^{II}(Tpe)(CN)_3]$. Suitable crystals for X-ray diffraction analysis were obtained by slow evaporation of a acetonitrile/water (3:1) mixture of $PPh_4[Fe^{II}(Tpe)(CN)_3]$.

Yield: 1.35 g (94.6%) as sodium salt.

Melting point: ~180°C (destruction)

^{1}H NMR (300.1 MHz, 298 K, methanol-d_4): δ = 5.55 (s, 2H, CH_2OH), 6.42 (bad resolved m, 3H, 4-pz-*CH*), 7.72 (m, 4H, PPh_4^+), 7.77 (m, 16H, PPh_4^+), 8.24 (s, 3H, 5-pz-*CH*), 8.33 (s, br, 3H, 3-pz-*CH*).

^{13}C NMR (75.5 MHz, 298 K, methanol-d_4): δ = 61.1 (s, 1C, CH_2OH), 84.6 (s, 1C, C^{apical}), 108.5 (s, 3C, 4-pz-*C*H), 119.3 (d, $^{1}J_{CP}$ = 90.6 Hz, 4C, PPh_4^+), 131.6 (d, $^{3}J_{CP}$ = 12.9 Hz, 8C, PPh_4^+), 135.8 (d, $^{2}J_{CP}$ = 10.2 Hz, 8C, PPh_4^+), 136.7 (d, $^{4}J_{CP}$ = 3.7 Hz, 4C, PPh_4^+), 149.2 (s, 3C, 5-pz-*C*H), 172.1 (s, 3C, *C*N).

^{15}N NMR (30.4 MHz, 298 K, methanol-d_4): δ = 212.9 (s, 3N, 1-pz-*N*), 253.5 (s, 3N, 2-pz-*N*).

Elemental analysis (%): calculated for $C_{38}H_{32}N_9OFeP \cdot 2\ H_2O$: C 60.57, H 4.82, N 16.73; found: C 60.27, H 4.46, N 17.05.

IR (ATR, ν, cm^{-1}): 620 (m), 691 (vs), 722 (vs), 756 (vs), 764 (vs), 865 (m), 924 (w), 962 (vw), 997 (w), 1028 (vw), 1051 (w), 1067 (w), 1094 (s), 1110 (vs), 1185 (vw), 1208 (w), 1218 (m), 1279 (w), 1318 (m), 1336 (w), 1396 (w), 1415 (w), 1440 (m), 1484 (vw), 1519 (vw), 1586 (vw), 1640 (vw), 2047 (vs), 2054 (vs), 2068 (m), 2870 (vw), 2998 (vw), 3058 (w), 3088 (w), 3109 (w), 3122 (w), 3149 (vw), 3456 (br, w).

$PPh_4[Fe^{III}(Tp^*)(CN)_3]$ (PPh_4[7])

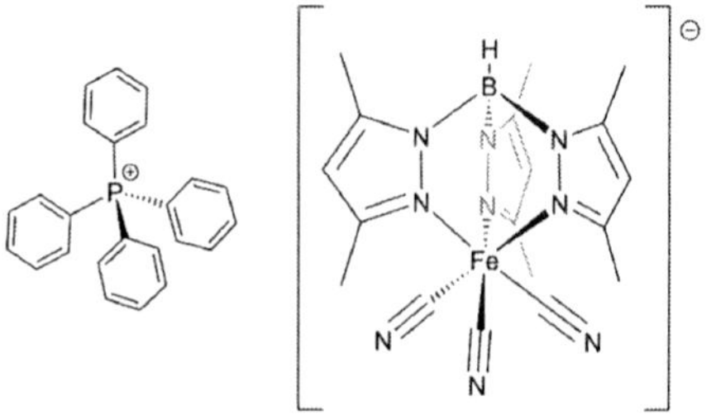

A degased solution of K[Tp*] (0.319 mg, 1.0 mmol) in 5 mL methanol was added dropwise to a methanolic solution of $FeCl_2 \cdot 4\,H_2O$ (1.0 mmol in 15 mL) The violet suspension was stirred for one hour, then added dropwise to a methanolic solution of NaCN (0.162 mg, 3.3 mmol). The resulting red suspension was stirred at room temperature overnight and the methanol was evacuated to dryness. The red resulting solid was redissolved in acetonitrile and filtered. Crystals of Na[**7**] were produced by slow evaporation of the acetonitrile solution. Crystals of PPh_4[7] suitable for X-ray diffraction analysis were produced in 1-2 weeks by slow evaporation of an acetonitrile solution of Na[7] in which was added one equivalent of tetraphenylphosphonium chloride and a small amount of water.

Yield: 0.144 mg (Na^+) (32%).

ESI-MS *m/z* (%): 431.14 (100) $[Fe^{III}(Tp^*)(CN)_3]^-$, 339.13 (100), $[PPh_4]^+$.

Elemental analysis (%): calculated for $C_{42}H_{42}BFeN_9P \cdot CH_3CN \cdot 0.5H_2O$: C64.41, H 5.65, N 17.07; found: C 64.71, H 5.43, N 16.79.

Melting point: ~190°C (decomposition).

IR (ATR, ν, cm^{-1}): 616 (vw), 646 (s), 689 (vs), 719 (vs), 755 (m), 789 (m), 804 (w), 817 (w), 867 (w), 884 (vw), 931 (vw), 996 (m), 1050 (m), 1063 (s), 1107 (vs), 1161 (vw), 1185 (m), 1205 (s), 1309 (vw), 1371 (m), 1388 (m), 1416 (s), 1434 (s), 1449 (m), 1483 (w), 1543 (s), 1585 (vw), 2119 (w), 2543 (vw), 2934 (vw), 2979 (vw), 3062 (vw), 3085 (vw).

$[Fe^{III}(Tpm^*)(CN)3]$ (8)

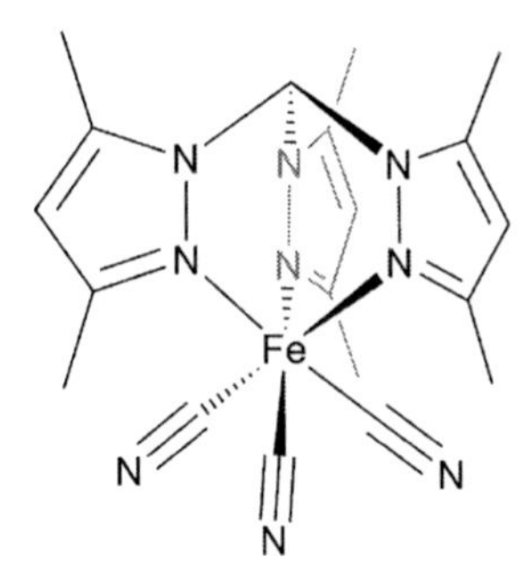

PPh_4[**3**] (0.348 g, 0.45 mmol) was dissolved in 45mL dry acetonitrile under inert conditions. [Fc][PF_6] (0.149 g, 0.45 mmol) was dissolved in 10 mL dry acetonitrile. The resulting midnight blue solution of ferrocenium was added dropwise to the yellow solution of PPh_4[**3**] under light exclusion conditions. The resulting brown suspension was further stirred 16 hours, before the solid was filtered, washed with about 5 mL acetonitrile to afford a golden powder. Crystals suitable for X-ray diffraction analysis were obtained by slow evaporation of an acetonitrile/water 4:1 solution within days.

Yield: 0.178 g (91.6%).

Elemental analysis (%): calculated for $C_{19}H_{22}N_9Fe \cdot H_2O \cdot 2\ CH_3CN$: C 51.89, H 5.679, N 28.94; found: C 51.55, H 4.81, N 28.85.

Melting point: ~245°C (decomposition).

^1H NMR (δ, 298 K, methanol-d_4): δ = 46.26 (s, 1 H), 38.50 (s, 9 H), 0.72 (s, 9 H), -2.80 (s, 3 H).

IR (ATR, ν, cm^{-1}): 205 (w), 213 (w), 227 (vw), 280 (vs), 322 (vw), 380 (vs), 414 (s), 471 (w), 509 (m), 532 (vw), 564 (vw), 595 (vw), 633 (vw), 701 (vs), 787 (s), 802 (m), 820 (m), 862 (vs), 921 (m), 987 (m), 1034 (w), 1050 (s), 1111 (w), 1140 (vw), 1154 (vw), 1254 (vs), 1299 (vs), 1382 (s), 1394 (s), 1405 (vs), 1456 (vs), 1557 (m), 1634 (vw), 2128 (w), 2882 (vw), 2927 (vw), 3143 (vw).

10.4 Syntheses of polynuclear complexes

{[Fe^{III}(Tp*)(CN)$_3$]$_2$[Co^{II}(bik)$_2$]$_2$}(ClO_4)$_2$ · 2 H_2O (10)

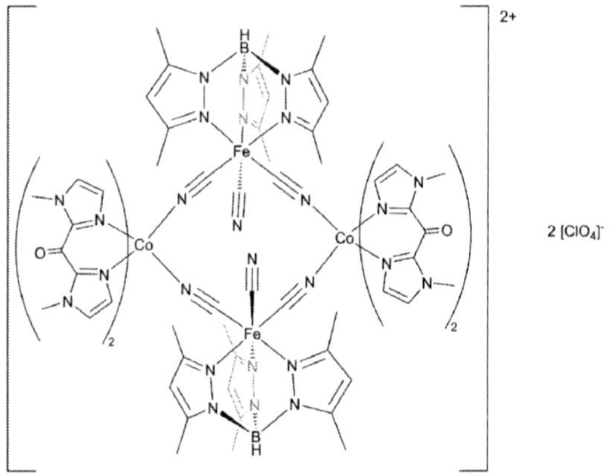

$Co^{II}(ClO_4)_2 \cdot 6\ H_2O$ (38 mg, 0.1 mmol) and bik ligand (38 mg, 0.2 mmol) were dissolved in 15 mL of a acetonitrile/water (4/1) mixture. The resulting yellow solution was added to a stirred red solution of Na[7] (46 mg, 0.1 mmol) in 15 mL of the same mixture of solvent. The red solution was further stirred about 40 minutes before being filtered. Slow evaporation of the reaction mixture produced red crystals suitable for X-ray diffraction analysis.

Yield: 28 mg (14.4%)

IR (ATR, ν, cm^{-1}): 215 (vs), 260(s), 282 (vs), 371 (s), 388 (s), 418 (s), 468 (s), 499 (s), 533 (m), 572 (m), 607 (s), 622 (vs), 642 (s), 651 (s), 691 (s), 724 (m), 774 (s), 788 (vs), 814 (s), 866 (m), 895 (vs), 931 (w), 950 (m), 982 (w), 1059 (vs), 1078 (s), 1102 (s), 1170 (w), 1203 (m), 1248 (vw), 1293 (m), 1373 (s), 1412 (vs), 1448 (m), 1484 (m), 1541 (w), 1639 (s), 2133 (vw), 2149 (w), 2159 (w), 2539 (vw), 2794 (vw), 2922 (vw), 2938 (vw), 2961 (vw), 2977 (vw), 2991 (vw), 3041 (vw), 3101 (vw), 3135 (w), 3162 (vw), 3261 (w), 3413 (m), 3572 (w).

$\{[Fe^{III}(Tp^*)(CN)_3]_2[Co^{II}(bik)_2]_2\}(BF_4)_2$ (11)

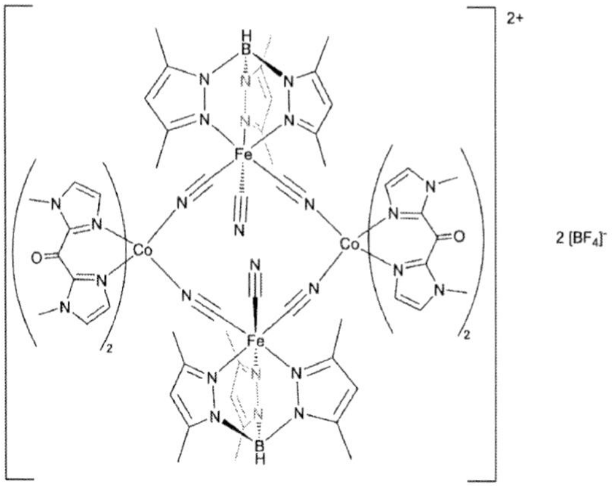

$Co^{II}(BF_4)_2 \cdot 6\,H_2O$ (34 mg, 0.1 mmol) and bik ligand (38 mg, 0.2 mmol) were dissolved in 15 mL of a acetonitrile/water (4:1) mixture. The resulting yellow solution was added to a stirred red solution of Na[7] (46 mg, 0.1 mmol) in 15 mL of the same mixture of solvent. The red solution was further stirred about 40 minutes before being filtered. Slow evaporation of the reaction mixture produced red crystals, but their quality was too low for X-ray diffraction analysis.

Yield: 13 mg (7%)

IR(ATR, ν, cm^{-1}): 218 (s), 266 (m), 283 (s), 336 (vw), 372 (m), 389 (m), 417 (s), 436 (s), 467 (m), 500 (m), 522 (m), 572 (w), 607 (m), 643 (m), 652 (s), 691 (m), 725 (m), 775 (s), 790 (vs), 813 (m), 869 (m), 895 (vs), 950 (m), 1050 (br, vs), 1059 (vs), 1088 (s), 1099 (s), 1134 (w), 1170 (m), 1204 (m), 1249 (vw), 1293 (m), 1374 (s), 1414 (vs), 1448 (m), 1484 (m), 1542 (w), 1639 (s), 2133 (vw), 2150 (w), 2160 (vw), 2539 (vw), 2806 (vw), 2861 (vw), 2935 (vw), 2962 (vw), 2981 (vw), 3040 (vw), 3057 (vw), 3108 (vw), 3140 (vw), 3160 (vw), 3261 (vw), 3420 (w), 3605 (vw).

Melting point: >400 K.

$\{[Fe^{III}(Tp^*)(CN)_3]_2[Fe^{II}(bik)_2]_2\}(ClO_4)_2 \cdot 2\ H_2O$ (12)

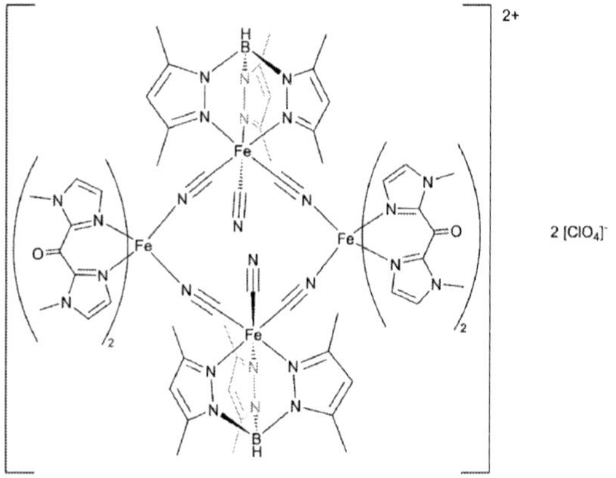

$Fe^{II}(ClO_4)_2 \cdot x\ H_2O$ (19 mg, 0.05 mmol) and bik ligand (19 mg, 0.1 mmol) were dissolved in 15 mL of a methanol/water (5/1) mixture. The resulting deep dark blue solution was added to a stirred red solution of Na[**7**] (23 mg, 0.05 mmol) in 15 mL of the same mixture of solvent. The purple solution was further stirred about 10 minutes before being filtered. Slow evaporation of the reaction mixture produced carmine red crystals suitable for X-ray diffraction analysis.

Yield: 40.1 mg (39.6%).

Elemental analysis (%): calculated for $C_{72}H_{84}N_{34}B_2Cl_2Fe_4O_{12} \cdot 5\ H_2O$: C 42.73, H 4.68, N 23.53; found: C 42.88, H 4.44, N 23.49.

IR (ATR, ν, cm^{-1}): 607 (s), 623 (s), 648 (m), 692 (m), 724 (w), 767 (m), 787 (s), 814 (m), 869 (w), 898 (vs), 950 (w), 987 (vw), 1054 (sh, s), 1064 (vs), 1093 (vs, br), 1171 (w), 1206 (m), 1253 (vw), 1292 (m), 1371 (s), 1385 (m), 1419 (vs, br), 1445 (m), 1488 (w), 1524 (vw), 1541 (w), 1634 (m), 2132 (vw), 2147 (vw), 2160 (vw), 2538 (vw), 2930 (vw), 2961 (vw), 3130 (vw), 3425 (br, vw), 3606 (br, vw).

Melting point: > 365 K

$\{\{[Fe^{III}(Tpm^*)(CN)_3]_2[Co^{II}(H_2O)_2]\}(ClO_4)_2 \cdot 2\,H_2O\}_\infty$ (13)

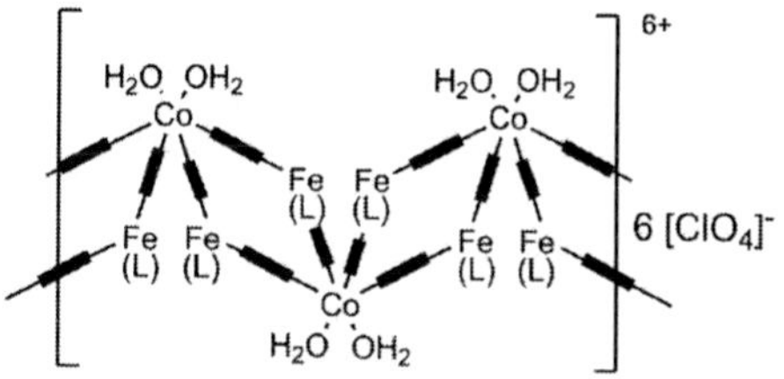

To a stirred orange reddish solution of **8** (0.1 mmol, 43 mg) in 10 mL of an acetonitrile/water (4:1) mixture, was added dropwise 2 mL of a pink solution of $Co^{II}(ClO_4)_2 \cdot 6\,H_2O$ (0.5 mmol, 183 mg) in the same solvent mixture. The resulting orange red solution was further stirred half an hour, before filtration. Slow evaporation of the solvent led after three weeks to yellow crystals of **8**, small red crystals of **13**, and a pink mother liquor. The reaction vessel was then covered with paraffin film to prevent further evaporation of the remaining solvent. After an overall two months, complete conversion of **8** into **13** was observed. Precipitation of a very small amount of low density green powder in suspension can be observed; in that case, it can be removed by filtration of the mother liquor.

Yield: 20 mg (34.6%)

IR(ATR, ν, cm^{-1}): 621 (vs), 701 (vs), 801 (s), 857 (m), 925 (m), 986 (w), 1029 (m), 1053 (s), 1090 (br, s), 1252 (m), 1262 (m), 1305 (m), 1379 (m), 1396 (m), 1411 (m), 1444 (sh, w), 1459 (m), 1562 (m), 1639 (br, w), 2126 (vw), 2177 (vw), 2932 (vw), 3017 (vw), 3250(br, m), 3351 (br, m).

Melting point: >400 K

$\{\{[Fe^{III}(Tpm^*)(CN)_3]_2[Mn^{II}(MeCN)_2]\}(ClO_4)_2 \cdot 2\ MeCN\}_\infty$ (14)

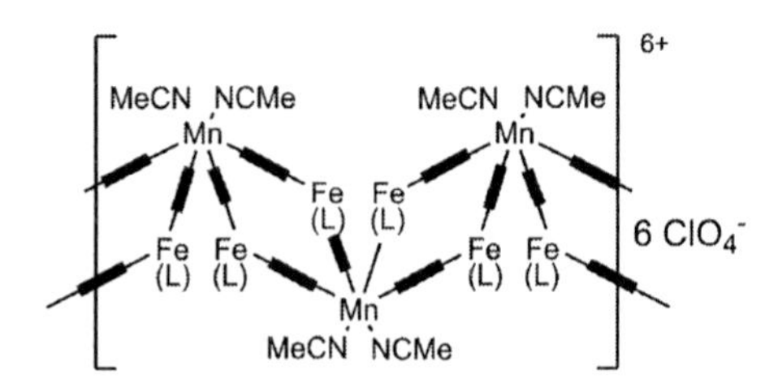

To a stirred orange reddish solution of **8** (0.1 mmol, 43 mg) in 10 mL of an acetonitrile/water (11:1) mixture, was added dropwise 4 mL of a colorless solution of $Mn^{II}(ClO_4)_2 \cdot 6\ H_2O$ (0.5 mmol, 186 mg) in the same solvent mixture. The resulting orange red solution was further stirred half an hour, before filtration. Slow evaporation of the solvent led after three weeks to yellow crystals of **8**, small red crystals of **14**, and a red mother liquor. The reaction vessel was then covered with paraffin film to prevent further evaporation of the remaining solvent. After an overall two months, complete conversion of **8** into **14** was observed.

Yield: 15 mg (25.0%)

IR(ATR, ν, cm^{-1}) with a 1 cm^{-1} resolution: 622 (vs), 661 (w), 700(s), 710 (m), 749 (w), 803 (s), 838 (w), 863 (m), 925 (m), 985 (w), 995 (w), 1031 (s), 1052 (vs), 1084 (br, vs), 1257 (m), 1304 (w), 1375 (m), 1391 (m), 1416 (m), 1462 (m), 1564 (m), 1636 (br, vw), 2158 (w), 2253 (vw), 2272 (vw), 2304 (vw), 2929(vw), 2940 (vw), 2993 (vw), 3138 (vw), 3456 (br, vw).

Melting point: > 400 K

$\{[Fe^{II}(Tpm^*)(CN)_3]_2[Co^{III}(bik)_2]_2\}(BF_4)_4 \cdot 7\ H_2O$ (15)

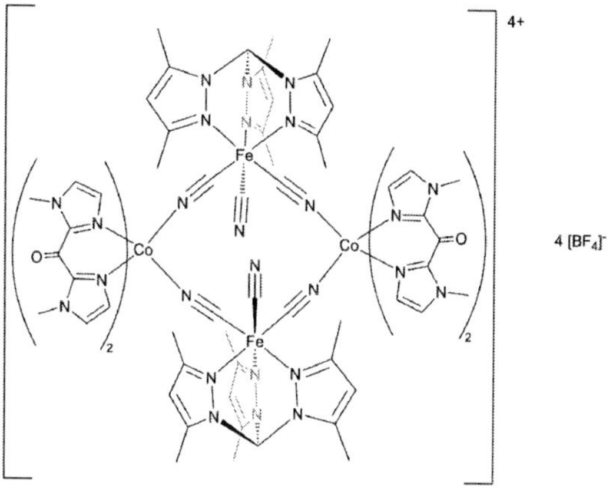

$Co^{II}(BF_4)_2 \cdot 6\ H_2O$ (35 mg, 0.1 mmol) and bik ligand (38 mg, 0.2 mmol) were dissolved in 15 mL of a mixture of acetonitrile/water (4/1). The yellow resulting solution was added to a stirred orange $[Fe^{III}(Tpm^*)(CN)_3]$ (**8**) solution (43 mg, 0.1 mmol in 15 mL of the same mixture of solvents). The resulting green solution was further stirred for 40 minutes before filtration. Slow evaporation provided deep dark green diamond-shaped crystals suitable for X-ray diffraction analysis after a few weeks.

Yield: 59.9 mg (56.3%).

Elemental Analysis (%): calculated for $C_{74}H_{84}B_4Co_2F_{16}Fe_2N_{34}O_4 \cdot 2H_2O$: C 41.80, H 4.17, N 22.40; found: C 41.76, H 4.19, N 22.27.

IR (ATR, ν, cm^{-1}): 610 (s), 628 (s), 655 (s), 689 (s), 703 (vs), 736 (s), 765 (s), 787 (s), 864 (s), 904 (vs), 979 (m), 1043 (br, vs), 1187 (m), 1264 (m), 1295 (m), 1308 (m), 1425 (vs), 1463 (m),1496 (m), 1542 (vw), 1568 (w), 1631(m), 1642 (m), 1672 (m), 2075 (m), 2114 (s), 2128 (m), 2876 (w), 2927 (w), 2960 (w), 3152 (w), 3408 (br, m).

Melting point: > 400 K.

$\{[Fe^{II}(Tpm^*)(CN)_3]_2[Co^{III}(bim)_2]_2\}(BF_4)_4 \cdot 12\ H_2O$ (16)

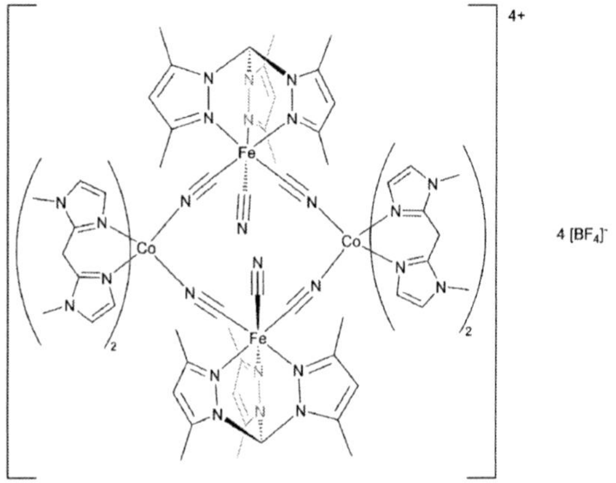

$Co^{II}(BF_4)_2 \cdot x\ H_2O$ (35 mg, 0.1 mmol) and bim ligand (38 mg, 0.2 mmol) were dissolved in 15 mL of a mixture of acetonitrile/water (4/1) and protected from light. The pale yellow resulting solution was added to a stirred orange $[Fe^{III}(Tpm^*)(CN)_3]$ solution (43 mg, 0.1 mmol in 15 mL of the same mixture of solvents). The resulting pink blackish solution was further stirred for 40 minutes before filtration. Slow evaporation provided dark-brown block-shaped crystals suitable for X-ray diffraction after a few weeks.

Yield: 25 mg (22.6%).

Elemental analysis (%): calculated for $C_{74}H_{92}N_{34}B_4Co_2F_{16}Fe_2 \cdot 10\ H_2O$: C 40.13, H 5.10, N 21.50; found: C 39.97, H 4.41, N 21.69.

IR (ATR, ν, cm^{-1}): 619 (w), 641 (m), 667 (m), 684 (w), 704 (s), 739 (m), 820 (m), 864 (m), 921 (w), 988 (s), 1032 (br, vs), 1155 (w), 1177 (w), 1228 (w), 1263 (m), 1292 (w), 1309 (m), 1413 (m), 1462 (m), 1519 (m), 1567 (m), 1631 (br, vw), 2077 (w), 2124 (s), 2134 (sh, m), 2249 (vw), 3143 (vw).

Melting point: >400 K

{[Fe^{III}(Tp)(CN)$_3$]$_2$[Co^{II}(Tpm*)(MeOH)]$_2$}(ClO$_4$)$_2$ · 2 MeOH (17)

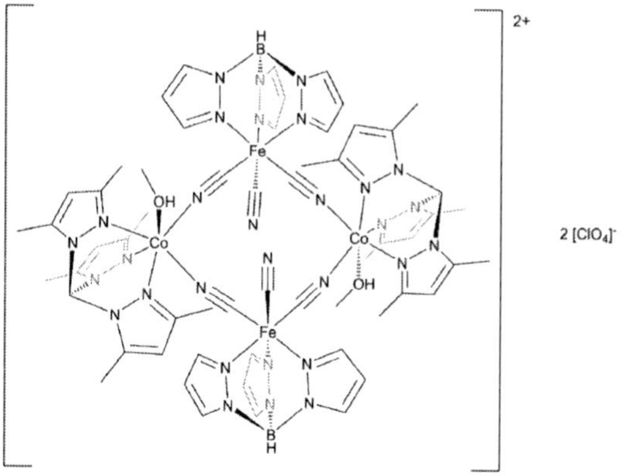

2+

2 [ClO$_4$]$^-$

A 5 mL methanolic solution of Co^{II}(ClO$_4$) · 6H$_2$O (37 mg, 0.1 mmol) was added dropwise to a solution of Tpm* (30 mg, 0.1 mmol) in 5 mL methanol. The resulting yellow solution was then added to a stirred red solution of Li[Fe^{III}(Tp)(CN)$_3$] (35 mg, 0.1 mmol) in 10 mL of the same solvent. The resulting red solution was stirred for 10 minutes before filtration. Slow evaporation of the filtrate produced red block-like crystals suitable for X-ray diffraction analysis within two weeks.

Yield: 30 mg (35.9%)

Elemental analysis (%): calculated for $C_{58}H_{72}B_2Cl_2Co_2Fe_2N_{30}O_{10}$ · 4H$_2$O: C 39.95, H 4.62, N 24.10; found: C 40.01, H 4.29, N 23.97.

IR (ATR, ν, cm^{-1}): 622 (vs), 655 (m), 667 (m), 700 (s), 709 (s), 765 (s), 778 (s), 794 (m), 805 (m), 820 (w), 855 (m), 913 (w), 988 (m), 1013 (vs), 1036 (vs), 1050 (vs), 1074 (s), 1093 (vs), 1193 (vw), 1212 (m), 1255 (w), 1266 (w), 1302 (m), 1314 (m), 1393 (m), 1408 (s), 1452 (w), 1501 (w), 1569 (w), 2143 (sh), 2149 (w), 2169 (vw), 2545 (vw), 2630 (vw), 2824 (w), 2939(vw), 3119 (w), 3147 (w), 3248 (w).

Melting point: >400 K

$\{[Fe^{III}(Tp)(CN)_3]_2[Mn^{II}(Tpm^*)(DMF)]_2\}(ClO_4)_2 \cdot 3\ DMF \cdot 2\ H_2O$ (18)

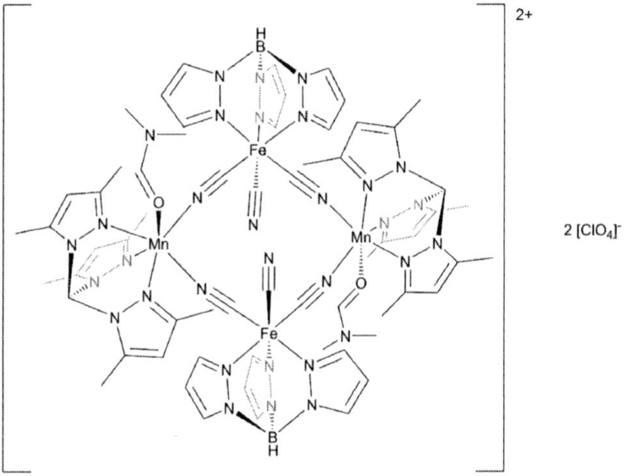

Treatment of a DMF (15 mL) solution of K[**1**] (0.386 g, 1.0 mmol) with $Mn(ClO_4)_2 \cdot 6\ H_2O$ (0.362 g, 1.0 mmol) in DMF (15 mL) rapidly afforded a blood-red mixture that was stirred for 2 h. A blood red oil was precipitated with 200 mL diethyl ether. 120 mg of this oil was redissolved in 8 mL DMF and a DMF solution of Tpm* (58 mg, 1.87 mmol) was added to it. The resulting red solution was layered with 20 mL diethylether to produce red blocks suitable for X-ray diffraction analysis.

Yield: 15 mg (9%)

IR (ATR, ν, cm^{-1}): 204 (s), 213 (s), 260 (m), 321 (s), 353 (m), 380 (m), 398 (m), 417 (m), 434 (m), 483 (m), 538 (vw), 620 (s), 636 (w), 659 (s), 678 (m), 706 (s), 767 (s), 785 (s), 821 (w), 859 (s), 905 (w), 988 (m), 1047 (vs), 1084 (br, vs), 1212 (m), 1258 (m), 1310 (s), 1376 (s), 1387 (s), 1408 (s), 1450 (w), 1502 (w), 1565 (w), 1651 (vs), 1673 (vs), 2122 (vw), 2148 (w), 2164 (vw), 2525 (vw), 2854 (vw), 2932 (vw), 3105 (vw), 3129 (vw), 3144 (vw).

$\{[Fe^{III}(Tp)(CN)_3]_4[Co(Tpe)]_2\} \cdot 4\ H_2O$ (19)

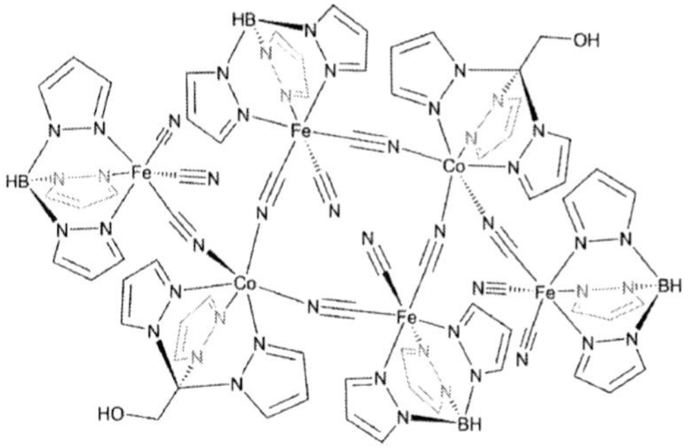

A solution of ^{H}Tpe (25 mg, 0.1 mmol) in 5 mL methanol/water (2:1) was added to a 5 mL solution of $Co^{II}(ClO_4)_2 \cdot 6\ H_2O$ (37 mg, 0.1 mmol) of the same solvent mixture. The resulting yellow solution was then added to a stirred red solution of Li[**1**] (35mg, 0.1 mmol) in 10 mL of the same solvent. The resulting red solution was further stirred 10 minutes before filtration. Slow evaporation of the reaction mixture gave X-ray diffraction suitable crystals within weeks.

Yield: 30 mg (60%).

Elemental analysis (%): calculated for $C_{70}H_{64}B_4Co_2Fe_4N_{48}O_2 \cdot 5\ H_2O$: C 40.69, H3.512, N 32.54; found: C 40.64, H 3.229, N 31.66.

IR (ATR, ν, cm^{-1}): 616 (s), 658 (s), 710 (s), 754 (vs), 775 (s), 796 (m), 822 (vw), 871 (m), 896 (vw), 922 (w), 969 (w), 990 (m), 1047 (vs), 1074 (m), 1091 (m), 1117 (m), 1212 (s), 1314 (s), 1338 (w), 1408 (s), 1501 (w), 1518 (sh, vw), 1612 (vw), 1649 (vw), 1664 (vw), 2122 (vw), 2132 (vw), 2149 (w), 2160 (w), 2516 (vw), 3118 (vw), 3153 (w), 3226 (vw), 3393 (vw), 3508 (vw), 3645 (vw).

Melting point: > 400 K (SQUID)

$\{[Fe^{III}(Tp)(CN)_3]_4[Mn(Tpe)]_2\} \cdot 4\ H_2O$ (20)

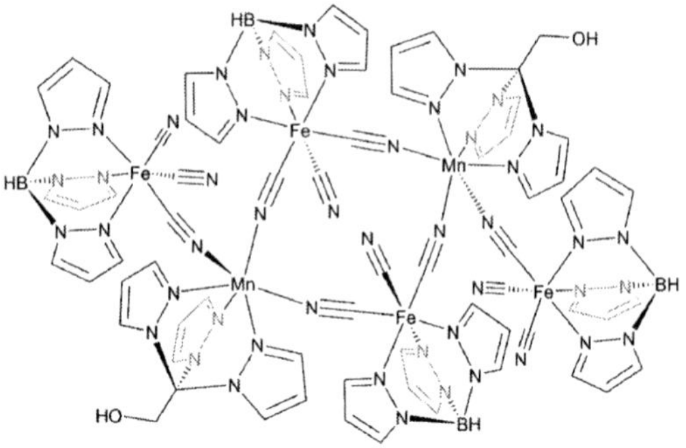

A solution of Tpe (25 mg, 0.1 mmol) in 5 mL methanol was added to a 5 mL solution of $M^{II}(ClO_4)_2 \cdot 6\ H_2O$ (37 mg, 0.1 mmol) of the same solvent. The resulting colourless solution was then added to a stirred red solution of Li[**1**] (35mg, 0.1 mmol) in 10 mL of methanol. The resulting red solution was further stirred 10 minutes before filtration. Slow evaporation of the reaction mixture provided a first species **20'** after 3-4 weeks, which was converted into crystals of **20** suitable for X-ray diffraction analysis if kept in the mother liquor for 4-6 supplementary weeks.

Yield: 20 mg (40%)

Melting point: >400 K (SQUID)

IR(ATR, ν, cm^{-1}): 616 (s), 658 (s), 710 (s), 754 (vs), 775 (s), 796 (m), 822 (vw), 871 (m), 896 (vw), 922 (w), 969 (w), 990 (m), 1047 (vs), 1074 (m), 1091 (m), 1117 (m), 1212 (s), 1314 (s), 1338 (w), 1408 (s), 1501 (w), 1518 (sh, vw), 1612 (vw), 1649 (vw), 1664 (vw), 2122 (vw), 2132 (vw), 2152 (w), 2162 (w), 3118 (vw), 3153 (w), 3226 (vw), 3393 (vw), 3508 (vw), 3645 (vw).

{[Fe^{III}(Tp)(CN)$_3$]$_4$[Co^{II}(Tpe)]$_4$}(ClO$_4$)$_4$ (21)

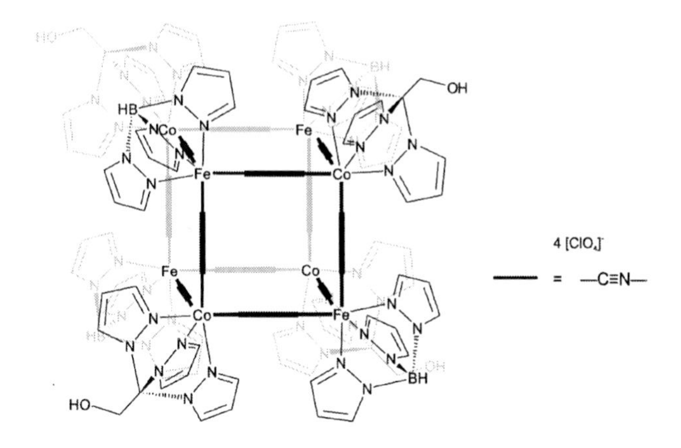

Solid K[**1**] (0.377 mg, 1.0 mmol) and solid Co^{II}(ClO$_4$)$_2$ · 6 H$_2$O (0.366 mg, 1.0 mg) were suspended in 10 mL DMF. The suspension was stirred 2 hours at room temperature, at which point the reaction mixture was a deep red solution. A deep red oil was precipitated with 60 mL diethylether and washed twice with a DMF/diethyl ether (8:1) to produce a brick red powder. It was redissolved in 25 mL CH$_2$Cl$_2$, filtered to remove K[ClO$_4$] and the solvent was evaporated. 101 mg of this powder (~0.05 mmol) was dissolved in 5 mL CH$_2$Cl$_2$ to produce a red solution. A colourless solution of Tpe ligand (84 mg, 0.4 mmol) was added dropwise to the latter, from which 75 mg of brick red solid precipitated. This powder was filtered off, and dissolved in 3 mL DMF. Layering of the DMF solution with 12 mL diethyl ether produced deep red crystals of **21**.

Yield: 75 mg (50.1%)

IR(ATR, ν, cm^{-1}): 214 (vs), 241 (s), 265 (m), 277 (m), 318 (s), 352 (s), 381 (m), 401 (m), 427 (s), 500 (m), 538 (w), 609 (sh, s), 618 (s), 659 (s), 711 (m), 767 (vs), 822 (vw), 872 (s), 901 (w), 922 (w), 965 (m), 986 (m), 1049 (vs), 1073 (vs), 1085 (vs), 1211 (m), 1229 (m), 1255 (w), 1315 (m), 1337 (s), 1386 (s), 1407 (w), 1500 (w), 1516 (sh, w), 1654 (br, vs), 2169 (w), 2522 (vw), 2866 (vw), 2885 (sh, vw), 2930 (vw), 2960 (vw), 2995 (vw), 3109 (vw), 3131 (vw), 3151 (vw), 3475 (br, vw).

K@{[Fe^{II}(Tp)(CN)$_3$]$_4$[Co^{III}(Ttp)]$_3$[Co^{II}(Ttp)]} (22)

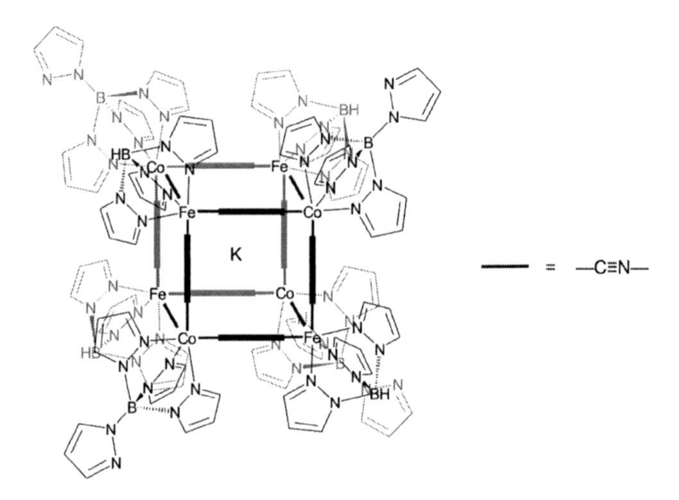

To a stirred yellow solution of [Et_4N][Fe^{III}(Tp)(CN)$_3$] (0.954 mg, 2.0 mmol) in 10 mL DMF was added solid $Co^{II}(ClO_4)_2 \cdot 6\ H_2O$ (732 mg, 2.0 mmol). The red resulting solution was stirred 20 minutes. It was precipitated with 100 mL Et_2O and the supernate was removed. 580 mg of this red powder was dissolved in 14 mL of DMF and about 6.3 equivalents of solid KTtp (400 mg, 1.26 mmol) were added. The stirred red solution immediately turned deep green and was further stirred overnight. The resulting suspension was centrifugated, and the yellow supernate removed. The Prussian blue coloured solid was washed several times with an Et_2O/DMF 8:1 mixture until the supernate was colourless. It was then dissolved in diethyl ether and filtrated to remove an off-white solid. Ether was evaporated and **22** was recrystallised either by layering a CH_2Cl_2 of **22** with *n*-pentane, or by slow evaporation of a CH_2Cl_2/DMF 4:1 solution of **22**.

Yield (as powder): 342 mg, (~62% of the brick red powder).

Elemental analysis (%): calculated for $C_{96}H_{88}N_{68}B_8Co_4Fe_4K \cdot 4\ H_2O \cdot 3\ C_3H_7NO$: C 41.08, H 3.84, N 32.39; found: C 40.92, H 3.55, N 32.44.

ESI-MS *m/z* (%) in CH_2Cl_2: 2779.4 (100) [**22**]$^+$; 2814 (100) [**22**]–Cl^-.

^{1}H NMR (400.1 MHz, 298 K, CD_2Cl_2): δ (ppm) = 94.08 (s, 3H), 37.57 (s, 3H), 18.46 (s, 6H), 17.93 (s, 3H), 15.93 (s, 3H), 11.41 (s, 3H), 10.46 (s, 3H), 9.71 (s, 3H+3H), 8.37 (s, 3H), 7.96 (s, 3H), 7.67 (s, 3H), 1.02 (s, 3H), -1.47 (s, 6H), -1.99 (s, 3H), -2.62 (s, 3H), -8.43 (s, 6H), -25.03 (s, 6H).

^{13}C NMR (100.6 MHz, 298 K, CD_2Cl_2): δ = 196.7 (s), 165.0 (s), 153.6 (br, s), 152.7 (br, s), 150.8 (br, s), 145.7 (s), 145.7 (s), 142.8 (s), 142.7 (s), 140.8 (s), 140.6 (s), 138.6 (s), 136.5 (s), 135.8 (s), 135.4 (s), 124.4 (br, s), 122.8 (br, s), 122.1 (s), 110.6 (s), 109.9 (s), 108.0 (s), 107.4 (s), 101.3 (s), 88.5 (s), 1.24 (s).

^{11}B NMR (96.29 MHz, 298 K, CD_2Cl_2): δ (ppm) = 196.5 (s, 1B, {Co^{II}(Ttp)}), 1.64 (s, 3B, {Co^{III}(Ttp)}, -13.9 (br, s, 4B, {Fe^{II}(Tp)}).

IR(ATR, ν, cm^{-1}, phase #2): 401 (w), 431 (w), 446 (w), 484 (m), 514 (s), 548 (w), 618 (vs), 658 (s), 715 (vs), 756 (vs), 801 (s), 825 (w), 850 (s), 860 (m), 926 (w), 976 (w), 1012 (w), 1041 (s), 1059 (s), 1094 (s), 1107 (m), 1152 (vw), 1178 (m – phase #1), 1207 (s), 1252 (w), 1296 (m), 1307 (s), 1386 (s), 1405 (m), 1430 (w), 1448 (vw), 1503 (w), 1669 (br, s), 2103 (br, m), 2480 (vw), 2845 (vw), 2932 (vw), 2960 (vw – phase #1), 3108 (vw), 3131 (vw), 3146 (vw).

Decomposition: > 400 K (SQUID, ATG)

11 Crystallographic data

Compound	PPh_4[**2**] · 2 H_2O
Molecular formula	$C_{37}H_{34}FeN_9O_2P$
M [g·mol^{-1}]	723.54
Crystal system	orthorhombic
Space group	*Pbcm*
a [Å]	7.3770(15)
b [Å]	16.106(3)
c [Å]	27.804(6)
α [°]	90.00
β [°]	90.00
γ [°]	90.00
V [Å^3]	3303.5(12)
Crystal size [mm]	0.10×0.10×0.05
μ [mm^{-1}]	0.555
$\rho_{calculated}$ [g·cm^3]	1.447
Z	4
T [K]	200
$2\theta_{max}$ [°]	58.392
Collected reflexions	34927
Unique reflexions	4541
Number of parameters/restraints	241/0
R1 [$I \geq 2\sigma(I)$]	0.1092
wR2 (all data)	0.1817
Max/min residual electron density [e×Å^{-3}]	1.59/-1.53
Radiation	Mo Kα (λ = 0.71073)
Diffractometer	STOE STADI 4

Compound	PPh_4[**3**] · 7 H_2O
Molecular formula	$C_{43}H_{56}FeN_9O_7P$
M [g·mol^{-1}]	897.80
Crystal system	triclinic
Space group	$P\bar{1}$
a [Å]	10.968(2)
b [Å]	11.206(2)
c [Å]	21.367(4)
α [°]	89.18(3)
β [°]	88.71(3)
γ [°]	61.17(3)
V [Å^3]	2300.1(10)
Crystal size	0.10×0.10×0.05
μ [mm^{-1}]	0.420
$\rho_{calculated}$ [g·cm^3]	1.296
Z	2
T [K]	200
$2\theta_{max}$ [°]	53.504
Collected reflexions	37143
Unique reflexions	9762
Number of parameters/restraints	551/0
R1 [I ≥ 2σ(I)]	0.1103
wR2 (all data)	0.1345
Max/min residual electron density [e×Å^{-3}]	1.73/-2.10
Radiation	Mo Kα (λ = 0.71073)
Diffractometer	STOE STADI 4

Compound	PPh_4[**3**] · 12 H_2O
Molecular formula	$C_{43}H_{66}FeN_9O_{12}P$
M [g·mol^{-1}]	963.68
Crystal system	triclinic
Space group	$P\bar{1}$
a [Å]	13.011(3)
b [Å]	13.487(3)
c [Å]	15.929(3)
α [°]	78.52(3)
β [°]	66.27(3)
γ [°]	78.69(3)
V [Å^3]	2486.7(11)
Crystal size	0.1×0.1×0.1
μ [mm^{-1}]	0.401
$\rho_{calculated}$ [g·cm^3]	1.287
Z	2
T [K]	200
$2\theta_{max}$ [°]	58.582
Collected reflexions	48522
Unique reflexions	13408
Number of parameters/restraints	596/0
R1 [$I \geq 2\sigma(I)$]	0.0555
$wR2$ (all data)	0.1062
Max/min residual electron density [e×Å^{-3}]	0.97/-0.96
Radiation	Mo Kα (λ = 0.71073)
Diffractometer	STOE STADI 4

Compound	PPh_4[**4**] · 2 H_2O
Molecular formula	$C_{38}H_{36}FeN_9O_3P$
M [g·mol^{-1}]	753.58
Crystal system	triclinic
Space group	$P\bar{1}$
a [Å]	8.9810(10)
b [Å]	13.821(3)
c [Å]	15.687(3)
α [°]	69.35(3)
β [°]	75.54(3)
γ [°]	86.34(3)
V [Å^3]	1763.7(7)
Crystal size	0.3×0.2×0.05
μ [mm^{-1}]	0.525
$\rho_{calculated}$ [g·cm^3]	1.419
Z	2
T [K]	200
$2\theta_{max}$ [°]	58.554
Collected reflexions	35915
Unique reflexions	9582
Number of parameters/restraints	476/0
R1 [$I \geq 2\sigma(I)$]	0.0899
$wR2$ (all data)	0.2950
Max/min residual electron density [e×Å^{-3}]	0.46/-1.49
Radiation	Mo Kα (λ = 0.71073)
Diffractometer	STOE IPDS II

Compound	$(PPh_4)_2$[**5**] · 2 MeCN · H_2O
Molecular formula	$C_{65}H_{57}FeN_{11}O_4P_2S$
M [g·mol^{-1}]	1206.10
Crystal system	triclinic
Space group	$P\bar{1}$
a [Å]	10.435(2)
b [Å]	15.852(3)
c [Å]	20.104(4)
α [°]	67.47(3)
β [°]	77.94(3)
γ [°]	77.22(3)
V [Å^3]	2967.0(12)
Crystal size	0.3×0.2×0.1
μ [mm^{-1}]	0.402
$\rho_{calculated}$ [g·cm^3]	1.348
Z	2
T [K]	200
$2\theta_{max}$ [°]	58.61
Collected reflexions	57874
Unique reflexions	16024
Number of parameters/restraints	767/0
R1 [$I \geq 2\sigma(I)$]	0.0618
$wR2$ (all data)	0.1151
Max/min residual electron density [e×Å^{-3}]	0.84/-1.52
Radiation	Mo Kα (λ = 0.71073)
Diffractometer	STOE IPDS II

Compound	PPh_4[**7**] · CH3CN
Molecular formula	$C_{44}H_{45}BFeN_{10}P$
M [g·mol^{-1}]	811.54
Crystal system	monoclinic
Space group	$P2_1/c$
a [Å]	16.3287(5)
b [Å]	9.8451(3)
c [Å]	26.4596(8)
α [°]	90.00
β [°]	103.2530(10)
γ [°]	90.00
V [Å^3]	4140.3(2)
Crystal size	0.160×0.120×0.100
μ [mm^{-1}]	0.448
$\rho_{calculated}$ [g·cm^3]	1.302
Z	4
T [K]	200
$2\theta_{max}$ [°]	60.23
Collected reflexions	41529
Unique reflexions	12087
Number of parameters/restraints	515/0
R1 [$I \geq 2\sigma(I)$]	0.0541
$wR2$ (all data)	0.1752
Max/min residual electron density [e×Å^{-3}]	0.84/-0.80
Radiation	Mo Kα (λ = 0.71073)
Diffractometer	Bruker Kappa Apex2

Compound	**8**
Molecular formula	$C_{19}H_{22}FeN_9$
M [g·mol^{-1}]	432.31
Crystal system	orthorhombic
Space group	*Pbca*
a [Å]	15.253(3)
b [Å]	15.858(3)
c [Å]	16.801(3)
α [°]	90.00
β [°]	90.00
γ [°]	90.00
V [Å^3]	4063.9(14)
Crystal size	0.30×0.20×0.20
μ [mm^{-1}]	0.767
$\rho_{calculated}$ [g·cm^3]	1.413
Z	8
T [K]	200
$2\theta_{max}$ [°]	50
Collected reflexions	26534
Unique reflexions	3549
Number of parameters/restraints	268/0
R1 [$I \geq 2\sigma(I)$]	0.0865
$wR2$ (all data)	0.2433
Max/min residual electron density [e×Å^{-3}]	0.74/-0.64
Radiation	Mo Kα (λ = 0.71073)
Diffractometer	STOE STADI 4

Compound	**8** · 2 CH_3CN
Molecular formula	$C_{21}H_{28}FeN_{11}$
M [g·mol^{-1}]	514.39
Crystal system	monoclinic
Space group	$P2_1/n$
a [Å]	9.1180(18)
b [Å]	16.429(3)
c [Å]	17.045(3)
α [°]	90.00
β [°]	94.98(3)
γ [°]	90.00
V [Å^3]	2543.7(9)
Crystal size	0.1×0.1×0.1
μ [mm^{-1}]	0.628
$\rho_{calculated}$ [g·cm^3]	1.348
Z	4
T [K]	200
2θ_{max} [°]	58.546
Collected reflexions	48858
Unique reflexions	6898
Number of parameters/restraints	317/0
R1 [I ≥ 2σ(I)]	0.0581
wR2 (all data)	0.1151
Max/min residual electron density [e×Å^{-3}]	0.79/-1.01
Radiation	Mo Kα (λ = 0.71073)
Diffractometer	STOE STADI 4

Compound	**8** · 0.5 HI_5 · H_2O
Molecular formula	$C_{19}H_{24.5}FeI_{2.5}N_9O$
M [g·mol^{-1}]	768.07
Crystal system	monoclinic
Space group	$C2/c$
a [Å]	21.299(4)
b [Å]	16.728(3)
c [Å]	18.217(4)
α [°]	90.00
β [°]	125.09(3)
γ [°]	90.00
V [Å^3]	5311(3)
Crystal size	0.4×0.15×0.2
μ [mm^{-1}]	3.504
$\rho_{calculated}$ [g·cm^3]	1.920
Z	8
T [K]	200
$2\theta_{max}$ [°]	58.544
Collected reflexions	50854
Unique reflexions	7194
Number of parameters/restraints	295/0
R1 [$I \geq 2\sigma(I)$]	0.0700
$wR2$ (all data)	0.1026
Max/min residual electron density [e×Å^{-3}]	1.88/-2.20
Radiation	Mo Kα (λ = 0.71073)
Diffractometer	STOE STADI 4

Compound	**10**
Molecular formula	$C_{72}H_{88}B_2Cl_2Co_2Fe_2N_{34}O_{14}$
M [g·mol^{-1}]	1975.80
Crystal system	triclinic
Space group	$P\bar{1}$
a [Å]	13.4512(6)
b [Å]	13.5064(6)
c [Å]	14.1015(7)
α [°]	108.700(3)
β [°]	102.335(3)
γ [°]	106.773(3)
V [Å^3]	2187.0(2)
Crystal size	0.160×0.07×0.040
μ [mm^{-1}]	6.768
$\rho_{calculated}$ [g·cm^3]	1.497
Z	1
T [K]	200
$2\theta_{max}$ [°]	119.45
Collected reflexions	14314
Unique reflexions	6330
Number of parameters/restraints	578/0
R1 [I ≥ 2σ(I)]	0.0532
wR2 (all data)	0.1043
Max/min residual electron density [e×Å^{-3}]	0.82/-0.53
Radiation	Cu Kα (λ = 1.54178)
Diffractometer	Bruker Kappa Apex2

Compound	**12**
Molecular formula	$C_{72}H_{88}B_2Cl_2Fe_4N_{34}O_{14}$
M [g·mol^{-1}]	1969.62
Crystal system	triclinic
Space group	$P\bar{1}$
a [Å]	13.430(3)
b [Å]	13.470(3)
c [Å]	13.907(3)
α [°]	102.34(3)
β [°]	108.67(3)
γ [°]	107.01(3)
V [Å^3]	2143.9(12)
Crystal size	0.2×0.2×0.05
μ [mm^{-1}]	0.808
$\rho_{calculated}$ [g·cm^3]	1.525
Z	1
T [K]	200
$2\theta_{max}$ [°]	56.564
Collected reflexions	36209
Unique reflexions	10640
Number of parameters/restraints	578/0
R1 [I ≥ 2σ(I)]	0.0877
$wR2$ (all data)	0.1819
Max/min residual electron density [e×Å^{-3}]	1.42/-2.82
Radiation	Mo Kα (λ = 0.71073)
Diffractometer	Stoe IPDS II

Compound	**13**
Molecular formula	$C_{38}H_{44}Cl_2CoFe_2N_{18}O_{14}$
M [g·mol^{-1}]	1218.44
Crystal system	monoclinic
Space group	*Cc*
a [Å]	25.898(5)
b [Å]	16.657(3)
c [Å]	13.278(3)
α [°]	90
β [°]	115.00(3)
γ [°]	90
V [Å^3]	5191(2)
Crystal size	0.3×0.1×0.1
μ [mm^{-1}]	1.049
$\rho_{calculated}$ [g·cm^3]	1.559
Z	4
T [K]	200
$2\theta_{max}$ [°]	58.524
Collected reflexions	65234
Unique reflexions	13927
Number of parameters/restraints	688/2
R1 [I ≥ 2σ(I)]	0.1297
wR2 (all data)	0.3683
Max/min residual electron density [e×Å^{-3}]	0.55/-1.35
Radiation	Mo Kα (λ = 0.71073)
Diffractometer	STOE STADI 4

Compound	**14**
Molecular formula	$C_{44}H_{56}Cl_2Fe_2MnN_{24}O_8$
M [g·mol^{-1}]	1286.66
Crystal system	monoclinic
Space group	$P2/c$
a [Å]	17.122(3)
b [Å]	12.199(2)
c [Å]	13.910(3)
α [°]	90
β [°]	97.68(3)
γ [°]	90
V [Å^3]	2879.3(10)
Crystal size	0.6×0.3×0.3
μ [mm^{-1}]	0.877
$\rho_{calculated}$ [g·cm^3]	1.484
Z	2
T [K]	200
$2\theta_{max}$ [°]	58.536
Collected reflexions	54754
Unique reflexions	7808
Number of parameters/restraints	374/0
R1 [$I \geq 2\sigma(I)$]	0.0576
$wR2$ (all data)	0.1751
Max/min residual electron density [e×Å^{-3}]	0.96/-0.97
Radiation	Mo Kα (λ = 0.71073)
Diffractometer	Stoe IPDS II

Compound	**15**
Molecular formula	$C_{74}H_{84}B_4Co_2F_{16}Fe_2N_{34}O_{11}$
M [g·mol^{-1}]	1101.24
Crystal system	monoclinic
Space group	$P2_1$
a [Å]	13.3881(5)
b [Å]	27.0794(10)
c [Å]	13.9341(5)
α [°]	90
β [°]	106.507(2)
γ [°]	90
V [Å^3]	4843.5(3)
Crystal size	0.090×0.050×0.040
μ [mm^{-1}]	5.899
$\rho_{calculated}$ [g·cm^3]	1.510
Z	4
T [K]	200
$2\theta_{max}$ [°]	132.344
Collected reflexions	22733
Unique reflexions	12715
Number of parameters/restraints	1154/1
R1 [I ≥ 2σ(I)]	0.0704
$wR2$ (all data)	0.1829
Max/min residual electron density [e×Å^{-3}]	0.70/-0.71
Radiation	Cu Kα (λ = 1.54180)
Diffractometer	Bruker Kappa Apex2

Compound	**16**
Molecular formula	$C_{74}H_{116}B_2Co_2F_{16}Fe_2N_{34}O_{12}$
M [g·mol^{-1}]	2250.62
Crystal system	monoclinic
Space group	$P2_1/n$
a [Å]	13.888(3)
b [Å]	13.505(3)
c [Å]	26.669(5)
α [°]	90
β [°]	103.32(3)
γ [°]	90
V [Å^3]	4867.3(17)
Crystal size	0.2×0.2×0.2
μ [mm^{-1}]	0.732
$\rho_{calculated}$ [g·cm^3]	1.519
Z	4
T [K]	200
$2\theta_{max}$ [°]	50.00
Collected reflexions	42307
Unique reflexions	8563
Number of parameters/restraints	644/19
R1 [I ≥ 2σ(I)]	0.0736
$wR2$ (all data)	0.2199
Max/min residual electron density [e×Å^{-3}]	1.65/-0.94
Radiation	Mo Kα (λ = 0.71073)
Diffractometer	Stoe IPDS II

Compound	**17**
Molecular formula	$C_{60}H_{80}B_2Cl_2Co_2Fe_2N_{30}O_{12}$
M [g·mol^{-1}]	1735.62
Crystal system	triclinic
Space group	$P\bar{1}$
a [Å]	11.3912(3)
b [Å]	13.3444(4)
c [Å]	13.7707(4)
α [°]	97.8170(10)
β [°]	105.3720(10)
γ [°]	92.7540(10)
V [Å^3]	1991.86(10)
Crystal size [mm]	0.3×0.1×0.1
μ [mm^{-1}]	0.908
$\rho_{calculated}$ [g·cm^3]	1.447
Z	1
T [K]	200
$2\theta_{max}$ [°]	61.16
Collected reflexions	46713
Unique reflexions	12194
Number of parameters/restraints	505/0
R1 [I ≥ 2σ(I)]	0.0284
$wR2$ (all data)	0.0790
Max/min residual electron density [e×Å^{-3}]	0.80/-0.45
Radiation	Mo Kα (λ = 0.71073)
Diffractometer	Bruker APEX-II CCD

Compound	**18**
Molecular formula	$C_{80}H_{124}B_2Cl_2Fe_2Mn_2N_{38}O_{18}$
M [g·mol^{-1}]	2220.19
Crystal system	triclinic
Space group	$P\bar{1}$
a [Å]	12.2707(5)
b [Å]	14.0882(5)
c [Å]	17.3515(7)
α [°]	90.813(2)
β [°]	104.002(2)
γ [°]	113.360(2)
V [Å^3]	2651.40(19)
Crystal size	0.15×0.10×0.05
μ [mm^{-1}]	5.170
$\rho_{calculated}$ [g·cm^3]	1.388
Z	1
T [K]	200
$2\theta_{max}$ [°]	133.19
Collected reflexions	22728
Unique reflexions	9098
Number of parameters/restraints	570/23
R1 [$I \geq 2\sigma(I)$]	0.0891
$wR2$ (all data)	0.3007
Max/min residual electron density [e×Å^{-3}]	2.24/-0.86
Radiation	Cu Kα (λ = 1.54180)
Diffractometer	Bruker Kappa Apex2

Compound	**19**
Molecular formula	$C_{70}H_{72}B_4Co_2Fe_4N_{48}O_6$
M [g·mol^{-1}]	2066.16
Crystal system	triclinic
Space group	$P\bar{1}$
a [Å]	12.642(3)
b [Å]	12.647(3)
c [Å]	15.966(3)
α [°]	71.20(3)
β [°]	67.16(3)
γ [°]	74.13(3)
V [Å^3]	2194.6(10)
Crystal size	0.6×0.3×0.1
μ [mm^{-1}]	1.093
$\rho_{calculated}$ [g·cm^3]	1.563
Z	1
T [K]	200
$2\theta_{max}$ [°]	58.684
Collected reflexions	47178
Unique reflexions	11900
Number of parameters/restraints	605/0
R1 [$I \geq 2\sigma(I)$]	0.0866
$wR2$ (all data)	0.2729
Max/min residual electron density [e×Å^{-3}]	1.22/-2.96
Radiation	Mo Kα (λ = 0.71073)
Diffractometer	STOE IPDS II

Compound	**20**
Molecular formula	$C_{70}H_{72}B_4Fe_4Mn_2N_{48}O_6$
M [g·mol^{-1}]	2058.28
Crystal system	triclinic
Space group	$P\bar{1}$
a [Å]	12.715(3)
b [Å]	12.735(3)
c [Å]	15.985(3)
α [°]	71.38(3)
β [°]	67.33(3)
γ [°]	74.17(3)
V [Å^3]	2230.0(10)
Crystal size	0.6×0.3×0.1
μ [mm^{-1}]	0.988
$\rho_{calculated}$ [g·cm^3]	1.536
Z	1
T [K]	200
$2\theta_{max}$ [°]	58.492
Collected reflexions	43480
Unique reflexions	12031
Number of parameters/restraints	619/0
R1 [$I \geq 2\sigma(I)$]	0.0513
$wR2$ (all data)	0.1673
Max/min residual electron density [e×Å^{-3}]	0.72/-1.17
Radiation	Mo Kα (λ = 0.71073)
Diffractometer	STOE IPDS II

Compound	**21**
Molecular formula	$C_{108}H_{72}Cl_4Co_4Fe_4N_{72}O_{21}$
M [g·mol^{-1}]	3315.12
Crystal system	trigonal
Space group	$R\bar{3}c$
a [Å]	23.181(3)
b [Å]	23.181(3)
c [Å]	63.948(13)
α [°]	90
β [°]	90
γ [°]	120
V [Å^3]	29759(10)
Crystal size	0.6×0.4×0.4
μ [mm^{-1}]	0.734
$\rho_{calculated}$ [g·cm^3]	1.071
Z	9
T [K]	260
$2\theta_{max}$ [°]	58.718
Collected reflexions	193345
Unique reflexions	9033
Number of parameters/restraints	339/0
R1 [$I \geq 2\sigma(I)$]	0.1308
$wR2$ (all data)	0.3968
Max/min residual electron density [e×Å^{-3}]	0.59/-1.66
Radiation	Mo Kα (λ = 0.71073)
Diffractometer	STOE IPDS II

Compound	**22**
Molecular formula	$B_8C_{108}Co_4Fe_4KN_{66}O_6$
M [g·mol^{-1}]	2902.30
Crystal system	trigonal
Space group	$R\bar{3}$
a [Å]	20.442(3)
b [Å]	20.442(3)
c [Å]	39.847(8)
α [°]	90
β [°]	90
γ [°]	120
V [Å^3]	14420(5)
Crystal size	0.4×0.3×0.2
μ [mm^{-1}]	0.699
$\rho_{calculated}$ [g·cm^3]	0.962
Z	3
T [K]	200
$2\theta_{max}$ [°]	56.626
Collected reflexions	87068
Unique reflexions	7989
Number of parameters/restraints	135/0
R1 [$I \geq 2\sigma(I)$]	0.1965
wR2 (all data)	0.5376
Max/min residual electron density [e×Å^{-3}]	2.24/-2.47
Radiation	Mo Kα (λ = 0.71073)
Diffractometer	STOE IPDS II

Compound	**22** · 12 CH_2Cl_2
Molecular formula	$C_{108}H_{112}B_8Cl_{24}Co_4Fe_4KN_{68}O$
M [$g \cdot mol^{-1}$]	3798.16
Crystal system	triclinic
Space group	$P\bar{1}$
a [Å]	16.406(3)
b [Å]	17.459(4)
c [Å]	29.831(6)
α [°]	84.34(3)
β [°]	81.26(3)
γ [°]	71.75(3)
V [$Å^3$]	8009(3)
Crystal size	0.433×0.351×0.142
μ [mm^{-1}]	1.249
$\rho_{calculated}$ [$g \cdot cm^3$]	1.575
Z	2
T [K]	200
$2\theta_{max}$ [°]	52.366
Collected reflexions	67971
Unique reflexions	31476
Number of parameters/restraints	1957/24
R1 [$I \geq 2\sigma(I)$]	0.0968
$wR2$ (all data)	0.2852
Max/min residual electron density [$e \times Å^{-3}$]	2.90/-1.68
Radiation	Mo Kα (λ = 0.71073)
Diffractometer	STOE STADI VARI

12 Annexes

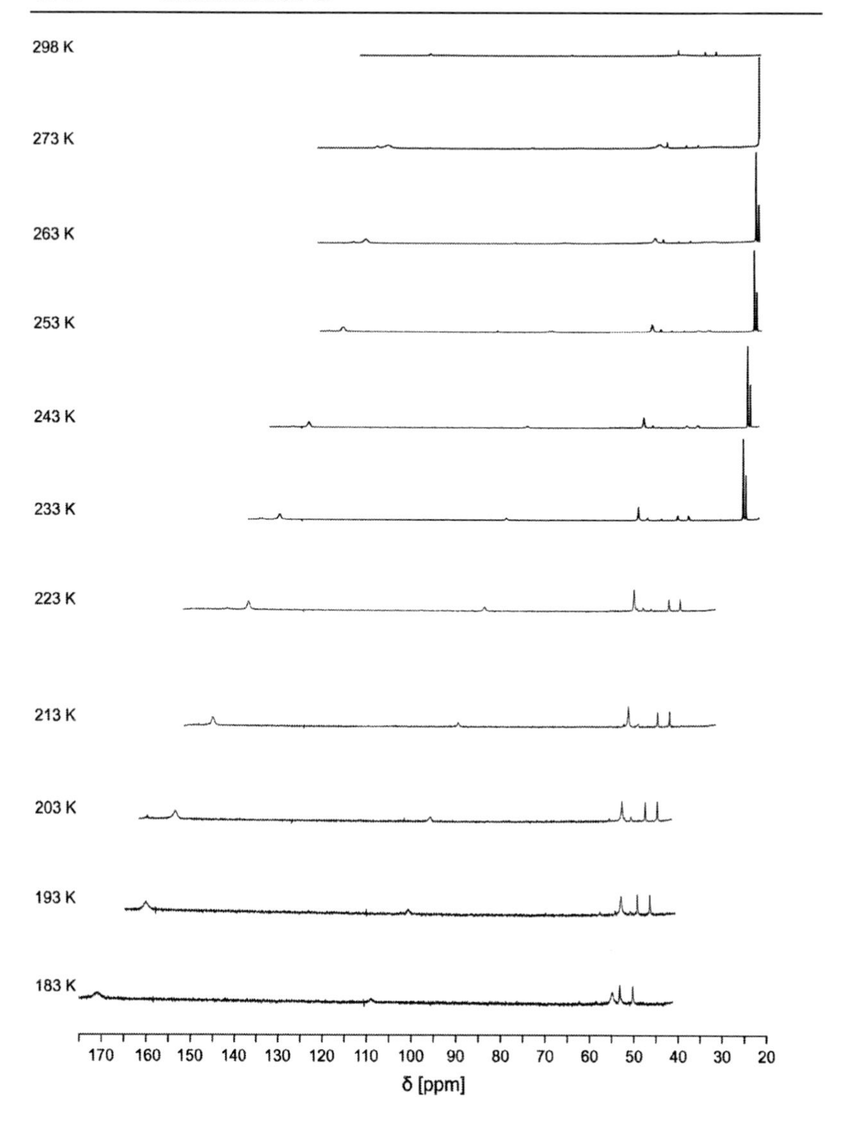

Figure 12.1 - Variable Temperature NMR of **22** (phase #2) in CD_2Cl_2 between 298 K and 183 K between 175 and 20 ppm.

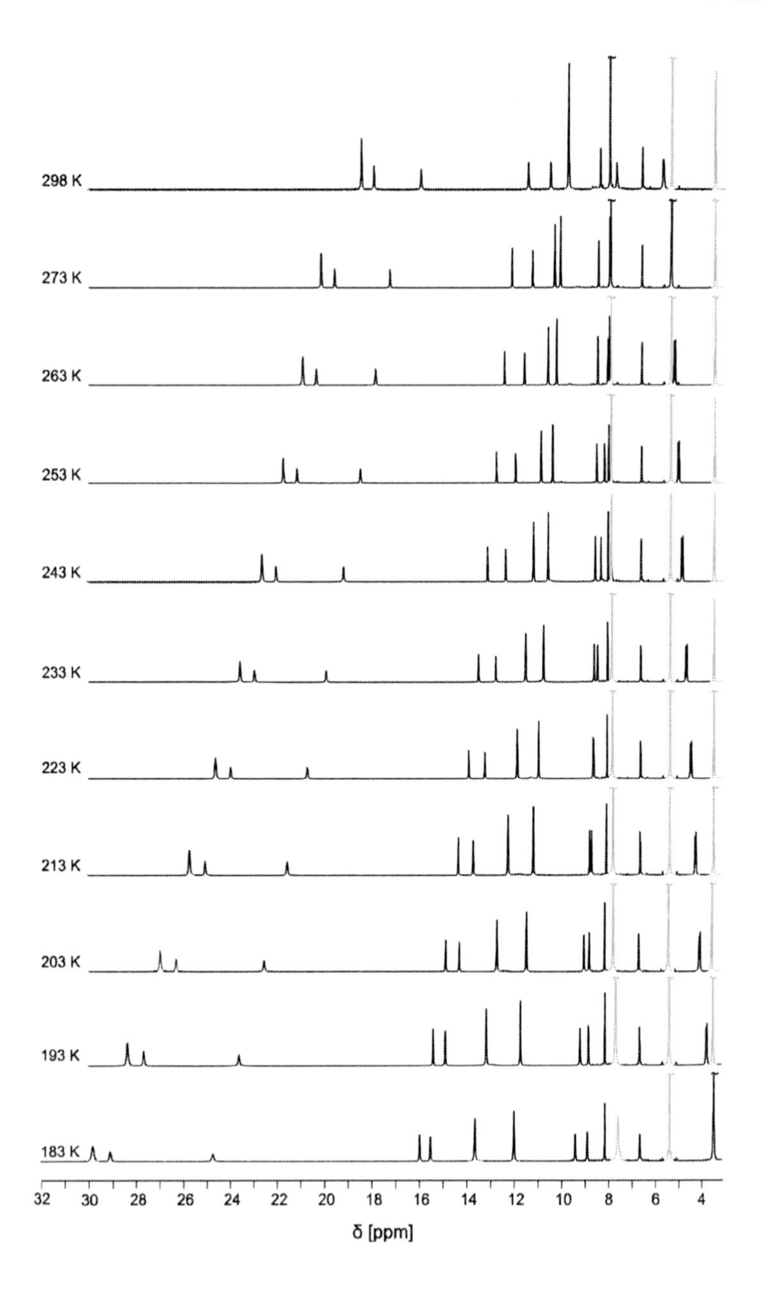

Figure 12.2 - Variable Temperature NMR of **22** (phase #2) in CD_2Cl_2 between 298 K and 183 K. Zoom between 3.1 and 30 ppm. The solvent peaks are marked as grey when no compound peak lies underneath.

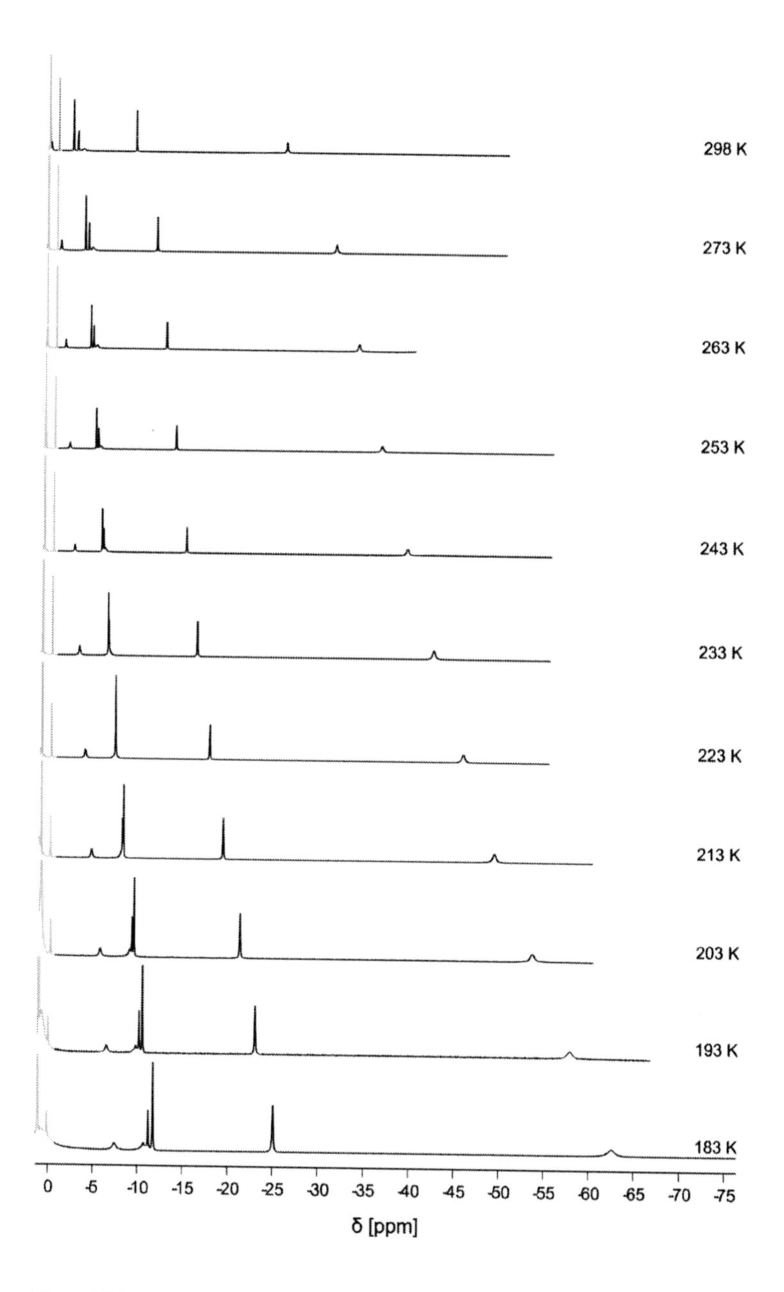

Figure 12.3 - Variable Temperature NMR of **22** (phase #2) in CD2Cl2 between 298 K and 183 K. Zoom between 1.4 and -76.4 ppm. The solvent peaks are marked as grey when no compound peak lies underneath.

13 List of abbreviations

χ_M	molar succeptibility
bik	bis(*N*-methylimidazolyl)ketone
bim	bis(*N*-methylimidazolyl)methane
CEA	Commissariat à l'Energie Atomique
CN	cyanide
Cp	cylclopentadienyl
Cp*	pentamethylcyclopentadienyl
CSA	Chemical Shift Anisotropy
δ	chemical shift
DFT	Density Functional Theory
DMF	dimethylformamide
dtbbpy	di*tert*butylbipyridine
EDX	Energy-Dispersive X-ray spectroscopy
EPR	Electron Paramagnetic Resonance
ETCST	Electron Transfer Coupled with Spin Transition
g	Landé factor
HMBC	Heteronuclear Multiple Bond Correlation
HMQC	Heteronuclear Multiple Quantum Coherence
HS	high-spin
Im	imidazolyl
IR	InfraRed
J	coupling constant
k_B	Boltzmann constant
LIESST	Light-Induced Spin State Trapping
LIETCST	Light-Induced Electron Transfer Coupled with a Spin Transition
LS	low-spin
μ_B, β	Bohr magneton
M	any transition metal
MAS	Magic Angle Spinning

Me	methyl
NMR	Nuclear Magnetic Resonance
NOE	Nuclear Overhauser Effect
Oe	Oestred
PBA	Prussian Blue Analogue
Ph	phenyl
PND	Polarised Neutron Diffraction
PPh_4	tetraphenylphosphonium
ptz	5-(pyrazinyl)tetrazolate
Py	pyridyl
pz	pyrazolyl
pz*	3,5-dimethylpyrazolyl
RT	Room Temperature
SCM	Single Chain Magnet
SMM	Single Molecule Magnet
SQUID	Superconducting Quantum Interference Device
T	temperature
Tacn	1,4,7-triazacyclononane
THF	TetraHydroFurane
tmphen	3,4,7,8-tetramethyl 1,10 phenanthroline
Tp	hydrotris(pyrazolyl)borate
Tp*	hydrotris(3,5-dimethylpyrazolyl)borate
Tpe	tris(pyrazolyl)ethanol
Tpm	tris(pyrazolyl)methane
Tpm*	tris(3,5-dimethylpyrazolyl)methane
Tpmd	tris(pyrazolyl)methanide
Tpms	tris(pyrazolyl)methanesulfonate
Triphos	bis(diphenylphosphinoethyl)phenylphosphine
Tt	hydrotris(1,2,4-triazolyl)borate
Ttp	tetrakispyrazolylborate

14 List of compounds

1 $[Fe^{III}(Tp)(CN)_3]^-$

2 $[Fe^{II}(Tpm)(CN)_3]^-$

3 $[Fe^{II}(Tpm*)(CN)_3]^-$

4 $[Fe^{II}(Tpe)(CN)_3]^-$

5 $[Fe^{II}(Tpms)(CN)_3]^{2-}$

6 $[Fe^{III}(Tt)(CN)_3]^-$

7 $[Fe^{III}(Tp*)(CN)_3]^-$

8 $[Fe^{III}(Tpm*)(CN)_3]$

9 $[Fe^{III}(Ttp)(CN)_3]^-$

10 $\{[Fe^{III}(Tp*)(CN)_3]_2[Co^{II}(bik)_2]_2\}(ClO4)_2 \cdot 2\ H_2O$

11 $\{[Fe^{III}(Tp*)(CN)_3]_2[Co^{II}(bik)_2]_2\}(BF_4)_2$

12 $\{[Fe^{III}(Tp*)(CN)_3]_2[Fe^{II}(bik)_2]_2\}(ClO_4)_2 \cdot 2\ H_2O$

13 $\{\{[Fe^{III}(Tpm*)(CN)_3]_2[Co^{II}(H_2O)_2]\}(ClO_4)_2 \cdot 2\ H_2O\}_\infty$

14 $\{\{[Fe^{III}(Tpm*)(CN)_3]_2[Mn^{II}(MeCN)_2]\}(ClO_4)_2 \cdot 2\ MeCN\}_\infty$

15 $\{[Fe^{II}(Tpm*)(CN)_3]_2[Co^{III}(bik)_2]_2\}(BF_4)_4 \cdot 7\ H_2O$

16 $\{[Fe^{II}(Tpm*)(CN)_3]_2[Co^{III}(bim)_2]_2\}(BF_4)_4 \cdot 12\ H_2O$

17 $\{[Fe^{III}(Tp)(CN)_3]_2[Co^{II}(Tpm*)(MeOH)]_2\}(ClO_4)_2 \cdot 2\ MeOH$

18 $\{[Fe^{III}(Tp)(CN)_3]_2[Mn^{II}(Tpm*)(DMF)]_2\}(ClO_4)_2 \cdot 3\ DMF \cdot 2\ H_2O$

19 $\{[Fe^{III}(Tp)(CN)_3]_4[Co(Tpe)]_2\} \cdot 4\ H_2O$

20 $\{[Fe^{III}(Tp)(CN)_3]_4[Mn(Tpe)]_2\} \cdot 4\ H_2O$

21 $\{[Fe^{III}(Tp)(CN)_3]_4[Co^{II}(Tpe)]_4\}(ClO_4)_4$

22 $K@\{[Fe^{II}(Tp)(CN)_3]_4[Co^{III}(Ttp)]_3[Co^{II}(Ttp)]\}$

15 Curriculum Vitae

Delphine Garnier

Nationality: French

Birth date: 22.01.1989 in Paris

Higher education

Oct. 2011- July 2015 — Split-site **PhD thesis** at the **Karlsruhe Institute of Technology** (KIT - Germany) and at the **University Pierre and Marie Curie** (UPMC - France).

Supervisors: Prof. Dr. Frank Breher and Prof. Dr. Rodrigue Lescouëzec.

Mobility grant from the DFH-UFA.

Subject: "*Open-shell Coordination Complexes based on Cyanide and Scorpionate Ligands*".

2008-2011 — **École Normale Supérieure** (ENS Paris), at the chemistry department.
University Pierre and Marie Curie (UPMC), at the chemistry department.

Degrees: Diploma of the ENS.
Master in Chemistry with high honours (UPMC).
Licence (≈ Bachelor) in chemistry with honours (UPMC).

2006-2008 — *Classes Préparatoires PC* (Physics and Chemistry) to the *Grandes Écoles* at **Lycée Saint-Louis** in Paris.

1999-2006 — High school at **Lycée Saint-Érembert**, Saint-Germain-en-Laye.

Degree: *Scientific Baccalauréat* (≈ A-Levels) with highest honours.

Internships

Feb. 2011- June 2011	5-month **Master thesis** in inorganic chemistry at the **Paris Institute of Molecular Chemistry** (IPCM), team MMMAX. Supervisors: Dr. Alexandrine Flambard and Dr. Rodrigue Lescouëzec Subject: "*Paramagnetic NMR: magnetic and structural probe for magnetic molecular materials*"
Feb. 2010- July 2010	6-month internship at the Karlsruhe Institute of Technology (KIT - Germany), team Breher. Supervisor: Prof. Dr. Frank Breher Subject: "*Coordination chemistry of ambidentate* N-*donor compounds*"
June 2009- July 2009	7-week **Bachelor internship** in inorganic chemistry at the **Paris Institute of Molecular Chemistry** (IPCM), team POM. Supervisor: Dr. Sébastien Blanchard Subject: "*Functionalisation of an electroactive organic molecule by a coordination site: radical-complex interaction and multifunctional molecular materials*"

Other

Oct. 2014	1-week NMR course on the acquisition of diffusional NMR data and their processing with the DiffAtOnce software package in Almería (Spain) in the work group of Ignacio Fernández de la Nieves.
2011-2014	Supervision of students during the inorganic chemistry practical courses for biology, applied chemistry and chemical biology students.
2009-2010	Private lessons for middle schoolers, high schoolers and students of *Classes Préparatoires*, scientific subjects.
Languages	French: mother tongue German: fluent English: fluent
Softwares	Pack Office (Outlook, Word, Powerpoint, Excel) Topspin, DiffAtOnce, MestreNova, NMRnotebook ChemBioDraw CorelDraw, OriginPro OPUS
2015	Last first-aid training

16 List of publications and conference communications

2015 "*Tris (3,5-dimethylpyrazolyl)methane-Based Heterobimetallic Complexes that Contain Zn- and Cd-Transition-Metal Bonds: Synthesis, Structures, and Quantum Chemical Calculations*" J. Meyer, S. González-Gallardo, S. Hohnstein, **D. Garnier**, M. K. Armbruster, K. Fink, W. Klopper and F. Breher, *Chem. Eur. J.* **2015**, *21*, 2905 – 2914.

2014 "*Synthesis and Structures of 1,3-Bis (organochalcogenyl) Distannatrisilabicyclo-[1.1.1]pentanes*" E. Moos, T. Augenstein, **D. Garnier**, and F. Breher, *Can. J. Chem.* **2014**, *92*, 574 – 579.

⋆This article is part of the GTL 2013 special issue.

"A metallacyclic alkyl-amido carbene complex (MCAAC)" I. Trapp, S. González-Gallardo, S. Hohnstein, **D. Garnier**, P. Oña-Burgos, and F. Breher, *Dalton Trans.* **2014**, *43*, 4313 – 4319.

⋆This article is part of the Dalton Transactions themed issue, New Talent: Europe

2012 Wöhlertagung, Göttingen – Poster on "*Towards Multimetallic Complexes of Tris(pyrazolyl)methanes and methanides*".

17 References

[1] S. Trofimenko, *J. Am. Chem. Soc.* **1966**, *88*, 1842–1844.

[2] S. Trofimenko, *J. Am. Chem. Soc.* **1967**, *89*, 3170–3177.

[3] S. Trofimenko, J. R. Long, T. Nappier, S. G. Shore, in *Inorg. Synth.* (Ed.: R.W. Parry), John Wiley & Sons, Inc., Hoboken, NJ, USA, **1970**, pp. 99–109.

[4] H. R. Bigmore, S. C. Lawrence, P. Mountford, C. S. Tredget, *Dalton Trans.* **2005**, 635–651.

[5] I. Kuzu, I. Krummenacher, J. Meyer, F. Armbruster, F. Breher, *Dalton Trans.* **2008**, 5836–5865.

[6] S. Trofimenko, *Chem. Rev.* **1993**, *93*, 943–980.

[7] A. Sella, S. E. Brown, J. W. Steed, D. A. Tocher, *Inorg Chem* **2007**, *46*, 1856–1864.

[8] S. Trofimenko, *Polyhedron* **2004**, *23*, 197–203.

[9] A. Otero, J. Fernández-Baeza, A. Antiñolo, J. Tejeda, A. Lara-Sánchez, *Dalton Trans.* **2004**, 1499–1510.

[10] I. Trapp, Untersuchungen zum Koordinationsverhalten neuerartiger tripodaler Liganden, Dissertation KIT, **2012**.

[11] S. Trofimenko, J. C. Calabrese, J. S. Thompson, *Inorg. Chem.* **1987**, *26*, 1507–1514.

[12] G. Parkin, in *Adv. Inorg. Chem.* (Ed.: A.G. Sykes), Academic Press, **1995**, pp. 291–393.

[13] S. Trofimenko, *J. Am. Chem. Soc.* **1970**, *92*, 5118–5126.

[14] D. L. Reger, T. C. Grattan, K. J. Brown, C. A. Little, J. J. S. Lamba, A. L. Rheingold, R. D. Sommer, *J. Organomet. Chem.* **2000**, *607*, 120–128.

[15] E. C. B. Alegria, L. M. D. R. S. Martins, M. Haukka, A. J. L. Pombeiro, *Dalton Trans.* **2006**, 4954–4961.

[16] T. F. S. Silva, E. C. B. A. Alegria, L. M. D. R. S. Martins, A. J. L. Pombeiro, *Adv. Synth. Catal.* **2008**, *350*, 706–716.

[17] M. G. Cushion, J. Meyer, A. Heath, A. D. Schwarz, I. Fernández, F. Breher, P. Mountford, *Organometallics* **2010**, *29*, 1174–1190.

[18] J. Meyer, S. González-Gallardo, S. Hohnstein, D. Garnier, M. K. Armbruster, K. Fink, W. Klopper, F. Breher, *Chem. – Eur. J.* **2014**, 2905–2914.

[19] W. Kläui, M. Berghahn, G. Rheinwald, H. Lang, *Angew. Chem.* **2000**, *112*, 2590–2592.

[20] W. Kläui, D. Schramm, W. Peters, G. Rheinwald, H. Lang, *Eur. J. Inorg. Chem. Eur. J. Inorg. Chem.* **2001**, *2001*, 1415–1424.

[21] W. Kläui, M. Berghahn, W. Frank, G. J. Reiß, T. Schönherr, G. Rheinwald, H. Lang, *Eur. J. Inorg. Chem.* **2003**, 2059–2070.

[22] P. Smoleński, C. Dinoi, M. F. C. Guedes da Silva, A. J. L. Pombeiro, *J. Organomet. Chem.* **2008**, *693*, 2338–2344.

[23] C. Dinoi, M. F. C. Guedes da Silva, E. C. B. A. Alegria, P. Smoleński, L. M. D. R. S. Martins, R. Poli, A. J. L. Pombeiro, *Eur. J. Inorg. Chem.* **2010**, 2415–2424.

[24] L. M. D. R. S. Martins, E. C. B. A. Alegria, P. Smoleński, M. L. Kuznetsov, A. J. L. Pombeiro, *Inorg. Chem.* **2013**, *52*, 4534–4546.

[25] B. G. M. Rocha, T. C. O. M. Leod, M. F. C. G. da Silva, K. V. Luzyanin, L. M. D. R. S. Martins, A. J. L. Pombeiro, *Dalton Trans.* **2014**, *43*, 15192–15200.

[26] D. L. Reger, T. C. Gratten, *Synthesis* **2003**, *2003*, 1306–1306.

[27] D. L. Reger, J. R. Gardinier, S. Bakbak, R. F. Semeniuc, U. H. F. Bunz, M. D. Smith, *New J. Chem.* **2005**, *29*, 1035–1043.

[28] D. L. Reger, K. J. Brown, M. D. Smith, *J. Organomet. Chem.* **2002**, *658*, 50–61.

[29] D. L. Reger, R. F. Semeniuc, V. Rassolov, M. D. Smith, *Inorg. Chem.* **2004**, *43*, 537–554.

[30] D. L. Reger, J. R. Gardinier, R. F. Semeniuc, M. D. Smith, *Dalton Trans.* **2003**, 1712–1718.

[31] D. L. Reger, R. F. Semeniuc, M. D. Smith, *Dalton Trans.* **2003**, 285–286.

[32] D. L. Reger, R. F. Semeniuc, M. D. Smith, *Dalton Trans.* **2008**, 2253–2260.

[33] D. L. Reger, J. R. Gardinier, S. Bakbak, R. F. Semeniuc, U. H. F. Bunz, M. D. Smith, *New J. Chem.* **2005**, *29*, 1035–1043.

[34] D. L. Reger, R. F. Semeniuc, M. D. Smith, *J. Chem. Soc. Dalton Trans.* **2002**, 476–477.

[35] D. L. Reger, R. F. Semeniuc, C. A. Little, M. D. Smith, *Inorg. Chem.* **2006**, *45*, 7758–7769.

[36] D. L. Reger, T. D. Wright, R. F. Semeniuc, T. C. Grattan, M. D. Smith, *Inorg. Chem.* **2001**, *40*, 6212–6219.

[37] T. F. S. Silva, M. F. Guedes da Silva, G. S. Mishra, L. M. D. R. S. Martins, A. J. L. Pombeiro, *J. Organomet. Chem.* **2011**, *696*, 1310–1318.

[38] R. Wanke, M. F. C. Guedes da Silva, S. Lancianesi, T. F. S. Silva, L. M. D. R. S. Martins, C. Pettinari, A. J. L. Pombeiro, *Inorg. Chem.* **2010**, *49*, 7941–7952.

[39] G. Dördelmann, H. Pfeiffer, A. Birkner, U. Schatzschneider, *Inorg. Chem.* **2011**, *50*, 4362–4367.

[40] I. Kuzu, D. Nied, F. Breher, *Eur. J. Inorg. Chem.* **2009**, 872–879.

[41] I. Kuzu, I. Krummenacher, I. J. Hewitt, Y. Lan, V. Mereacre, A. K. Powell, P. Höfer, J. Harmer, F. Breher, *Chem. - Eur. J. Chem. - Eur. J.* **2009**, *15*, 4350–4365.

[42] I. Kuzu, Metallorganische Chemie Und Koordinationschemie von Anionischen Tris(pyrazolyl)methaniden, Dissertation TH Karlsruhe, **2009**.

[43] I. Krummenacher, H. Rüegger, F. Breher, *Dalton Trans.* **2006**, 1073–1081.

[44] J. Meyer, Koordinationsverhalten von Ambidenten Tris(pyrazolyl)methaniden: Untersuchungen an Gruppe 2 Und 12 Komplexen, Dissertation KIT, **2010**.

[45] D. Kratzert, D. Leusser, D. Stern, J. Meyer, F. Breher, D. Stalke, *Chem. Commun.* **2011**, *47*, 2931–2933.

[46] J.-P. Launay, M. Verdaguer, *Electrons in Molecules: From Basic Principles to Molecular Electronics*, Oxford University Press, **2013**.

[47] M. Emmelius, G. Pawlowski, H. W. Vollmann, *Angew. Chem. Int. Ed. Engl.* **1989**, *28*, 1445–1471.

[48] O. Sato, J. Tao, Y.-Z. Zhang, *Angew. Chem. Int. Ed.* **2007**, *46*, 2152–2187.

[49] O. Sato, *Acc. Chem. Res.* **2003**, *36*, 692–700.

[50] H. Paulsen, L. Duelund, H. Winkler, H. Toftlund, A. X. Trautwein, *Inorg. Chem.* **2001**, *40*, 2201–2203.

[51] P. Gütlich, *Eur. J. Inorg. Chem.* **2013**, 581–591.

[52] J. A. Real, A. B. Gaspar, M. C. Muñoz, *Dalton Trans.* **2005**, 2062–2079.

[53] A. B. Gaspar, M. Seredyuk, P. Gütlich, *J. Mol. Struct.* **2009**, *924–926*, 9–19.

[54] P. Gütlich, J. Jung, *J. Mol. Struct.* **1995**, *347*, 21–38.

[55] P. Sonar, C. M. Grunert, Y.-L. Wei, J. Kusz, P. Gütlich, A. D. Schlüter, *Eur. J. Inorg. Chem.* **2008**, 1613–1622.

[56] Y. Wei, P. Sonar, M. Grunert, J. Kusz, A. D. Schlüter, P. Gütlich, *Eur. J. Inorg. Chem.* **2010**, 3930–3941.

[57] I. Boldog, F. J. Muñoz-Lara, A. B. Gaspar, M. C. Muñoz, M. Seredyuk, J. A. Real, *Inorg. Chem.* **2009**, *48*, 3710–3719.

[58] A. B. Gaspar, V. Ksenofontov, M. Seredyuk, P. Gütlich, *Coord. Chem. Rev.* **2005**, *249*, 2661–2676.

[59] T. Liu, Y.-J. Zhang, S. Kanegawa, O. Sato, *Angew. Chem. Int. Ed.* **2010**, *49*, 8645–8648.

[60] L. Cambi, L. Szegö, *Berichte Dtsch. Chem. Ges. B Ser.* **1931**, *64*, 2591–2598.

[61] L. Cambi, L. Szegö, *Berichte Dtsch. Chem. Ges. B Ser.* **1933**, *66*, 656–661.

[62] E. K. Barefield, D. H. Busch, S. M. Nelson, *Q. Rev. Chem. Soc.* **1968**, *22*, 457–498.

[63] L. Sacconi, *Pure Appl Chem* **n.d.**, *27*, 161.

[64] D. M. Halepoto, D. G. L. Holt, L. F. Larkworthy, G. J. Leigh, D. C. Povey, G. W. Smith, *J. Chem. Soc. Chem. Commun.* **1989**, 1322–1323.

[65] S. R. Batten, J. Bjernemose, P. Jensen, B. A. Leita, K. S. Murray, B. Moubaraki, J. P. Smith, H. Toftlund, *Dalton Trans.* **2004**, 3370–3375.

[66] D. L. Reger, J. D. Elgin, M. D. Smith, F. Grandjean, L. Rebbouh, G. J. Long, *Eur. J. Inorg. Chem. Eur. J. Inorg. Chem.* **2004**, *2004*, 3345–3352.

[67] D. L. Reger, J. R. Gardinier, J. D. Elgin, M. D. Smith, D. Hautot, G. J. Long, F. Grandjean, *Inorg. Chem.* **2006**, *45*, 8862–8875.

[68] C. Piquer, F. Grandjean, O. Mathon, S. Pascarelli, D. L. Reger, C. A. Little, G. J. Long, *Inorg. Chem.* **2003**, *42*, 982–985.

[69] D. L. Reger, C. A. Little, A. L. Rheingold, M. Lam, L. M. Liable-Sands, B. Rhagitan, T. Concolino, A. Mohan, G. J. Long, V. Briois, et al., *Inorg. Chem.* **2001**, *40*, 1508–1520.

[70] O. Kahn, *Science* **1998**, *279*, 44–48.

[71] K. S. Murray, *Eur. J. Inorg. Chem.* **2008**, 3101–3121.

[72] S. Decurtins, P. Gütlich, C. P. Köhler, H. Spiering, A. Hauser, *Chem. Phys. Lett.* **1984**, *105*, 1–4.

[73] A. Paquirissamy, A. R. Ruyack, A. Mondal, Y. Li, R. Lescouëzec, C. Chanéac, B. Fleury, *J. Mater. Chem. C* **2015**, *3*, 3350–3355.

[74] C. M. Quintero, G. Félix, I. Suleimanov, J. Sánchez Costa, G. Molnár, L. Salmon, W. Nicolazzi, A. Bousseksou, *Beilstein J. Nanotechnol.* **2014**, *5*, 2230–2239.

[75] T. Miyamachi, M. Gruber, V. Davesne, M. Bowen, S. Boukari, L. Joly, F. Scheurer, G. Rogez, T. K. Yamada, P. Ohresser, et al., *Nat. Commun.* **2012**, *3*, 938.

[76] P. Poganiuch, S. Decurtins, P. Guetlich, *J. Am. Chem. Soc.* **1990**, *112*, 3270–3278.

[77] O. Sato, T. Iyoda, A. Fujishima, K. Hashimoto, *Science* **1996**, *272*, 704–705.

[78] O. Sato, Y. Einaga, T. Iyoda, A. Fujishima, K. Hashimoto, *J. Electrochem. Soc.* **1997**, *144*, L11–L13.

[79] H. J. Buser, D. Schwarzenbach, W. Petter, A. Ludi, *Inorg. Chem.* **1977**, *16*, 2704–2710.

[80] F. Herren, P. Fischer, A. Ludi, W. Haelg, *Inorg. Chem.* **1980**, *19*, 956–959.

[81] M. Verdaguer, *Science* **1996**, *272*, 698–699.

[82] A. Flambard, F. H. Köhler, R. Lescouëzec, *Angew. Chem. Int. Ed.* **2009**, *48*, 1673–1676.

[83] A. Bleuzen, V. Marvaud, C. Mathoniere, B. Sieklucka, M. Verdaguer, *Inorg. Chem.* **2009**, *48*, 3453–3466.

[84] J.-D. Cafun, G. Champion, M.-A. Arrio, C. C. dit Moulin, A. Bleuzen, *J. Am. Chem. Soc.* **2010**, *132*, 11552–11559.

[85] N. Shimamoto, S. Ohkoshi, O. Sato, K. Hashimoto, *Inorg. Chem.* **2002**, *41*, 678–684.

[86] O. Kahn, *Molecular Magnetism*, Wiley-VCH, New York, **1993**.

[87] S. Wang, X.-H. Ding, J.-L. Zuo, X.-Z. You, W. Huang, *Coord. Chem. Rev.* **2011**, *255*, 1713–1732.

[88] E. Pardo, M. Verdaguer, P. Herson, H. Rousselière, J. Cano, M. Julve, F. Lloret, R. Lescouëzec, *Inorg. Chem.* **2011**, *50*, 6250–6262.

[89] C. P. Berlinguette, A. Dragulescu-Andrasi, A. Sieber, J. R. Galán-Mascarós, H.-U. Güdel, C. Achim, K. R. Dunbar, *J. Am. Chem. Soc.* **2004**, *126*, 6222–6223.

[90] C. P. Berlinguette, A. Dragulescu-Andrasi, A. Sieber, H.-U. Güdel, C. Achim, K. R. Dunbar, *J. Am. Chem. Soc.* **2005**, *127*, 6766–6779.

[91] M. Shatruk, A. Dragulescu-Andrasi, K. E. Chambers, S. A. Stoian, E. L. Bominaar, C. Achim, K. R. Dunbar, *J. Am. Chem. Soc.* **2007**, *129*, 6104–6116.

[92] K. E. Funck, M. G. Hilfiger, C. P. Berlinguette, M. Shatruk, W. Wernsdorfer, K. R. Dunbar, *Inorg. Chem.* **2009**, *48*, 3438–3452.

[93] D. Li, R. Clérac, O. Roubeau, E. Harté, C. Mathonière, R. Le Bris, S. M. Holmes, *J. Am. Chem. Soc.* **2008**, *130*, 252–258.

[94] M. Nihei, Y. Sekine, N. Suganami, K. Nakazawa, A. Nakao, H. Nakao, Y. Murakami, H. Oshio, *J. Am. Chem. Soc.* **2011**, *133*, 3592–3600.

[95] D. Siretanu, D. Li, L. Buisson, D. M. Bassani, S. M. Holmes, C. Mathonière, R. Clérac, *Chem. – Eur. J.* **2011**, *17*, 11704–11708.

[96] J. Mercurol, Y. Li, E. Pardo, O. Risset, M. Seuleiman, H. Rousselière, R. Lescouëzec, M. Julve, *Chem. Commun.* **2010**, *46*, 8995–8997.

[97] A. Mondal, Y. Li, M. Seuleiman, M. Julve, L. Toupet, M. Buron-Le Cointe, R. Lescouëzec, *J. Am. Chem. Soc.* **2013**, *135*, 1653–1656.

[98] A. Mondal, Switchable Molecular Magnetic Materials, Dissertation UPMC, **2013**.

[99] E. S. Koumousi, I.-R. Jeon, Q. Gao, P. Dechambenoit, D. N. Woodruff, P. Merzeau, L. Buisson, X. Jia, D. Li, F. Volatron, et al., *J. Am. Chem. Soc.* **2014**, *136*, 15461–15464.

[100] Y. Zhang, D. Li, R. Clérac, M. Kalisz, C. Mathonière, S. M. Holmes, *Angew. Chem. Int. Ed.* **2010**, *49*, 3752–3756.

[101] Y.-Z. Zhang, P. Ferko, D. Siretanu, R. Ababei, N. P. Rath, M. J. Shaw, R. Clérac, C. Mathonière, S. M. Holmes, *J. Am. Chem. Soc.* **2014**, *136*, 16854–16864.

[102] G. N. Newton, M. Nihei, H. Oshio, *Eur. J. Inorg. Chem.* **2011**, 3031–3042.

[103] R. Lescouëzec, J. Vaissermann, C. Ruiz-Pérez, F. Lloret, R. Carrasco, M. Julve, M. Verdaguer, Y. Dromzee, D. Gatteschi, W. Wernsdorfer, *Angew. Chem. Int. Ed.* **2003**, *42*, 1483–1486.

[104] D.-P. Dong, T. Liu, S. Kanegawa, S. Kang, O. Sato, C. He, C.-Y. Duan, *Angew. Chem. Int. Ed.* **2012**, *51*, 5119–5123.

[105] N. Hoshino, F. Iijima, G. N. Newton, N. Yoshida, T. Shiga, H. Nojiri, A. Nakao, R. Kumai, Y. Murakami, H. Oshio, *Nat. Chem.* **2012**, *4*, 921–926.

[106] M. Verdaguer, *Nat. Chem.* **2012**, *4*, 871–872.

[107] A. Mondal, Y. Li, P. Herson, M. Seuleiman, M.-L. Boillot, E. Rivière, M. Julve, L. Rechignat, A. Bousseksou, R. Lescouëzec, *Chem. Commun.* **2012**, *48*, 5653–5655.

[108] M. Nihei, M. Ui, H. Oshio, *Polyhedron* **2009**, *28*, 1718–1721.

[109] M. Nihei, M. Ui, M. Yokota, L. Han, A. Maeda, H. Kishida, H. Okamoto, H. Oshio, *Angew. Chem. Int. Ed.* **2005**, *44*, 6484–6487.

[110] J. M. Herrera, V. Marvaud, M. Verdaguer, J. Marrot, M. Kalisz, C. Mathonière, *Angew. Chem. Int. Ed.* **2004**, *43*, 5468–5471.

[111] G. Rombaut, M. Verelst, S. Golhen, L. Ouahab, C. Mathonière, O. Kahn, *Inorg. Chem.* **2001**, *40*, 1151–1159.

[112] S. Ohkoshi, H. Tokoro, K. Hashimoto, *Coord. Chem. Rev.* **2005**, *249*, 1830–1840.

[113] S. Ohkoshi, H. Tokoro, *Acc. Chem. Res.* **2012**, *45*, 1749–1758.

[114] R. Lescouëzec, J. Vaissermann, F. Lloret, M. Julve, M. Verdaguer, *Inorg. Chem.* **2002**, *41*, 5943–5945.

[115] D. Li, S. Parkin, G. Wang, G. T. Yee, S. M. Holmes, *Inorg. Chem.* **2006**, *45*, 1951–1959.

[116] Z.-G. Gu, J.-L. Zuo, Y. Song, C.-H. Li, Y.-Z. Li, X.-Z. You, *Inorganica Chim. Acta* **2005**, *358*, 4057–4061.

[117] J. Kim, S. Han, I.-K. Cho, K. Y. Choi, M. Heu, S. Yoon, B. J. Suh, *Polyhedron* **2004**, *23*, 1333–1339.

[118] P. G. Edwards, A. Harrison, P. D. Newman, W. Zhang, *Inorganica Chim. Acta* **2006**, *359*, 3549–3556.

[119] S. Liang, H. Wang, T. Deb, J. L. Petersen, G. T. Yee, M. P. Jensen, *Inorg. Chem.* **2012**, DOI 10.1021/ic301409s.

[120] C.-C. Shi, C.-S. Chen, S. C. N. Hsu, W.-Y. Yeh, M. Y. Chiang, T.-S. Kuo, *Inorg. Chem. Commun.* **2008**, *11*, 1264–1266.

[121] I. Trapp, S. González-Gallardo, S. Hohnstein, D. Garnier, P. Oña-Burgos, F. Breher, *Dalton Trans.* **2014**, *43*, 4313–4319.

[122] W. Kläui, M. Berghahn, G. Rheinwald, H. Lang, *Angew. Chem. Int. Ed.* **2000**, *39*, 2464–2466.

[123] S. Miguel, J. Diez, M. P. Gamasa, M. E. Lastra, *Eur. J. Inorg. Chem.* **2011**, *2011*, 4745–4755.

[124] D. V. Fomitchev, C. C. McLauchlan, R. H. Holm, *Inorg. Chem.* **2002**, *41*, 958–966.

[125] R. S. Herrick, T. J. Brunker, C. Maus, K. Crandall, A. Cetin, C. J. Ziegler, *Chem. Commun.* **2006**, 4330–4331.

[126] C. C. McLauchlan, M. P. Weberski Jr., B. A. Greiner, *Inorganica Chim. Acta* **2009**, *362*, 2662–2666.

[127] C. Santini, M. Pellei, G. G. Lobbia, A. Cingolani, R. Spagna, M. Camalli, *Inorg. Chem. Commun.* **2002**, *5*, 430–433.

[128] E. T. Papish, M. T. Taylor, F. E. Jernigan, M. J. Rodig, R. R. Shawhan, G. P. A. Yap, F. A. Jové, *Inorg. Chem.* **2006**, *45*, 2242–2250.

[129] C. M. Nagaraja, M. Nethaji, B. R. Jagirdar, *Organometallics* **2007**, *26*, 6307–6311.

[130] F. Marchetti, C. Pettinari, R. Pettinari, A. Cerquetella, L. M. D. R. S. Martins, M. F. C. Guedes da Silva, T. F. S. Silva, A. J. L. Pombeiro, *Organometallics* **2011**, *30*, 6180–6188.

[131] P. Smoleński, C. Pettinari, F. Marchetti, M. F. C. Guedes da Silva, G. Lupidi, G. V. Badillo Patzmay, D. Petrelli, L. A. Vitali, A. J. L. Pombeiro, *Inorg. Chem.* **2015**, *54*, 434–440.

[132] R. M. Silverstein, F. X. Webster, D. J. Kiemle, *Identification spectrométrique de composés organiques*, De Boeck, Bruxelles, **2007**.

[133] K. Nakamoto, *Infrared and Raman Spectra of Inorganic and Coordination Compounds*, John Wiley & Sons, Inc., Hoboken, NJ, USA, **2008**.

[134] Y.-Z. Zhang, P. Ferko, D. Siretanu, R. Ababei, N. P. Rath, M. J. Shaw, R. Clérac, C. Mathonière, S. M. Holmes, *J. Am. Chem. Soc.* **2014**, *136*, 16854–16864.

[135] N. G. Connelly, W. E. Geiger, *Chem. Rev.* **1996**, *96*, 877–910.

[136] M. E. Lines, *J. Chem. Phys.* **1971**, *55*, 2977–2984.

[137] D. G. Davis, *J. Chem. Phys.* **1967**, *46*, 388.

[138] C. P. Grey, N. Dupré, *Chem. Rev.* **2004**, *104*, 4493–4512.

[139] R. J. Clément, A. J. Pell, D. S. Middlemiss, F. C. Strobridge, J. K. Miller, M. S. Whittingham, L. Emsley, C. P. Grey, G. Pintacuda, *J. Am. Chem. Soc.* **2012**, *134*, 17178–17185.

[140] N. Trease, T. K.-J. Köster, C. P. Grey, *J Electrochem Soc* **2011**, *20*, 69.

[141] N. Dupré, M. Cuisinier, D. Guyomard, *J Electrochem Soc* **2011**, *20*, 61.

[142] M. Schnellbach, F. H. Köhler, J. Blümel, *J. Organomet. Chem.* **1996**, *520*, 227–230.

[143] G. Kervern, G. Pintacuda, Y. Zhang, E. Oldfield, C. Roukoss, E. Kuntz, E. Herdtweck, J.-M. Basset, S. Cadars, A. Lesage, et al., *J. Am. Chem. Soc.* **2006**, *128*, 13545–13552.

[144] G. Otting, *Annu. Rev. Biophys.* **2010**, *39*, 387–405.

[145] J. Iwahara, G. M. Clore, *Nature* **2006**, *440*, 1227–1230.

[146] G. M. Clore, J. Iwahara, *Chem. Rev.* **2009**, *109*, 4108–4139.

[147] F. H. Köhler, R. Lescouëzec, *Angew. Chem. Int. Ed.* **2004**, *43*, 2571–2573.

[148] N. Baumgärtel, A. Flambard, F. H. Köhler, R. Lescouëzec, *Inorg. Chem.* **2013**, *52*, 12634–12644.

[149] C. Belle, C. Bougault, M. T. Averbuch, A. Durif, J. L. Pierre, J. M. Latour, L. Le Pape, *J. Am. Chem. Soc.* **2001**, *123*, 8053–8066.

[150] F. H. Köhler, in *eMagRes*, John Wiley & Sons, Ltd, **2007**.

[151] R. J. Kurland, B. R. McGarvey, *J. Magn. Reson. 1969* **1970**, *2*, 286–301.

[152] A. K. Koh, D. J. Miller, *At. Data Nucl. Data Tables* **1985**, *33*, 235–253.

[153] J. R. Morton, K. F. Preston, *J. Magn. Reson. 1969* **1978**, *30*, 577–582.

[154] H. Heise, Dissertation TU München, **1999**.

[155] M. Kaupp, F. H. Köhler, *Coord. Chem. Rev.* **2009**, *253*, 2376–2386.

[156] S. De, Unpublished Results, **2015**.

[157] R. K. Harris, E. D. Becker, S. M. Cabral de Menezes, P. Granger, R. E. Hoffman, K. W. Zilm, *Pure Appl. Chem.* **2008**, *80*, 59–84.

[158] J. Cano, E. Ruiz, S. Alvarez, M. Verdaguer, *Comments Inorg. Chem.* **1998**, *20*, 27–56.

[159] E. Ruiz, J. Cirera, S. Alvarez, *Coord. Chem. Rev.* **2005**, *249*, 2649–2660.

[160] H. Paulsen, H. Grünsteudel, W. Meyer-Klaucke, M. Gerdan, H. F. Grünsteudel, A. I. Chumakov, R. Rüffer, H. Winkler, H. Toftlund, A. X. Trautwein, *Eur. Phys. J. B - Condens. Matter Complex Syst.* **2001**, *23*, 463–472.

[161] B. Li, R.-J. Wei, J. Tao, R.-B. Huang, L.-S. Zheng, Z. Zheng, *J. Am. Chem. Soc.* **2010**, *132*, 1558–1566.

[162] R.-J. Wei, Q. Huo, J. Tao, R.-B. Huang, L.-S. Zheng, *Angew. Chem. Int. Ed.* **2011**, *50*, 8940–8943.

[163] B. N. Figgis, M. Gerloch, J. Lewis, F. E. Mabbs, G. A. Webb, *J. Chem. Soc. Inorg. Phys. Theor.* **1968**, 2086–2093.

[164] H. Sakiyama, R. Ito, H. Kumagai, K. Inoue, M. Sakamoto, Y. Nishida, M. Yamasaki, *Eur. J. Inorg. Chem.* **2001**, 2027–2032.

[165] F. Lloret, M. Julve, J. Cano, R. Ruiz-García, E. Pardo, *Inorganica Chim. Acta* **2008**, *361*, 3432–3445.

[166] A. Mondal, Y. Li, L.-M. Chamoreau, M. Seuleiman, L. Rechignat, A. Bousseksou, M.-L. Boillot, R. Lescouëzec, *Chem. Commun.* **2014**, *50*, 2893–2895.

[167] K. Mitsumoto, M. Ui, M. Nihei, H. Nishikawa, H. Oshio, *CrystEngComm* **2010**, *12*, 2697–2699.

[168] R. Lescouëzec, J. Vaissermann, L. M. Toma, R. Carrasco, F. Lloret, M. Julve, *Inorg. Chem.* **2004**, *43*, 2234–2236.

[169] H.-R. Wen, C.-F. Wang, J.-L. Zuo, Y. Song, X.-R. Zeng, X.-Z. You, *Inorg. Chem.* **2006**, *45*, 582–590.

[170] Y.-J. Zhang, T. Liu, S. Kanegawa, O. Sato, *J. Am. Chem. Soc.* **2009**, *131*, 7942–7943.

[171] L. Jiang, X.-L. Feng, T.-B. Lu, S. Gao, *Inorg. Chem.* **2006**, *45*, 5018–5026.

[172] D. Li, S. Parkin, G. Wang, G. T. Yee, A. V. Prosvirin, S. M. Holmes, *Inorg. Chem.* **2005**, *44*, 4903–4905.

[173] Y. Zhang, D. Li, R. Clérac, M. Kalisz, C. Mathonière, S. M. Holmes, *Angew. Chem. Int. Ed.* **2010**, *49*, 3752–3756.

[174] P. Ferko, S. Holmes, *Curr. Inorg. Chem.* **2013**, *3*, 172–193.

[175] M. Nihei, M. Ui, N. Hoshino, H. Oshio, *Inorg. Chem.* **2008**, *47*, 6106–6108.

[176] K. Mitsumoto, H. Nishikawa, G. N. Newton, H. Oshio, *Dalton Trans.* **2012**, *41*, 13601–13608.

[177] A. Mondal, S. Durdevic, L.-M. Chamoreau, Y. Journaux, M. Julve, L. Lisnard, R. Lescouëzec, *Chem. Commun.* **2013**, *49*, 1181–1183.

[178] E. S. Koumousi, I.-R. Jeon, Q. Gao, P. Dechambenoit, D. N. Woodruff, P. Merzeau, L. Buisson, X. Jia, D. Li, F. Volatron, et al., *J. Am. Chem. Soc.* **2014**, *136*, 15461–15464.

[179] D. Li, S. Parkin, R. Clérac, S. M. Holmes, *Inorg. Chem.* **2006**, *45*, 7569–7571.

[180] D. Li, S. Parkin, G. Wang, G. T. Yee, R. Clérac, W. Wernsdorfer, S. M. Holmes, *J. Am. Chem. Soc.* **2006**, *128*, 4214–4215.

[181] J. Kim, S. Han, K. I. Pokhodnya, J. M. Migliori, J. S. Miller, *Inorg. Chem.* **2005**, *44*, 6983–6988.

[182] J. L. Heinrich, P. A. Berseth, J. R. Long, *Chem. Commun.* **1998**, 1231–1232.

[183] J. Y. Yang, M. P. Shores, J. J. Sokol, J. R. Long, *Inorg. Chem.* **2003**, *42*, 1403–1419.

[184] Y. Zhang, U. P. Mallik, N. Rath, G. T. Yee, R. Clérac, S. M. Holmes, *Chem. Commun.* **2010**, *46*, 4953–4955.

[185] E. J. Schelter, F. Karadas, C. Avendano, A. V. Prosvirin, W. Wernsdorfer, K. R. Dunbar, *J. Am. Chem. Soc.* **2007**, *129*, 8139–8149.

[186] J. L. Boyer, H. Yao, M. L. Kuhlman, T. B. Rauchfuss, S. Wilson, *Eur. J. Inorg. Chem.* **2007**, *2007*, 2721–2728.

[187] J. L. Boyer, M. L. Kuhlman, T. B. Rauchfuss, *Acc. Chem. Res.* **2007**, *40*, 233–242.

[188] K. K. Klausmeyer, S. R. Wilson, T. B. Rauchfuss, *J. Am. Chem. Soc.* **1999**, *121*, 2705–2711.

[189] "WHO | WHO Model Lists of Essential Medicines," can be found under http://www.who.int/medicines/publications/essentialmedicines/en/.

[190] J. R. Sheets, F. A. Schultz, *Polyhedron* **2004**, *23*, 1037–1043.

[191] S. E. H. Nasab, A. H. Nasab, S. A. Ataei, N. Amiri, *Int. J. Electrochem. Sci.* **2013**, *8*, 8800–8811.

[192] C. Chai, W. L. F. Armarego, *Purification of Laboratory Chemicals, Fifth Edition*, Butterworth-Heinemann, **2003**.

[193] S. Trofimenko, in *Prog. Inorg. Chem.* (Ed.: S.J. Lippard), John Wiley & Sons, Inc., **1986**, pp. 115–210.

[194] C. Janiak, *Chem. Ber.* **1994**, *127*, 1379–1385.

[195] W. Kläui, M. Berghahn, G. Rheinwald, H. Lang, *Angew. Chem. Angew. Chem.* **2000**, *112*, 2590, 2590–2592, 2592.

[196] N. Braussaud, T. Rüther, K. J. Cavell, B. W. Skelton, A. H. White, *Synthesis* **2001**, *2001*, 0626–0632.

[197] G. R. Fulmer, A. J. M. Miller, N. H. Sherden, H. E. Gottlieb, A. Nudelman, B. M. Stoltz, J. E. Bercaw, K. I. Goldberg, *Organometallics* **2010**, *29*, 2176–2179.

[198] H. E. Gottlieb, V. Kotlyar, A. Nudelman, *J. Org. Chem.* **1997**, *62*, 7512–7515.

18 Acknowledgement

First of all, this PhD thesis would never have been born without Prof. Dr. Frank Breher. Apart for the financial support for which I'm and will ever be endless grateful, his consent, efforts and engagement to make my dream of a French-German joint-PhD, a so-called "Co-tutelle", come true earn him one of the biggest thanks of this acknowledgement section. France and Germany are not lands reputed for their easy-going bureaucraty: I know it must have been a real nightmare, but it worked, and I think it was worth it. Moreover, I got free reins and means from him to make the chemistry I like, without restraints, and without pressure. I worked these three and half years in an optimal research environment, and I felt there "at home". Many thanks for that, and for all the things I'm also very grateful for (the Almería stay for instance) I can't describe here in detail if I don't want this section to be bigger than my PhD thesis itself.

There was Karlsruhe, but there was also Paris. No joint-PhD without two working groups. I would like to express my sincere gratitude to Prof. Dr. Rodrigue Lescouëzec, first for accepting me as a delocalised Ph.D student, but also for the fruitful discussions, guidance on my PhD subject and great humour.

I'd also like to express my gratitude to the DFH-UFA for their mobility financial support.

On the Parisian side, my deepest thanks to Dr. Yanling Li for her kindness and valuable work. She performed most of my magnetic measurements, and spent a huge amount of energy used up to keep the SQUID magnetometers working over the last years. I also owe her many thanks for the help with the magnetic data processing.

Many thanks to Prof. Dr. Yves Journaux for his patient explanations, and above all, for the countless hours his spent simulating the magnetic behaviour of my compounds.

I sincerely thank Lise-Marie Chamoreau, Geoffrey Gontard and especially Patrick Herson for the XRD analysis done in Paris, but also the reviewing some difficulties on the crystal structures from Karlsruhe.

I'd like to address a special thanks to our MAS-NMR specialist Dr. Alexandrine Flambard for the patient teaching and reviewing of the analysis of the paramagnetic spectra.

I want to thank Dr. Benoît Fleury, Dr Laurent Lisnard, Dr. Mannan Seuleiman and of course, my fellow PhD students: Dr. Abhishake Mondal, Siddhartha De and Juan Ramón Jiménez Gallego for the pleasant atmosphere in the ERMMES group.

Of course, my thanks goes to all members of the 4th floor of the IPCM (Institut Parisien de Chimie Moléculaire), but especially to my former internship supervisor Dr. Sébastien Blanchard for his nice reference letter for my master 1 internship application: I don't know what he did wrote and probably never will, but without it, I surely woudn't have done my Master 1 internship in Karlsruhe, and without internship here, no PhD thesis. He was also the one to tell me: "go at the end of the corridor and meet Dr. Rodrigue Lescouëzec, he might be interested by your joint-PhD idea".

On the Karlsruhe side, and to archieve a good transition, I'd like to thank the nice French company of this institute: Dr. Marie Fustier-Boutignon and Dr. Nicolas Zill. Marie also owes my thanks for lending me some KTp* for my last syntheses, after a painful mishap with my own organic synthesis.

Dr. Timo Augenstein owes my thanks for help by skilful cyclovoltammetric and EPR measurements. He was a great first lab mate, and we had tons of fun together.

Eva Deck owes a "big thank you" for the nice company in the Almeria stay (but not just!) and her eagerness to learn NMR at my side.

A very warm thank you goes to Silvia Hohnstein for company and interesting NMR and chemistry talks. All the best for your project.

Thanks to Alexander Feyrer and Dr. Steffen Styra for the nice coffee pauses and nice compagny.

To Eric Moos goes one of my greatest "thank you" of this section. With upmost patience, he measured a whole bunch of "very bad" to "could be far better" crystals, and was of greatest help for the resolution of the (somewhat) disordered structures. I still wonder how we got a crystal structure for some of them. On the same note, I'd like to thank Dr.

Michael Gamer for XRD time on the Roesky group microsource: getting a crystal of the ordered phase of my cube (with twelve dichloromethane per cube), on the diffractometer without it getting destroyed in the process is an exploit per se.

I'd like to thank Florian Walz for the always opened ears for problems of all kind. Good luck with the writing.

Even though they are new in the lab, I also would like to thank Martin Simon and Hanna Wagner for their nice company.

My gratitude goes also to all former members of the DeBrehers working group: Dr. Sandra Gallardo-González, Dr. Matthias Löble, Dr. Felix Armbruster, Dr. Dominik Nied, Dr. Tanja Wolfer and Dr. Ina Trapp.

It also goes to all permanent members of the institute: Gabi Leichle, Herr Munchi, the workshop team in a whole and every one of them; of course, Frau Baust, Frau Kayas, Frau Pendl and Frau Pollich. Thanks also to Petra Smie (ESI), Sybille Schneider (XRD), Dr. Silke Wolf (für ATG), Dr. Ralf Köppe (for IR) and Frau Maurer (Praktika) and Frau Klaassen (EA).

I'd like to highlight the help of Dr. Ana Kuzmanoski and Lennart Brütsch from the Feldmann group. I've be meowing for almost four years at their doors to get access to their lab IR spectrometer. Not simple to bring aqueous samples in our glovebox! Thanks also to Lennart for his help with the EDX qualitative analyses.

One of the biggest thanks goes to Frau Berberich, for her patience by NMR teaching, for her trust with the device and always nice welcome downstairs. And of course, for her busy work for the whole Institute. I know exactly how much work it is, and I'm very grateful she is there to do it for us.

Of course, I can't forget to thank my actual lab mate, Dr. E. Sattler. His presence makes my lab the best lab room of this institute. Always a wonder to observe, I learnt a lot by his side. Thank you so much for the cake and tea afternoon breaks, for the nice discussions over nearly everything, for listening for my chemical joys and deceptions and sharing yours with me. I'm really going to miss you.

I would like to thank Dr Pascual Oña-Burgos, Prof. Dr. Ignacio "Nacho" Fernández de las Nieves and Francisco "Curro" Manuel Arrabal Campos for the opportunity and the nice welcome in Almeria.

Many thanks to all my students: I may have been complaining a lot over you during all those years, but I really enjoyed the time I spent supervising you in the lab. It gave me good stories to tell, and I wish you all the best for your future.

This thesis would have never been finished without the people correcting it. On that note I'd like to thanks all my collegues of Karlsruhe, my two supervisors, as well as Sabrina and Michael Capper. All the best with Annelie! You were quicker with her than I am with this work! A special "thank you" goes to my friend Mélanie Démeraux, for moral support as well as skillful English translational help. Je croise les doigts pour toi!

Many thanks to my Joël. It has been tough years, but you were always there for me when I needed it. Merci de tout coeur.

Last but not least: though 600 km away, my family have been the best support I would ever dream. They did not understand everything, but they did their best to encourage me to go on. Mamie, Mimie, Jo, Maman, Papa, merci pour tout.